Food Industry Quality Control Systems

Food Industry Quality Control Systems

Edited by **Hugh Brennan**

SYRAWOOD
PUBLISHING HOUSE
New York

Published by Syrawood Publishing House,
750 Third Avenue, 9th Floor,
New York, NY 10017, USA
www.syrawoodpublishinghouse.com

Food Industry Quality Control Systems
Edited by Hugh Brennan

International Standard Book Number: 978-1-68286-333-6 (Hardback)

Printed in China.

Contents

Preface

Quality control systems are of immense importance in food processing industries in order to ensure the safety and quality of processed food products. This book studies, analyses and upholds the pillars of food quality control systems and its utmost significance in modern times. Food engineering and production, toxicology, food safety analysis, etc. are some of the topics discussed in this book that will provide innovative insights to the readers. From theories to research to practical applications, case studies related to all contemporary topics of relevance to this field have been included in this book.

This book unites the global concepts and researches in an organized manner for a comprehensive understanding of the subject. It is a ripe text for all researchers, students, scientists or anyone else who is interested in acquiring a better knowledge of this dynamic field.

I extend my sincere thanks to the contributors for such eloquent research chapters. Finally, I thank my family for being a source of support and help.

Editor

1

The "Dark Side" of Food Stuff Proteomics: The CPLL-Marshals Investigate

Pier Giorgio Righetti [1,*], Elisa Fasoli [1], Alfonsina D'Amato [1] and Egisto Boschetti [2]

[1] Department of Chemistry, Materials and Chemical Engineering "Giulio Natta", Politecnico di Milano, Via Mancinelli 7, Milano 20131, Italy; E-Mails: elisa.fasoli@polimi.it (E.F.); alfonsina.damato@chem.polimi.it (A.D.)

[2] EB Consulting, Paris, France; E-Mail: egisto.boschetti@gmail.com

* Author to whom correspondence should be addressed; E-Mail: piergiorgio.righetti@polimi.it

---

Abstract: The present review deals with analysis of the proteome of animal and plant-derived food stuff, as well as of non-alcoholic and alcoholic beverages. The survey is limited to those systems investigated with the help of combinatorial peptide ligand libraries, a most powerful technique allowing access to low- to very-low-abundance proteins, i.e., to those proteins that might characterize univocally a given biological system and, in the case of commercial food preparations, attest their genuineness or adulteration. Among animal foods the analysis of cow's and donkey's milk is reported, together with the proteomic composition of egg white and yolk, as well as of honey, considered as a hybrid between floral and animal origin. In terms of plant and fruits, a survey is offered of spinach, artichoke, banana, avocado, mango and lemon proteomics, considered as recalcitrant tissues in that small amounts of proteins are dispersed into a large body of plant polymers and metabolites. As examples of non-alcoholic beverages, ginger ale, coconut milk, a cola drink, almond milk and orgeat syrup are analyzed. Finally, the trace proteome of white and red wines, beer and aperitifs is reported, with the aim of tracing the industrial manipulations and herbal usage prior to their commercialization.

Keywords: animal foods; plant foods; alcoholic beverages; non-alcoholic beverages; aperitifs; proteomics; combinatorial peptide libraries

---

1. Introduction

The present review aims at offering a panorama of what modern pre-fractionation technologies can achieve in detecting the low- to very-low abundance proteome in food stuff and beverages, an analyte fraction that is quite invisible even to the most sophisticated modern mass spectrometers (MS), whose sensitivity spans in general five orders of magnitude in relative concentrations of proteins present in a sample. Yet, in human biological fluids, such as plasma, such dynamic range can cover up to 12 orders of magnitude [1] and, in living cells, can span at least seven orders of magnitude. Thus, it is quite obvious that MS alone cannot cover efficiently the grounds and thus additional techniques would be needed to achieve the goal of (hopefully) a global coverage of any proteome. One such technique, which will be reviewed here with examples of what can be achieved in analysis of animal and "botanical" food stuff, as well as of non-alcoholic and alcoholic beverages, is definitely the combinatorial peptide ligand library (CPLL) technology that we have developed over the years, and which has now been taken at a level of maximum performance, as summarized in a recent book [2].

In a way it sounds preposterous that one should tackle such a detection of trace species in food stuff when, in the field of human nutrition, the problems we are facing are enormous, as today we have to feed a global population that has reached seven billion and is projected to reach nine billion by the year 2050 [3]. In the book here quoted [3], Brown depicts an almost apocalyptic scenario of land erosion, desertification, water shortage, dramatic increments of dangerous gases such as CO_2 and NO and yields of cereals (the basic staple food for humans) rapidly reaching a limit. Were this not enough, there would be a dramatic imbalance of food distribution in the world population. About one third suffers from malnutrition, although in a peculiar way: 1.1 billion face starvation whereas 1.3 billion are either overweight or obese. One might think that the overweight population would be typical of the rich countries, such as USA (where indeed the phenomenon is reaching alarming proportions) but in reality it is widespread also in poor countries, since obese conditions can be attained by eating "junk food". Yet, in well-to-do countries, the focus today is not on food quantity (the supermarkets seem to be bursting with food stuff) but on food quality, i.e., on how food can affect health and wellbeing of people, as illustrated below.

2. Nutraceutical and Functional Foods

The focus today is on detrimental to beneficial constituents in foods, as discussed in a recent issue of *Journal of Agricultural and Food Chemistry* [4–6]. This has led to a new science, called "nutraceutical", a term that is a fusion of the words nutrition and pharmaceutical, thus emphasizing the potential therapeutic properties of foods. A new category of foods has thus emerged, the "functional food", i.e., a food stuff that provides health or medical benefits beyond the basic nutrients it contains, including not just metabolites but proteins and peptides as well [7–9]. It is now well ingrained, in fact, that, whereas bad feeding habits are associated with, e.g., cardiovascular diseases, diabetes and even cancer, regular ingestion of "healthy foods" may prevent such diseases and/or mitigate them during onset and development [10]. The global nutraceutical market, defined as the cumulative sales of nutraceutical foods, beverages and supplements fortified with bio-active components, is now reaching a substantial segment of the food market. For instance, this market, which was worth $117.3 billion in

2007, has increased to 176.7 billion in 2013. A rather large proportion of these nutraceuticals is based on health claims, i.e., statements used on labels, in marketing or in advertising that health benefits can result from consuming a given food or from one of its components such as vitamins and minerals, fibre, and "probiotic" bacteria (which would ultimately form the population of gut microbiota) [11]. There are different types of health claims. For instance, statements that a food can help reinforce the body's natural defenses or enhance learning ability are called "general function" claims. Examples also include claims on the reduction of disease risk and other substances that may improve or modify the normal functions of the body. Other types of claims are "nutrition claims", which state or suggest that a food has particular beneficial nutritional properties. Examples comprise "low fat", "source of omega-3 fatty acids" or "high in fibre". In Europe alone, nutraceuticals have produced a stampede of applications to the regulatory agencies, for approval of their products, by mini- and start-up companies active (or entering) in this field. EFSA (European Food Safety Agency) has been flooded, only in the last few years, with 44,000 applications for approval of "potential" nutraceuticals, claiming a rainbow of biological activities. These applications have been gathered, analyzed and consolidated into about 4600 claims, of which, in 2012, only 222 were approved [12]. One wonders what the reason is for such a low promotion score. The vast majority of claims, indeed, lack the following basic information: (1) most of them do not even list or know which ones are the active and health beneficial ingredients, since they propose simply ground botanical extracts as such or only partially purified, whose chemical composition remains largely unknown; (2) by the same token, they lack *in vivo* demonstration of such health benefits, not just on humans, but not even on animal models. It is clear that tools for fully exploring their nature are sorely needed, especially those able to detect trace peptide/protein components, whose biological activities have to be demonstrated.

3. Separation Science at Work: Combinatorial Peptide Ligand Libraries (CPLL)

As stated above, CPLLs appear to be a unique tool for exploring the "dark side" of any proteome, due to their unique property of providing millions of affinity ligands able to find a partner in any protein species present in biological materials. It must be emphasized that in just about any biological specimen a small set of proteins (often as few as 20–30) are present in a large excess and could constitute, like in human sera, as much as 99% of the total protein mass. This would leave little room for sampling (and thus detecting) all other species present therein. A solution proposed already in 2002 by the Anderson's lab is immuno-subtraction, i.e., preparation of affinity resins containing antibodies against the six most abundant proteins in sera (first extended to 12 and now to 20) [13]. Believed to permit access to low-abundance species [14], in reality it did not quite live up to expectations [15,16], for several reasons, among them the major issue being that much too little sample volume can be processed in a single sweep (barely 100 μL serum/plasma). Application of this methodology to plant proteomes, such as the immuno-depletion of RuBisCO, also did not lead to any major improvement [17].

The CPLL technique we have developed is immune from such drawbacks. Here is a brief description of the chemistry and mechanism of action of CPLLs. The material for the treatment of the biological sample is constituted of a mixture of discrete sorbents each of them carrying a different ligand (a hexapeptide). The number of individual species is extremely large. Each of them represents an item of a full mathematical combination of 15 amino acids built to a length of hexapeptides and obtained by combinatorial synthesis. Each discrete peptide is a potential ligand for one or more proteins from the sample. By loading the combined sorbent mix with a biological sample exceeding the binding capacity, each single affinity ligand will be saturated with corresponding protein partners and the excess will be found in the so-called flow-through. Abundant proteins will rapidly saturate the corresponding ligand while the rare proteins will continue to be adsorbed as long as the loading is increased, this process leading to their enrichment onto the beads. Once the sorbent mix is extensively washed in order to remove the excess proteins, the captured species are harvested by means of appropriate desorbing agents. All proteins are potentially present but their relative concentration is deeply changed with a strong compression of dynamic concentration range. The consequence of such a situation is the detection of many novel proteins (the low-abundance ones) as a result of (1) the annihilation of the signal suppression due to concentrated species (e.g., albumin in serum) and (2) the detection of very low concentration proteins that were below the detectability level prior to sample treatment. These two distinct phenomena, co-existing during the CPLL treatments, are very powerful when the loading is very large and when all captured proteins are harvested [2,18–20]. Figure 1A is a schematic representation of this sequence of steps and *modus operandi* of the beads, while Figure 1B is an artistic rendering of a bead suspension docking onto the respective partners (with a proviso, though: the beads do not have just two ligands but are coated with millions of ligands, all of the same hexapeptide type). We have applied the CPLL technique to analysis of food stuff and beverages for at least three main reasons:

1. to detect trace proteins/peptides exhibiting negative effects on health (e.g., allergens);
2. to detect trace proteins/peptides displaying positive effects on health (e.g., anti-microbial, anti-hypertensive and anti-oxidant activities);
3. to expose frauds in commercial food products and provide a proof of genuineness for "correct" commercial foods, as found in supermarkets.

The examples listed below are limited to analyses of foods we have performed over the years via the CPLL methodology. Moreover, we cannot possibly cover here the vast literature on plant and food proteomics, but we just mention two recent reviews (and references therein) covering this field quite extensively [21,22]. Special issues of different journals appear from time to time covering plant proteomics, including analysis of food and beverages. A recent one has been released by *Journal of Proteomics* [23]. Within this issue see, for instance, Boggess *et al.* [22], Nakamura *et al.* [24], Uvackova *et al.* [25], Agrawal *et al.* [26] and Ribeiro Demartini *et al.* [27].

Figure 1. (**A**) Mechanism of action of combinatorial peptide ligand libraries. (**B**) Artistic rendering of combinatorial peptide ligand library (CPLL) beads with proteins docked on the hexapeptide baits.

A

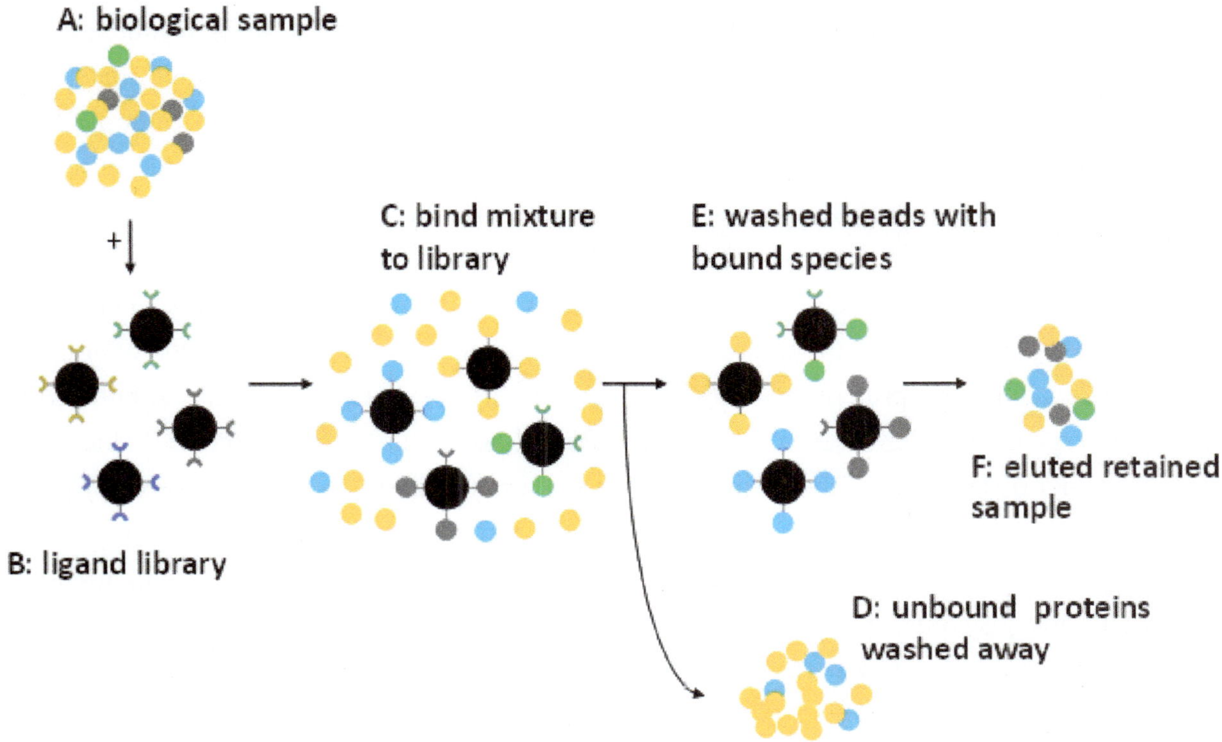

B

4. Analysis of Animal Food Stuff

Already in 2008 we applied the CPLL technique to the analysis of the egg white proteome [28], soon followed by the exploration of egg yolk [29], the reason being that, in both cases, barely a handful of proteins was known, those of very high abundance. The first major exploration, though, performed via the most sophisticated MS equipment, had just described, prior to our investigations, 78 unique gene products, a quantum jump in respect to what was known up to that time [30]. Yet, in our case, even by using less powerful MS tools, we could almost double that value to 148 unique protein species, a non-negligible increment. Basically, the same events were repeated in regard to egg yolk. Just prior to our report, Mann and Mann [31] reported 115 unique gene products in the yolk plasma, whereas our findings have more than doubled this value to 255 unique protein species. These two manuscripts formed the core of an extensive review on the egg white and yolk interactomes [32]. Thus, these very first investigations proved that CPLLs had a lot to offer in terms of food analysis. Soon after we tackled the proteome of cow's whey, not only for an in depth analysis, but also to detect "hidden" allergens that seemed to be responsible for unknown allergies suffered by babies that at that time were under observation at the Hôpital Cochin in Paris. Whereas comprehensive whey protein lists progressively increased in the last six years from 17 unique gene products to more than 100, our findings considerably expanded this list to a total of 149 unique protein species, of which 100 were not described in any previous proteomics study [33]. As an additional interesting result, a polymorphic alkaline protein was observed with a strong positive signal when blotted from an isoelectric focusing separation in gel and tested with sera of allergic patients. This polymorphic protein, found only after treatment with the peptide library, was identified as an immunoglobulin (Ig), a minor allergen that had been largely amplified. In terms of allergy to milk proteins, often quite deleterious for small babies, it had been known for quite some time that donkey's milk was an excellent substitute for cow's milk, allowing a normal and healthy growth for allergic children. So much so that two Italian pharmacists from Sicily, having observed that, opened a farm on the slopes of the Etna and started offering this milk with the brand name "Asilat" (asina is the Italian name of the she donkey), which has been a life-saver for thousands of Italian affected children. Yet, its composition remained largely unknown till our study with CPLLs [34]. By treating large volumes (up to 300 mL) of defatted, de-caseinized (whey) milk, we have been able to identify 106 unique gene products, by far the largest description so far of this precious nutrient. Due to poor knowledge of the donkey's genetic asset, only 10% of the proteins could be identified by consulting the data base of *Equus asinus*; the largest proportion (70%) could be identified by homology with the proteins of *Equus caballus*. Among the species identified we could recognized also some epidermal growth factors, suggesting that Poppea's (emperor Nero's wife) habit of bathing in donkey's milk, notwithstanding her scarce knowledge of proteomics, had some scientific merits! Our work on different types of milks deserves some further comments. In 2011, Le *et al.* [35] in an extensive investigation of both cow's whey and colostrum, requiring a rather complex experimental set-up and analysis of several eluted fractions, reported a grand total of 293 unique proteins and thus stated that "the 149 characterized proteins may be a superficial representation of the proteome of bovine milk whey". Soon after, Nissen *et al.* [36] delved again onto bovine colostrum and reported identification of 402 unique species, suggesting that the data of Le *et al.* [35] did not quite represent an in-depth exploration. Moreover, it just so happens that we are now analyzing cow's colostrum by our (quite simple) CPLL methodology in collaboration with Prof. G. Aldini (University of Milano) in order to increase the knowledge of pharmaceutical activities related to low abundance proteins, of which we find a large number not reported before. So, at this point we are forced to state that their investigation [35] (notwithstanding the quite complex and elaborated experimental set-up) might be a "really superficial representation of this proteome"! All these collective data have helped Bislev *et al.* [37] to construct a Bovine Peptide Atlas comprising 1921 proteins at 1.2% false discovery rate (FDR) and 8559 distinct peptides at 0.29% FDR (with a caveat, though: this atlas comprises not only milk and colostrum, but a variety of bovine tissues, including hoof specimens).

As an additional example of the application of CPLLs to food analysis, Di Girolamo *et al.* [38] recently investigated the proteome content of different unifloral honeys (from chestnut, acacia, sunflower, eucalyptus and orange) in order to see if any plant proteins present would allow the proteo-typing of these different varieties. It turned out that all proteins identified (except one, the enzyme glyceraldehyde-3-phosphate dehydrogenase) were not of plant origin but belonged to the *Apis mellifera* proteome. Among the total proteins detected (eight, but only seven as basic constituents of all types of honey), five belonged to the family of major royal jelly proteins 1–5, and were also the most abundant ones in any type of honey, together with α-glucosidase and defensin-1. It thus appears that honey has a proteome resembling the royal jelly proteome (but with considerably fewer species), except that its protein concentration is lower by three to four orders of magnitude as compared to royal jelly. The only other report available on the honey proteome could identify barely one protein [39]. In a way, this food stuff should be listed in the section below, but in reality it is a hybrid: although the origin is floral, the proteome is essentially of animal origin. We have also analyzed honeydew from *Abies alba*, produced by bees towards the end of August, when no more flowers are available, from fir resin that had been already pre-digested by insects. Even in honeydew (in Italian "melata"), only five bee's proteins out of the seven of all other types of honeys could be detected (unpublished experiments with Dr. M.M. Staver, University of Rijeka).

5. Analysis of Vegetable and Fruit Food Stuff

One of the first nourishment of plant origin we have studied is spinach, possibly under the influence of the Popeye saga. Fasoli *et al.* [40] mapped the cytoplasmic proteome of spinach leaves by capturing it at three pH values (4.0, 7.0 and 9.3) and eluting in 4% boiling SDS, 20 mM DTT. The number of gene products found exclusively thanks to the use of CPLLs and classified as low-abundance was particularly high, with 208 proteins *versus* 114 found in the initial extract with an overall gene product count of 322. Two main phenomena allowed detecting so many more proteins: the massive reduction of the concentration of RuBisCO and the enrichment of many rare proteins initially present in trace amounts made possible by the use of a relatively large initial volume of spinach extract. Among the enriched cellular components were ribosomal proteins, such as 30S, 40S, 50S and 60S. Other low-abundance proteins found in spinach leaves were those related to chloroplast category and those operating in the photosynthetic organelles, e.g., inorganic pyrophosphatase-1 and phosphoglycerate kinase. From the same category, but involved in the CO_2 fixation and carbohydrate metabolism, were glutamate decarboxylase 2, glucose-1-phosphate adenylyltransferase, phosphoglucomutase, fructose bisphosphate aldolase 2, ribulose-phosphate 3-epimerase and sucrose-phosphate synthase.

The next proteome of importance in human nutrition we have analyzed is the one of olive pulp and seed (*Olea europaea* L.) due to the importance of this fruit in production of olive oil. Esteve *et al.* [41] have tackled this task via the CPLL approach. Thanks to its use, a quite large number of compounds has been indeed identified: 61 in the seed (*vs.* only four reported in current literature) and 231 in the pulp (*vs.* 56 described so far) [42]. In the seed, it highlights the presence of seed storage proteins, oleosins and histones (of which as many as 14 different unique species have been listed). In the pulp, the allergenic thaumatin-like protein (Ole e 13) was confirmed, among the other 231, as the most abundant protein in the olive fruit.

Although the results on both systems were exciting, due to large increments of discovery as compared to published literature, yet we were annoyed by the fact that, in plant specimens, we did seem to be unable to break the barrier of 300 detected proteins, a much too low proteome harvest, considering that several thousand gene products should be present in these specimens too (although it should be noted that, up the these investigations, we only analyzed cytoplasmic soluble species, captured under native conditions). We thus turned to "recalcitrant" tissues, in search of ways and means to break these barriers. Two such tissues would certainly be the banana and avocado proteomes, since here tiny amounts of proteins (*ca.* 1%) would be dispersed into huge amounts of interfering plant matrices (polysaccharides and solid oils, respectively). These would be quite challenging projects and it was felt that one needed extra efforts to study these tissues and in general any plant proteome, efforts that have materialized in more than one extraction protocol, i.e., under native and denaturing conditions, in both cases followed by capture with CPLLs of the extracted proteins. The extraction under denaturing conditions was performed in 3% SDS, an anathema in CPLL treatments, since it would completely inhibit the capture. The SDS had to be removed either via the classical acetone/methanol precipitation, or by dilution from 3% to 0.1% in the presence of another compatible surfactant, like 1% CHAPS, conditions that would be compatible with the CPLL technology. The results in the case of avocado were outstanding: about 1300 unique gene products were detected, a four-fold increment as compared to the findings in olive pulp proteome. Figure 2 details the various discoveries obtained with the treatments: the unique species identified in the control were 236 *vs.* 796 in the global CPLL treatment, 250 being common in the two sets. The Venn diagram to the right shows how much more the denaturing conditions' extraction contributed to the total discovery as compared to the native protocol: the unique catch has been more than double. The bar graph of GO analysis (lower panel in Figure 2) shows also some outstanding data: not only all the various Go categories are considerably more populated in the CPLL-captured samples, but also five new metabolic processes, not discernible in the control, could be described [43].

Figure 2. Upper panel: Venn diagrams showing the total identifications of avocado proteins in either the control or in the CPLL treated samples (left diagrams). The Venn diagrams to the right display the relative contributions to the total IDs of the denatured *vs.* the native sample extractions. Lower panel: GO analysis of the proteins detected in various metabolic processes in the control *vs.* CPLL-treated samples. In the latter case, five additional metabolic processes can be described, which are not visible in the control sample and most likely represent low- to very-low abundance proteins, whose visibility has been substantially enhanced via the CPLL technology (modified from Esteve *et al.* [43]).

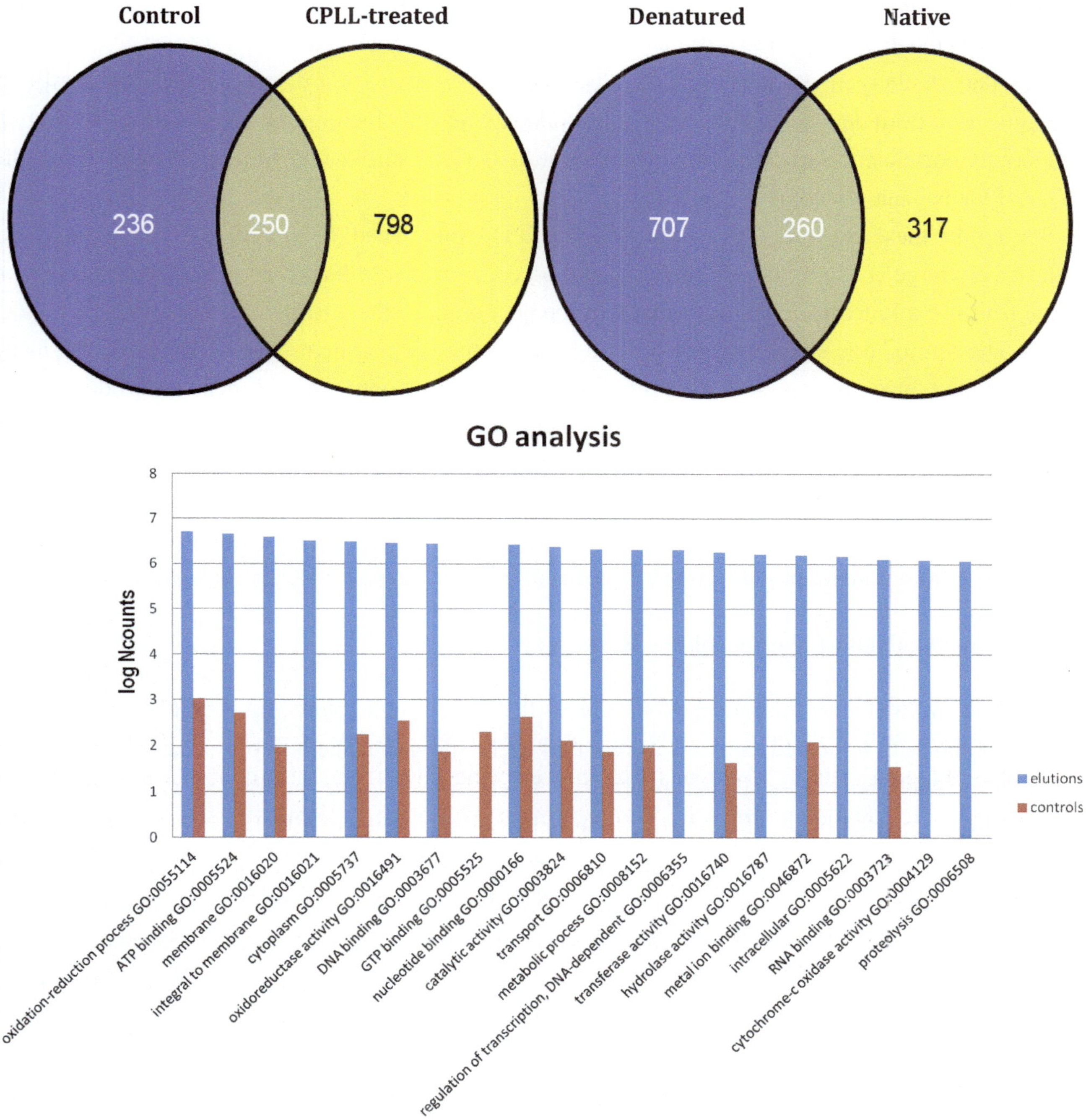

The next fruit investigated was the banana. *Musa* spp., comprising banana and plantain, is grown extensively in many developing countries and is considered to be one of the most important sources of energy in the diet of people living in tropical humid regions. Due to its antioxidant and cell antiproliferative activities, the consumption of banana has been associated with reduced risk of chronic diseases such as cardiovascular diseases and cancer [44]. Up to date, no in depth work has been focused on identifying the banana fruit proteome; since fresh banana pulp contains approximately 20% of carbohydrates and only 1% of proteins, this fruit has been traditionally considered as a difficult matrix for protein extraction, being a target in studies of optimization of protein extraction methodologies [45]. Fruits, as every biological source, contain highly-abundant proteins, which are often of limited interest for proteome analysis, whereas other proteins may be orders-of-magnitude less abundant, although still of high importance. Here too, by using CPLLs, advanced mass spectrometry techniques and *musa* mRNAs database in combination with Uniprot_viridiplantae database, we could identify, for the first time, 1131 proteins [46]. Among this large amount of species found, several already known allergens such as *musa a 1*, pectinesterase, superoxide dismutase and potentially new allergens have been detected. Additionally, several enzymes involved in degradation of starch granules and strictly correlated to ripening stage were identified. These results constitute the largest description so far of the banana proteome.

Spurred by these encouraging results, we decided to explore also the proteome of mango, a tropical fruit of very large consumption. In this case, we opted for analysis also of its peel (since also this part of the fruit is exploited for cooking) as well of the peptidome content (the latter as captured with a C_{18} resin). The aim of this study was not only to perform the deepest investigation so far of the mango proteome, but also to assess the potential presence of allergens and of peptides endowed with biological activities. The proteins of peel and pulp were captured under both native and denaturing extraction techniques and the harvest was outstanding. A total of 334 unique protein species have been identified in the peel *vs.* 2855 in the pulp, via capture with CPLLs at different pH values (2.2 and 7.2). In this very large pulp protein list, we could also identify some well-known allergens, in particular the non-specific lipid transfer protein, superoxide dismutase, germin-like protein and profilin [47].

6. Proteome Analysis of Non-Alcoholic Beverages

Here too, we will limit this survey to beverages that have been explored by the CPLL methodology, in comparison with untreated samples, as reported in the literature. We will only deal with proteomes, although in such beverages also a peptidome screening could be performed with CPLLs. The investigations reported below demonstrated the unique performance of CPLLs, such as: (1) the ability to handle large sample volumes (one liter and more) in an easy and user-friendly protocol; (2) sample enhancement factors up to four orders of magnitude; (3) extremely high detection sensitivities, reaching as little as 1 μg protein/L of sample, a sensitivity rarely obtained by present-day methodologies. Needless to say, these results were much superior than those obtained with conventional techniques, due to their inability to properly concentrate and "normalize" the relative concentration of the proteomes under analysis.

There are at least two reasons for exploring the trace proteome of non-alcoholic beverages (whose proliferation in the shelves of supermarkets is appalling, with so many brands and fancy appellations as to stun the customers): (1) in order to certify the genuineness of such products and find out if they contain proteins of the vegetable extracts they have been prepared from; (2) in order to screen for the presence of any potential allergen in such beverages. Indeed, such analyses have been made on almond's milk and orgeat syrup [48], coconut milk [49], as well as on a cola drink [50], ginger ale [51] and even white-wine vinegar [52]. In the case of almond's milk, 137 unique protein species were identified, whereas in the case of orgeat syrup, a handful of proteins (just 13) were detected, belonging to a bitter almond extract. In both cases, the genuineness of such products was verified. On the contrary, cheap orgeat syrups produced by local supermarkets and sold as their own brands were found not to contain any residual proteins, suggesting that they were likely produced only with synthetic aromas and no natural plant extracts. Figure 3 displays the SDS-PAGE profiles of orgeat syrup (two tracks to the left) and of almond's milk. In the case of orgeat, it can be appreciated that no single protein band is visible in the control, untreated sample; as for almond's milk, whereas the control lane shows two major bands at 23 and 35 kDa, plus a few other zones, the two eluates after capture at pH 7.0 and 9.3 exhibit a very extensive band profile from 10 to 100 kDa. In the case of coconut milk, a grand total of 307 unique gene products could be listed, 200 discovered via CPLL capture, 137 detected in the control, untreated material and 30 species in common between the two sets of data. This is by far the most extensive mapping of this nutritious beverage, in which, up to the present, only a dozen proteins were known, those belonging to the high- to very-high abundance class.

An interesting detective story can here be told about the "invisible" proteome of a cola drink (nothing to do with Coca Cola, this one is an English beverage!) [50], stated to be produced with a cola nut extract. Indeed, a few proteins in the molecular mass (Mr) 15–20 kDa range could be identified by treating large beverage volumes (one liter) and performing the capture with CPLLs at very acidic pH values (pH 2.2) under conditions mimicking reverse-phase adsorption. It turned out that the proteins identified were barely three: one belonging indeed to the cola nut, the other two identified as agave species. Indeed, in this beverage label, it was stated that the cola drink had been generously sprinkled with 6% agave syrup, honor saved, all around. Conversely, things did not go so well (for the producer) in the case of a ginger beverage [51]. Although in traces, the presence of five grape proteins and one apple protein could be confirmed, but not even the faintest trace of any ginger root proteins. The first two findings are correct, as the producer stated that this beverage had been reinforced with 12% grape juice and 6% apple juice, but the absence of even traces of ginger proteins did not permit the classification of this beverage as a ginger extract on a proteomics scale. It was thus concluded that either the *Zingiber officinalis* was present in traces, or only its flavors had been added and not any root extract.

Figure 3. SDS-PAGE profiles of orgeat syrup (two tracks to the left) and of almond's milk (right panel). Orgeat: CTRL = 100 µL of control orgeat solution before CPLL treatment. 2.2 E = 50 µL 2% SDS eluate of orgeat solution treated with a home-made-CPLL library at pH = 2.2. Almond's milk: control (A-CTRL) and eluates from a 1:1 mixed library of PM and home-made-CPLLs at pH 7.2 (7A) and pH 9.3 (9A) values (modified from Fasoli *et al.* [48]).

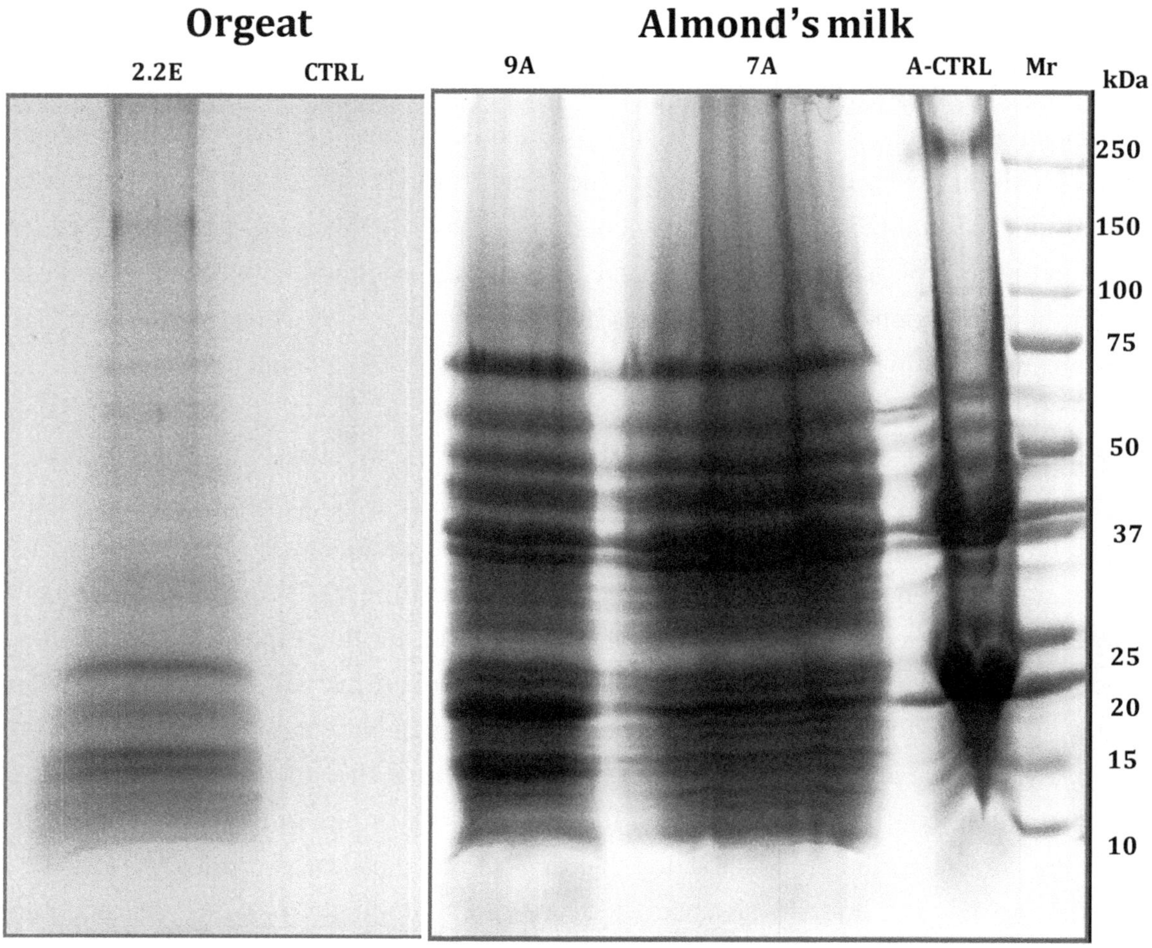

In another investigation on non-alcoholic beverages, the trace proteome of white-wine vinegar was identified again via CPLLs but under conditions mimicking reverse-phase capture, i.e., at pH 2.2 in presence of 0.1% trifluoroacetic acid. A total of 27 unique gene products could be tabulated, of which 10 specific of the database *Vitis vinifera*, 13 found in the general database Uniprot_viridiplantae and four in Swiss Prot_all entries. The most abundant species detected, on the basis of spectral counts, appears to be the whole genome shotgun sequence of line PN40024, scaffold_22 (a protein of the glycosyl hydrolase family, indicated by an arrow in Figure 4) [52]. The hypothesis set forward regarding these 27 surviving proteins to the strongly acidic, pH 2.2, environment of vinegar is that they could be the only ones resistant to denaturation, a phenomenon that typically occurs with proteins which have a tightly packed hydrophobic core and are thus insensitive to pH denaturation. The other interesting aspect of this set of 27 survivors is the fact that they seem to be resistant also to proteolytic attack, since their apparent Mr values (see Figure 4) are well distributed in the 70–120 kDa range. Conversely, when the ultra-trace proteome of a cola drink was explored, the minute traces of the only three surviving proteins were found solely in the leading/terminating boundary of the Laemmli discontinuous buffer in the SDS-PAGE gel, where typically only protein fragments are confined, suggesting that they were all degraded species.

Figure 4. SDS-PAGE of white-vine vinegar. Mr: molecular mass ladder. Laemmli discontinuous buffer. Gel: 8%–18% polyacrylamide gradient. Run: three hours at 300 V. Sample load: 25 μL containing *ca.* 50 μg protein. Staining with micellar Coomassie Brilliant Blue. The major protein (the whole genome shotgun sequence of line PN40024, scaffold_22, a protein belonging to the glycosyl hydrolase 32 family) is indicated with an arrow (modified from Di Girolamo *et al.* [52]).

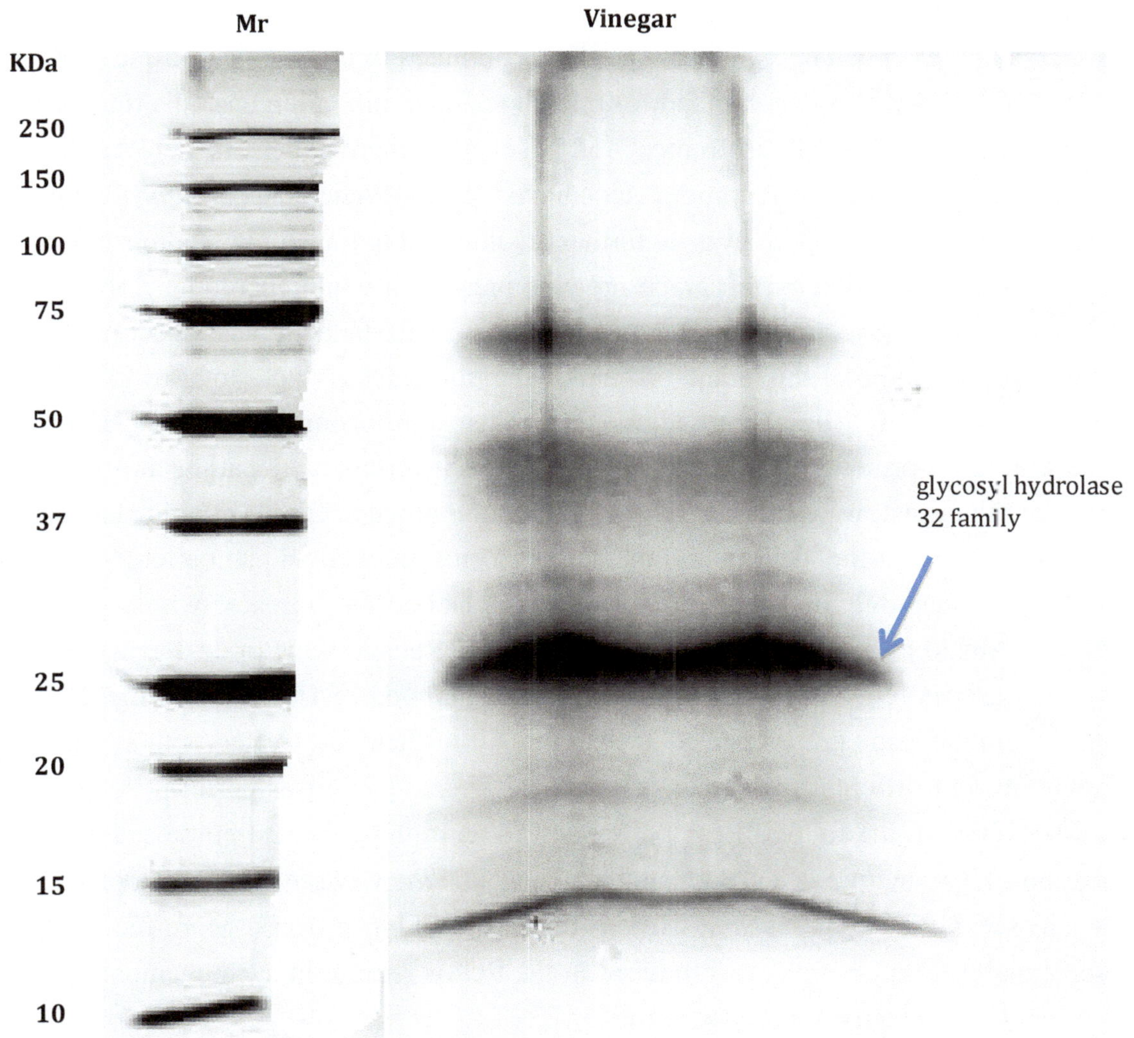

7. Proteome Analysis of Alcoholic Beverages

Due to its extensive consumption worldwide, an investigation on wine proteome was surely due. To complicate the matters, though, modern wines might be quite different from those drunk by our ancestors. One of the main reasons is that the residual grape proteins, that survived the fermentation process, slowly aggregate leading to amorphous sediments or flocculates, causing turbidity. A haze or deposit in bottled wine indicates that the product is unstable, has a low commercial value and is therefore unacceptable for sale. For these reasons, it has become customary to remove the residual proteins remaining in the finished product, so as to prevent haze formation and sediment in the bottled wines available for sale. Among the fining agents, one of the most popular is casein derived from bovine milk, which, however, is a known allergen, whose presence should be declared on the label (something that no wine producer has ever done!). In fact, when an entire bottle of white wine (750 mL) was treated with CPLLs at pH 3.3, the results were quite exceptional, in that as little as 1 μg casein could be assessed as a residue after the treatment [53]. The major result of the above investigation was that it was proven that the CPLL technique had a sensitivity 200 times higher than the current ELISA test, a non-negligible accomplishment. In a competing study, Monaci *et al.* [54] stated "when fined wine samples were considered, the lowest added concentration for which the peptide marker could be detected was 50 μg/mL" (the peptide marker referring to casein digests, as identified by mass spectrometry), which seems to be a terribly low detection sensitivity, 50,000 times lower than that reported by Cereda *et al.* [53]. These authors, actually, continued in their search of traces of caseins in white wines and reported yet another method for tracing residual milk allergens, this time based on the use of a single-stage Orbitrap MS instrument [55]. Yet, the improvement in detectability was not spectacular, in their own words: "minimum detectable added caseinate concentrations were estimated between 39 and 51 μg/mL". There was a further evolution on this topic. Palmisano *et al.* [56] published an extensive investigation on grape proteins present in white chardonnay. They adopted a multiplexed glycopeptide enrichment strategy in combination with tandem mass spectrometry in order to analyze the glycoproteome of this brand of white wine, thus identifying a total of 28 glycoproteins and 44 glycosylation sites. The identified glycoproteins were from grape and yeast origin. In particular, several glycoproteins derived from grape, like invertase and pathogenesis-related (PR) proteins, and from the yeast, were found after the vinification process. Bioinformatic analysis revealed sequence similarity between the identified grape glycoproteins and known plant allergens. These data, though, could be obtained only by analyzing a wine which had not been treated with any fining agent. Aware of that, D'Amato *et al.* [57] analyzed some bottles of Recioto wine (a dessert wine made from partly dehydrated grapes left for longer periods on the plants) and of Garganega wine, a white table wine produced from the same grapes picked at ripening, all of them not treated with any fining agent. A CPPL capture at four different pH values (pH 2.2, 3.8, 7.2 and 9.3) was performed. The combined data on the discoveries in the four CPLL eluates, as well as in the collected bottle sediment, allowed the identification of 106 unique gene products belonging to *Vitis vinifera*, as well as of an additional 11 proteins released by the *S. cerevisiae* used in the fermentation process. More recently, we have continued such analyses on champagne, produced in the Reims region: in aged champagne, 40 unique gene products could be detected for the first time [58].

The next logical step was to extend these investigations to red wines, here too either to detect traces of fining agents or the entire grape-proteomic asset in non-treated samples. In the case of red wines, one would expect to find ovalbumin or entire egg white, since these are the customary fining agents adopted. However, in an extensive investigation of Italian red wines, D'Amato *et al.* [59] found them to not contain traces of egg albumins but again of bovine caseins, especially α- and κ-caseins, with essentially no residual grape proteins, except for traces of thaumatin fragments. With another surprising finding: the wines analyzed (in North Italy, around the lake of Garda, 2009 harvest) contained (albeit in traces) abnormal amounts of proteins originating from fungal infection, such as from *Botryotinia fuckeliana*, *Sclerotinia sclerotiorum*, *Aspergillus aculeatus*, suggesting adverse meteorological conditions at harvest in that year. Traces of egg-white proteins in red wines were indeed detected Tolin *et al.* [60] but in wines that they had treated with different amounts of ovalbumin just to determine the detection limit of the MS analyses.

Following these extensive investigations of white and red wines, a study on the beer proteome was definitely due. This came from Fasoli *et al.* [61] who assessed this proteome again via CPLL capture at three different pH (4.0, 7.0 and 9.3) values. Via mass spectrometry analysis of the recovered fractions, after elution of the captured populations in 4% boiling SDS, they could categorize such species in 20 different barley protein families and two maize proteins, the only ones that had survived the brewing process (the most abundant ones being *Z*-serpins and lipid transfer proteins). In addition to those, they could identify no less than 40 unique gene products from *Saccharomyces cerevisiae*, as routinely used in the malting process. These latter species must represent trace components, as in previous proteome investigations barely two such yeast proteins could be detected.

The natural extension of these investigations on the proteome of alcoholic beverages was to explore the trace proteome of aperitifs, i.e., alcoholic beverages half a way in between wines (with a 11%–13% alcohol content) and hard liquors (e.g., cognac, whiskey, grappa and the like, typically with more than 40% alcohol). Aperitifs generally have an alcohol content around 20%–22% (although their alcohol content can be lower, down to *ca.* 15%) and are in general made with herbal infusions. Their widespread popularity stemmed from their large consumption in pubs and social gatherings and from the medical observation that they would favor food intake and stimulate digestion (these bitter liquors were classified as *cholagogues*, i.e., medicinal agents promoting the discharge of bile from the system). Since in general all producers declare that their aperitifs are made with herbal infusions, it was of interest to investigate their trace proteome, in order to verify the genuineness of such products. With this in mind, Fasoli *et al.* [62] decided to analyze a popular aperitif in North Italy, Amaro Braulio, in search of residual proteins, if any. This aperitif is made with an infusion of 13 mountain herbs and berries, among which four are officially indicated in the label: *Achillea moschata*, juniper (*Juniperus communis* subsp. *alpina*) berries, absinthe (*Artemisia absinthium*) and gentian (*Gentiana alpina*) roots. Via capture with solid-phase combinatorial peptide ligand library at pH 7.0 and 2.2 these authors were able to identify 72 unique gene products, among which the PR5 (parasite resistance) allergen *Jun r 3.2*, a 25 kDa species from *Juniperus rigida*. Due to the paucity of data on these alpine herbs, it was difficult to attribute these proteins to the specific plant extracts presumably present in this beverage; however, most of the species identified indeed belong to alpine herbs and plants, living in a habitat between 1000 and 2000 m of elevation. Most of them are enzymes, spanning a Mr range from 10 to 65 kDa. Thus, the genuineness of this product was amply proven. Things did not go so well, though, with another popular aperitif sold the world over, Cynar, stated to be made with an infusion of artichoke leaves reinforced too with an extract of 13 secret herbs. To err on the safe side, Saez *et al.* [63] adopted a three-pronged attack to tackle the issue. First, different extraction techniques were used for the characterization of the artichoke's proteome, secondly a home-made infusion was analyzed and finally the proteome of the commercial drink was checked. The artichoke proteome was evaluated via prior capture with CPLLs at four different pH (2.2, 4.0, 7.2 and 9.3) values. Via mass spectrometry analysis of the recovered fractions, after elution of the captured populations in 4% boiling SDS, these authors could identify a total of 876 unique gene products in the artichoke extracts, 18 in the home-made infusion and no proteins at all in the Italian Cynar liqueur, casting severe doubts on the procedure stated by the manufacturer (that should be made by an infusion of artichoke leaves plus 13 different herbs).

It would thus appear that CPLLs could be a formidable tool for investigating the genuineness and natural origin of all commercial drinks in order to protect consumers from adulterated products.

As we are Italians, we could not possibly terminate this review without investigating the proteome of another very popular digestive, namely Limoncello, stated to be an infusion of the outer lemon peel, the flavedo. We took this occasion to investigate too the proteomes of lemon peels and pulp, of a home-made alcoholic infusion and finally, to close the circle, of commercial liqueurs. The aim of this study was not only to perform the deepest investigation so far of the lemon peel proteome but also to assess the genuineness of the commercial liqueur via a three-pronged attack. First, different extraction techniques have been used for the characterization of the peel (and additionally of the pulp) proteome, secondly a home-made infusion was analyzed and finally the proteome of the commercial drink was checked. The peel (the flavedo, not the underlying layer called albedo) proteome was evaluated via prior capture with CPLLs at different pH values (2.2 and 7.2). Via mass spectrometry analysis of the recovered fractions, after elution of the captured populations in 4% boiling SDS, we could identify a total of 1011 unique gene products in the peel extracts and 674 in the pulp, 264 proteins in the home-made infusion and just eight proteins (and protein fragments), together with 12 peptides, in one Italian Limoncello produced in the Sorrento Region, thus proving the genuineness of this product [64]. On the contrary, cheaper Limoncellos were devoid of any protein/peptide, casting doubts on their production from vegetable extracts. Thus, it would appear that CPLLs could prove a powerful technique for investigating the genuineness and natural origin of commercial drinks in order to protect consumers from adulterated products.

8. Conclusions: Mammalian *versus* Plant Proteomics

When surveying the deeds of scientists working with mammalian proteomics, one can find that today they can explore to an incredible extent the proteome of any living cell line, as shown in Figure 5. When analyzing 11 human cell lines, Geiger *et al.* [65] could identify a total of 11,731 proteins and on average 10,361 ± 120 proteins in each cell line, an outstanding catch, indeed. Interestingly, a very large number of them represent a common set shared by all cell lines, amounting to 8522 unique species. Each individual cell then displays from 200 to 500 proteins specific of each line. The latter probably represent proteins that characterize each individual line and ensure its specific biological activity. Likely, they could also be low-abundance species. How can there be such a discrepancy with plant proteomics, when in this last domain we are lucky if we can find a few hundred species in a single run? It should be noted that, in a way, these mammalian cell lines grown in *in vitro* cultures are rather "easy" samples, in that they are not embedded in, e.g., fibrous tissues, muscles and other body compartments that would represent a complex matrix from which such cells would have to be extracted. On the contrary, in plant proteomics, most of the time the proteins to be identified are dispersed into a very complex matrix and often are present in low amounts as compared to the plant biopolymers (e.g., polysaccharides, polyphenols) and metabolites constituting the specimen mass. This is the reason why, most of the time, it has been difficult to detect more than a few hundred species in any plant tissue. Thus, the CPLL methodology here described, allowing access to an order of magnitude more so (up to 3000 unique gene products) represents an important advance in the field.

Clearly, however, more efforts should be dedicated to improving extraction technologies so as to match what today is obtained in mammalian proteomics.

Figure 5. Venn diagrams of the proteins identified in 10 different human cell lines. Each cell exhibited an average of 10,361 species, of which 8522 proteins represented a common set and the remaining a set specific for each cell line, as indicated (modified from Geiger *et al.* [65]).

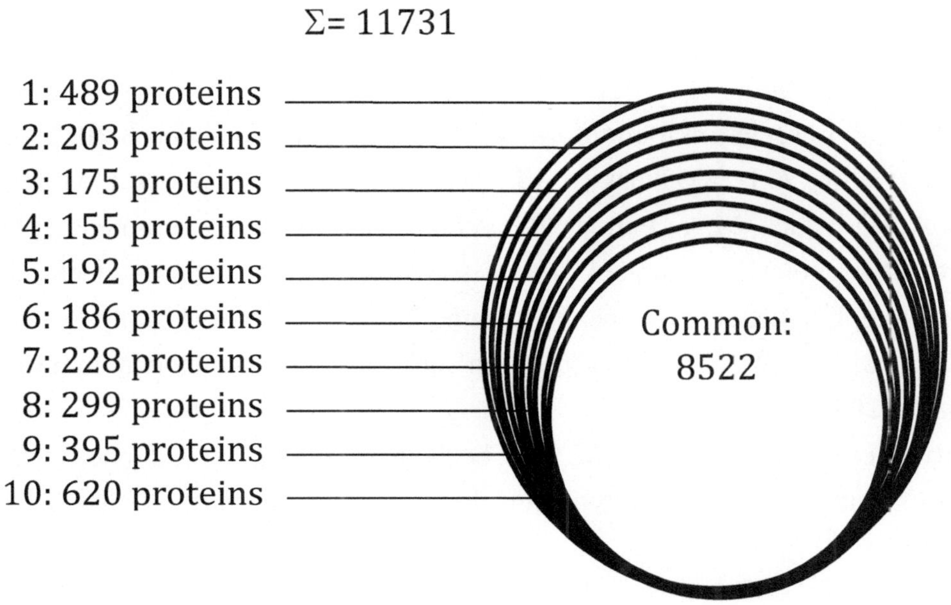

Conflicts of Interest

The authors declare no conflict of interest.

References

1. Anderson, N.L.; Anderson, N.G. The human plasma proteome: History, character, and diagnostic prospects. *Mol. Cell. Proteomics* **2002**, *1*, 845–867.
2. Boschetti, E.; Righetti, P.G. *Low-Abundance Protein Discovery: State of the Art and Protocols*; Elsevier: Amsterdam, The Netherlands, 2013.
3. Brown, L.R. *Full Planet, Empty Plates. The New Geopolitics of Food Scarcity*; Earth Policy Institute: Washington, DC, USA, 2012.
4. Wu, X.; Schuss, A.G. Mitigation of inflammation with foods. *J. Agric. Food Chem.* **2012**, *60*, 6703–6917.
5. Seiber, J.N.; Kleinschmidt, L. From detrimental to beneficial constituents of foods. *J. Agric. Food Chem.* **2012**, *60*, 6644–6647.
6. Tomàs-Barberàn, F.A.; Somoza, V.; Finley, J. Food bioactives research and the *Journal of Agricultural and Food Chemistry*. *J. Agric. Food Chem.* **2012**, *60*, 6641–6643.
7. Bagchi, D.; Lau, F.C.; Bagchi, M. *Genomics, Proteomics and Metabolomics in Nutraceutical and Functional Foods*; Wiley-Blackwell: Ames, IA, USA, 2010.

8. Ganesh, V.; Hettiarachchy, N.S. Nutriproteomics: A promising tool to link diet and diseases in nutritional research. *Biochim. Biophys. Acta* **2012**, *1824*, 1107–1117.
9. Kussmann, M.; Panchaud, A.; Affolter, M. Proteomics in nutrition: Status quo and outlook for biomarkers and bioactives. *J. Proteome Res.* **2010**, *9*, 4876–4887.
10. Gorduza, E.V.; Indrei, L.L.; Gorduza, V.M. Nutrigenomics in the post-genomic era. *Rev. Med. Chir. Soc. Med. Nat. Isai* **2008**, *112*, 152–164.
11. Del Chierico, F.; Vernocchi, P.; Bonizzi, L.; Carsetti, R.; Castellazzi, A.M.; Dallapiccola, B.; de Vos, W.; Guerzoni, M.E.; Manco, M.; Marseglia, G.L.; *et al.* Early-life gut microbiota under physiological and pathological conditions: The central role of combined meta-omics-based approaches. *J. Proteomics* **2012**, 75, 4580–4587.
12. Testori Coggi, P. EU rules on food information to consumers. *Nutrafoods* **2012**, *11*, 75–80.
13. Pieper, R.; Gatlin, C.L.; Makusky, A.J.; Russo, P.S.; Schatz, C.R.; Miller, S.S.; Su, Q.; McGrath, A.M.; Estock, M.A.; Parmar, P.P.; *et al.* The human serum proteome: Display of nearly 3700 chromatographically separated protein spots on two-dimensional electrophoresis gels and identification of 325 distinct proteins. *Proteomics* **2003**, *3*, 1345–1364.
14. Fang, X.; Zhang, W.W. Affinity separation and enrichment methods in proteomic analysis. *J. Proteomics* **2008**, *71*, 208–303.
15. Tu, C.; Rudnick, P.A.; Martinez, M.Y.; Cheek, K.L.; Stein, S.E.; Slebos, R.J.; Liebler, D.C. Depletion of abundant plasma proteins and limitations of plasma proteomics. *J. Proteome Res.* **2010**, *9*, 4982–4991.
16. Shen, Y.; Kim, J.; Strittmatter, E.F.; Jacobs, J.M.; Camp, D.G., 2nd; Fang, R.; Tolié, N.; Moore, R.J.; Smith, R.D. Characterization of the human blood plasma proteome. *Proteomics* **2005**, *5*, 4034–4045.
17. Widjaja, I.; Naumann, K.; Roth, U.; Wolf, N.; Mackey, D.; Dangl, J.L.; Scheel, D.; Lee, J. Combining subproteome enrichment and RuBisCO depletion enables identification of low abundance proteins differentially regulated during plant defense. *Proteomics* **2009**, *9*, 138–147.
18. Boschetti, E.; Lomas, L.; Citterio, A.; Righetti, P.G. Romancing the "hidden proteome", Anno Domini two zero zero six. *J. Chromatogr. A* **2007**, *1153*, 277–290.
19. Righetti, P.G.; Boschetti, E.; Lomas, L.; Citterio, A. Protein Equalizer technology: The quest for a "democratic proteome". *Proteomics* **2006**, *6*, 3980–3992.
20. Righetti, P.G.; Boschetti, E.; Kravchuk, A.V.; Fasoli, E. The proteome buccaneers: How to unearth your treasure chest via combinatorial peptide ligand libraries. *Expert Rev. Proteomics* **2010**, *7*, 373–385.
21. Agrawal, G.K.; Sarkar, A.; Righetti, P.G.; Pedreschi, R.; Carpentier, S.; Wang, T.; Barkla, B.J.; Kohli, A.; Ndimba, B.K.; Bykova, N.V.; *et al.* A decade of plant proteomics and mass spectrometry: Translation of technical advancements to food security and safety issues. *Mass Spectrom. Rev.* **2013**, *32*, 335–365.
22. Boggess, M.V.; Lippolis, J.D.; Hurkman, W.J.; Fagerquist, C.K.; Briggs, S.P.; Gomes, A.V.; Righetti, P.G.; Bala, K. The need for agriculture phenotyping: "Moving from genotype to phenotype". *J. Proteomics* **2013**, *93*, 20–39.

23. Jorrín-Novo, J.; Valledor Gonzalez, L. Special Issue: Translational Plant Proteomics. *J. Proteomics* **2013**, *93*, pp. 1–368. Available online: http://www.sciencedirect.com/science/journal/18743919/93 (accessed on 4 February 2014).
24. Nakamura, R.; Teshima, R. Proteomics-based allergen analysis in plants. *J. Proteomics* **2013**, *93*, 40–49.
25. Lubica Uvackova, L.; Skultety, L.; Bekesova, S.; McClain, S.; Hajduch, M. The MS^E-proteomic analysis of gliadins and glutenins in wheat grain identifies and quantifies proteins associated with celiac disease and baker's asthma. *J. Proteomics* **2013**, *93*, 65–73.
26. Agrawal, G.K.; Timperio, A.M.; Zolla, L.; Bansal, V.; Shukla, R.; Rakwal, R. Biomarker discovery and applications for foods and beverages: Proteomics to nanoproteomics. *J. Proteomics* **2013**, *93*, 74–92.
27. Ribeiro Demartini, D.; Pasquali, G.; Regina Carlini, C. An overview of proteomics approaches applied to biopharmaceuticals and cyclotides research. *J. Proteomics* **2013**, *93*, 224–233.
28. D'Ambrosio, C.; Arena, S.; Scaloni, A.; Guerrier, L.; Boschetti, E.; Mendieta, M.E.; Citterio, A.; Righetti, P.G. Exploring the chicken egg white proteome with combinatorial peptide ligand libraries. *J. Proteome Res.* **2008**, *7*, 3461–3474.
29. Farinazzo, A.; Restuccia, U.; Bachi, A.; Guerrier, L.; Fortis, F.; Boschetti, E.; Fasoli, E.; Citterio, A.; Righetti, P.G. Chicken egg yolk cytoplasmic proteome, mined via combinatorial peptide ligand libraries. *J. Chromatogr. A* **2009**, *1216*, 1241–1252.
30. Mann, K. The chicken egg white proteome. *Proteomics* **2007**, *7*, 3558–3568.
31. Mann, K.; Mann, M. The chicken egg yolk plasma and granule proteomes. *Proteomics* **2008**, *8*, 178–191.
32. D'Alessandro, A.; Righetti, P.G.; Fasoli, E.; Zolla, L. The egg white and yolk interactomes as gleaned from extensive proteomic data. *J. Proteomics* **2010**, *73*, 1028–1042.
33. D'Amato, A.; Bachi, A.; Fasoli, E.; Boschetti, E.; Peltre, G.; Sénéchal, H.; Righetti, P.G. In-depth exploration of cows whey proteome via combinatorial peptide ligand libraries. *J. Proteome Res.* **2009**, *8*, 3925–3936.
34. Cunsolo, V.; Muccilli, V.; Fasoli, E.; Saletti, R.; Righetti, P.G.; Foti, S. Poppea's bath liquor: The secret proteome of she-donkey's milk. *J. Proteomics* **2011**, *74*, 2083–2099.
35. Le, A.; Barton, L.D.; Sanders, J.T.; Zhang, Q. Exploration of bovine milk proteome in colostral and mature whey using an ion-exchange approach. *J. Proteome Res.* **2011**, *10*, 692–704.
36. Nissen, A.; Bendixen, E.; Ingvartsen, K.L.; Røntved, C.M. In-depth analysis of low abundant proteins in bovine colostrum using different fractionation techniques. *Proteomics* **2012**, *12*, 2866–2878.
37. Bislev, S.L.; Deutsch, E.W.; Sun, Z.; Farrah, T.; Aebersold, R.; Moritz, R.L.; Bendixen, E.; Codrea, M.C. A Bovine Peptide Atlas of milk and mammary gland proteomes. *Proteomics* **2012**, *12*, 2895–2899.
38. Di Girolamo, F.; D'Amato, A.; Righetti, P.G. Assessment of the floral origin of honey via proteomic tools. *J. Proteomics* **2012**, *75*, 3688–3693.
39. Won, S.R.; Lee, D.C.; Ko, S.H.; Kim, J.W.; Rhee, H.I. Honey major protein characterization and its application to adulteration detection. *Food Res. Int.* **2008**, *41*, 952–956.

40. Fasoli, E.; D'Amato, A.; Kravchuk, A.V.; Boschetti, E.; Bachi, A.; Righetti P.G. Popeye strikes again: The deep proteome of spinach leaves. *J. Proteomics* **2011**, *74*, 127–136.
41. Esteve, C.; D'Amato, A.; Marina, M.L.; García, M.C.; Citterio, A.; Righetti, P.G. Identification of olive (*Olea europaea*) seed and pulp proteins by *n*LC-MS/MS via combinatorial peptide ligand libraries. *J. Proteomics* **2012**, *75*, 2396–2403.
42. Esteve, C.; Cañas, B.; Moreno-Gordaliza, E.; Del Rio, C.; García, M.C.; Marina, M.L. Identification of olive (*Olea europaea*) pulp proteins by matrix-assisted laser desorption/ionization time-of-flight mass spectrometry and nano-liquid chromatography tandem mass spectrometry. *J. Agric. Food Chem.* **2011**, *59*, 12093–12101.
43. Esteve, C.; D'Amato, A.; Marina, M.L.; Concepción García, M.; Righetti, P.G. Identification of avocado (*Persea americana*) pulp proteins by nanoLC-MS/MS via combinational peptide ligand libraries. *Electrophoresis* **2012**, *33*, 2799–2805.
44. Sun, J.; Chu, Y.F.; Wu, X.; Liu, R.H. Antioxidant and antiproliferative activities of common fruits. *J. Agric. Food Chem.* **2002**, *50*, 7449–7454.
45. Saravanan, R.S.; Rose, J.K.C. A critical evaluation of sample extraction techniques for enhanced proteomic analysis of recalcitrant plant tissues. *Proteomics* **2004**, *4*, 2522–2532.
46. Esteve, C.; D'Amato, A.; Marina, M.L.; Concepción García, M.; Righetti, P.G. In-depth proteomic analysis of banana (*Musa* spp.) fruit with combinatorial peptide ligand libraries. *Electrophoresis* **2013**, *34*, 207–214.
47. Fasoli, E.; Righetti, P.G. The peel and pulp of mango fruit: A proteomic samba. *Biochim. Biophys. Acta* **2013**, *1834*, 2539–2545.
48. Fasoli, E.; D'Amato, A.; Kravchuk, A.V.; Citterio, A.; Righetti, P.G. In-depth proteomic analysis of non-alcoholic beverages with peptide ligand libraries. I: Almond milk and orgeat syrup. *J. Proteomics* **2011**, *74*, 1080–1090.
49. D'Amato, A.; Fasoli, E.; Righetti, P.G. Harry Belafonte and the secret proteome of coconut milk. *J. Proteomics* **2012**, *75*, 914–920.
50. D'Amato, A.; Fasoli, E.; Kravchuk, A.V.; Righetti, P.G. going nuts for nuts? The trace proteome of a cola drink, as detected via combinatorial peptide ligand libraries. *J. Proteome Res.* **2011**, *10*, 2684–2686.
51. Fasoli, E.; D'Amato, A.; Citterio, A.; Righetti, P.G. Ginger rogers? No, Ginger Ale and its invisible proteome. *J. Proteomics* **2012**, *75*, 1960–1965.
52. Di Girolamo, F.; D'Amato, A.; Righetti P.G. *Horam nonam exclamavit: Sitio*. The trace proteome of your daily vinegar. *J. Proteomics* **2011**, *75*, 718–724.
53. Cereda, A.; Kravchuk, A.V.; D'Amato, A.; Bachi, A.; Righetti, P.G. Proteomics of wine additives: Mining for the invisible via combinatorial peptide ligand libraries. *J. Proteomics* **2010**, *73*, 1732–1739.
54. Monaci, L.; Losito, I.; Palmisano, F.; Visconti, A. Identification of allergenic milk proteins markers in fined white wines by capillary liquid chromatography-electrospray ionization-tandem mass spectrometry. *J. Chromatogr. A* **2010**, *1217*, 4300–4305.

55. Monaci, L.; Losito, I.; Palmisano, F.; Godula, M.; Visconti, A. Towards the quantification of residual milk allergens in caseinate-fined white wines using HPLC coupled with single-stage Orbitrap mass spectrometry. *Food Addit. Contam. Part A Chem. Anal. Control Expo. Risk Assess.* **2011**, *28*, 1304–1314.
56. Palmisano, G.; Antonacci, D.; Larsen, M.R. Glycoproteomic profile in wine: A "sweet" molecular renaissance *J. Proteome Res.* **2010**, *9*, 6148–6159.
57. D'Amato, A.; Fasoli, E.; Kravchuk, A.V.; Righetti, P.G. *Mehercules, adhuc bacchus*! The debate on wine proteomics continues. *J. Proteome Res.* **2011**, *10*, 3789–3801.
58. Cilindre, C.; Fasoli, E.; D'Amato, A.; Liger-Belair, G.; Righetti, P.G. It's time to pop a cork on Champagne's proteome! *J. Proteomics* 2014, submitted for publication.
59. D'Amato, A.; Kravchuk, A.V.; Bachi, A.; Righetti, P.G. Noah's nectar: The proteome content of a glass of red wine. *J. Proteomics* **2010**, *73*, 2370–2377.
60. Tolin, S.; Pasini, G.; Curioni, A.; Arrigoni, G.; Masi, A.; Mainente, F.; Simonato, B. Mass spectrometry detection of egg proteins in red wines treated with egg white. *Food Control* **2012**, 23, 87–94.
61. Fasoli, E.; Aldini, G.; Regazzoni, L.; Kravchuk, A.V.; Citterio, A.; Righetti, P.G. Les Maîtres de l'Orge: The proteome content of your beer mug. *J. Proteome Res.* **2010**, *9*, 5262–5269.
62. Fasoli, E.; D'Amato, A.; Citterio, A.; Righetti, P.G. Anyone for an aperitif? Yes, but only a Braulio DOC with its certified proteome. *J. Proteomics* **2012**, *75*, 3374–3379.
63. Saez, V.; Fasoli, E.; D'Amato, A.; Simó-Alfonso, E.; Righetti, P.G. Artichoke and Cynar liqueur: Two (not quite) entangled proteomes. *Biochim. Biophys. Acta* **2013**, *1834*, 119–126.
64. Fasoli, E.; Colzani, M.; Aldini, G.; Citterio, A.; Righetti, P.G. Lemon peel and Limoncello liqueur: A proteomic duet. *Biochim. Biophys. Acta* **2013**, *1834*, 1484–1491.
65. Geiger, T.; Wehner, A.; Schaab, C.; Cox, J.; Mann, M. Comparative proteomic analysis of eleven common cell lines reveals ubiquitous but varying expression of most proteins. *Mol. Cell. Proteomics* **2012**, *11*, M111.014050.

2

The Effect of Oral Intake of Low-Temperature-Processed Whey Protein Concentrate on Colitis and Gene Expression Profiles in Mice

Sharmila Jayatilake [1,2], Katsuhito Arai [1,2], Nanami Kumada [2], Yoshiko Ishida [2], Ichiro Tanaka [2], Satoru Iwatsuki [3,†], Takuji Ohwada [1,2], Masao Ohnishi [1,2], Yoshihiko Tokuji [1,2,*] and Mikio Kinoshita [1,2]

[1] United Graduate School of Agricultural Sciences, Iwate University, Ueda 3-18-8, Morioka, Iwate 020-8550, Japan; E-Mails: sharmilaeas@gmail.com (S.J.); s17046@st.obihiro.ac.jp (K.A.); taku@obihiro.ac.jp (T.O.); mohnishi@fujijoshi.ac.jp (M.O.); kinosita@obihiro.ac.jp (M.K.)

[2] Department of Food Science, Obihiro University of Agriculture and Veterinary Medicine, Inada-cho W2-11, Obihiro, Hokkaido 080-8555 Japan; E-Mails: gekogeko773@yahoo.co.jp (N.K.); y-ishida@gh-assoc.ne.jp (Y.I.); tanaka-ichiro@outlook.jp (I.T.)

[3] Asama Chemical Co. LTD., Nihonbashi-kodenma-cho 20-3, Chuo-ku, Tokyo 103-0001, Japan

† Deceased on 17 November 2012.

* Author to whom correspondence should be addressed; E-Mail: tokuji@obihiro.ac.jp

---

Abstract: Inflammatory bowel disease (IBD) is an autoimmune disease of unknown etiology and can lead to inflammation and cancer. Whey proteins contain many bioactive peptides with potential health benefits against IBD. We investigated the effect of low-temperature-processed whey protein concentrate (LWPC) on the suppression of IBD by using a dextran sodium sulfate (DSS)-induced colitis model in BALB/c mice. Oral intake of LWPC resulted in improved recovery of body weight in mice. Histological analysis showed that the epithelium cells of LWPC-treated mice were healthier and that lymphocyte infiltration was reduced. The increase in mucin due to the LWPC also reflected reduced inflammation in the colon. Transcriptome analysis of the colon by DNA microarrays revealed marked downregulation of genes related to immune responses in LWPC-fed mice. In particular, the expression of interferon gamma receptor 2 (Ifngr2) and guanylate-binding proteins (GBPs) was increased by DSS treatment and decreased in

LWPC-fed mice. These findings suggest that LWPCs suppress DSS-induced inflammation in the colon by suppressing the signaling of these cytokines. Our findings suggest that LWPCs would be an effective food resource for suppressing IBD symptoms.

Keywords: colitis; anti-inflammation; whey protein; DNA microarray; immunomodulation

1. Introduction

Inflammatory bowel disease (IBD), including Crohn's disease (CD) and ulcerative colitis (UC), is a chronic autoimmune disease condition, the pathogenesis of which is unknown [1]. IBD is considered to occur due to a combination of genetic, environmental, microbial and immunological factors characterized by the pronounced destruction of the gastrointestinal mucosa and the epithelial cell barrier. Subsequent infiltration of luminal antigens activate macrophages and T lymphocytes, which stimulate the production of proinflammatory cytokines, such as interleukin 1 (IL-1), interferons and tumor necrosis factor α (TNF-α), giving rise to a cascade of inflammatory reactions [2]. Manifestations of IBD include diarrhea, cramping, abdominal pain and fever [3]. Prolonged exposure to IBD is known to lead to colorectal cancer, which is one of the most common cancers, accounting for about 100,000 hospitalizations per year in the U.S. alone, and the incidence of IBD has continued to increase in recent decades in both developed and developing countries [4]. Current treatments for IBD typically include standard drug treatments, such as anti-inflammatory drugs, immune modifiers and antibiotics, and are frequently associated with severe side effects [5], which indicate the need for new and safe medicaments to contribute to the prevention and control of this disease.

Recent research reveals that food-derived peptides from various sources, including seafood, milk and plants, contain bioactive functional properties, especially in the suppression of colitis. Young *et al.* [6] showed that soy-derived peptides have the ability to suppress dextran sodium sulfate (DSS)-induced colitis through reducing the anti-inflammatory mediators. A novel purified peptide from pacific oyster (*Crassostrea gigas*) has been shown to reduce inflammation through immunomodulation [7]. Milk is known to be a rich source of nutrients coupled with bioactive properties that supports the growth and development of infants. Whey protein, a by-product of the cheese-making process, represents approximately 20% of the total protein content in bovine milk. The major protein components in whey are β-lactoglobulin, α-lactalbumin, immunoglobins, serum albumin, lactoferrin and lactoperoxidase [8]. Bioactive peptides are released from native protein during the course of gastric digestion. Most of such whey peptides have shown interactions directly with the immune cells, thereby modulating the immune system [9], or indirectly acting upon stimulating beneficial bacteria in the gut, cardiovascular health [10], improved muscle strength, inhibition of osteoporosis, anti-inflammatory, antibacterial and anticancer effects [11,12]. Both β-lactoglobulin and α-lactalbumin have been shown to inhibit colon carcinoma [13–15]. Whey peptides, such as β-casomorphins and α-lactorphins, derived from milk enhance the secretion of mucin from goblet cells via upregulation of the mucin gene [16], showing protective effects through maintaining gut homeostasis and suppressing intestinal inflammation. Recent work reveals a milk-derived glycomacropeptide, which shows anti-inflammatory functions in colitis mouse via modulation of macrophages and Th1 and Th17 lymphocytes. This modulatory effect was observed to arrest the signal transducers and activators of transcription (STAT) pathway and to suppress interferon gamma (IFNγ) expression [17]. These studies indicate the importance of whey proteins as an alternative medicament for the suppression of IBD. However, the mechanism of action or exact therapeutic use of whey proteins against IBD has not been clearly evaluated to date.

The functional properties of cheese whey protein are known to differ from the properties of raw milk based on the sterilization method applied during processing. The milk pasteurization process is known to cause a low degree of protein denaturation; thus, whey generated by pasteurized milk processing has been reported to have more bioactive functions when compared to whey processed at higher temperatures. Li *et al.* [18] showed that a minimally processed whey protein concentrate (heat treated, <40 °C) had positive effects on intestinal structure, function and integrity compared with whey proteins treated at higher temperatures. Elfstrand *et al.* [19] found the recovery of immunoglobins to be increased by low-temperature-treated colostrum compared to the findings for colostrum prepared by standard higher temperature treatments.

Therefore, the present study was conducted to evaluate the effects of low-temperature-treated whey proteins in relation to their suppression of colon inflammation in the dextran sulfate sodium (DSS) mouse model of experimental colitis. The possible mechanisms by which whey protein may exert its action were studied via DNA microarrays followed by a comparison of the gene expression levels.

2. Experimental Section

2.1. Preparation of Whey Protein

The low-temperature-processed whey concentrate (LWPC) powder was a commercial product kindly gifted by Asama Chemical Co. Ltd. (Tokyo, Japan) The LWPC was dissolved in distilled water and heated at 70 °C for 2 h, and this solution was concentrated by freeze-drying. The resulting powder is named high-temperature-processed whey protein concentrate (HWPC). The protein profiles of HWPC and LWPC were analyzed by SDS-polyacrylamide gel electrophoresis [20] with 5%–20% gradient gels, followed by Coomassie brilliant blue staining.

2.2. Animals and Diets

The two treatment diets were prepared based on the AIN-76 diet (the composition is given in the Supplementary Materials) [21,22], where 50% of the casein in the AIN-76 diet was replaced separately with each of the above-described processed whey protein concentrates, HWPC or LWPC. Normal AIN-76 was used as a control diet.

Female BALB/c mice (4 weeks old) were obtained from CREA Japan Inc. (Tokyo, Japan) and housed in isolated cages at 20 °C under a 12 h light/dark cycle. After 10 days of acclimatization with the AIN-76 diet and water provided *ad libitum*, the mice were randomly divided into 4 experimental groups (5 mice per group). Colitis was induced in Groups 1 to 3 through the administration of 2.5% dextran sulfate sodium (DSS) in the drinking water [23], and each group was separately fed the AIN-76 basal diet, AIN-76 + HWPC or AIN-76 + LWPC in parallel with the induction of colitis. Group 4 was untreated. After a period of 14 days, DSS administration was arrested, and the experiment was continued until Day 21. On Day 21, the mice were sacrificed under ether anesthesia, and the portion of the large intestine from the cecum to the vent was removed and rinsed in cold saline. All protocols involving animals were approved by the Institutional Animal Care and Use Committee (20-101) and conducted according to the guidelines of Obihiro University of Agriculture and Veterinary Medicine.

2.3. Histopathological Experiments

Five colon tissue samples were obtained from each treatment group representing all 5 mice. Each tissue sample was taken from a different site so as to represent the whole column within a treatment group. The colon tissues were fixed overnight in 4% paraformaldehyde in phosphate-buffered saline, embedded in paraffin wax and sectioned (10 μm) using a microtome, followed by staining with hematoxylin and eosin (HE). Twenty-five slide spots from each treatment group were used for microscopical examination with 5 spots representing each tissue section. The slides were viewed under a light microscope. The slides were further evaluated with regard to histological damage to the colon using a semi-quantitative scoring system (Table 1) with minor alterations to the method used by McCafferty *et al.* [24], with a maximum possible total damage score of 18 when summed. Three semi-trained panelists performed the scoring under the guidance of a trained pathologist. Scores obtained for different treatments were analyzed using a non-parametric one-way ANOVA in STATA version 11.1 followed by Tukey HSD pairwise comparisons.

Table 1. Parameters and scores used to assess damage in the colon tissues of mice.

Parameters	**Scores**			
	0	**1**	**2**	**3**
Extent of damage to colon structure	normal	mild	moderate	extensive
Crypt atrophy	not present	mild	moderate	extensive
Degree of cellular infiltration	normal	mild	moderate	extensive
Extent of muscle thickening	normal	mild	moderate	extensive
Extent of crypt abscess	not present	mild	moderate	extensive
Goblet cell depletion	normal	mild	moderate	extensive

The final score allocated is the average value from 25 slide spots from each different treatment.

2.4. Western Blotting

The mouse colon was homogenized, and proteins were extracted with general SDS-PAGE sample buffer. The protein concentration was assessed using the Bio-Rad protein assay method. Equal amounts of each extract were then subjected to protein separation on SDS-PAGE (15% gel) and then electro-transferred onto a nitrocellulose membrane. The membrane was blocked with Tris-buffered saline and 6% milk protein and then incubated overnight at 4 °C with primary rabbit antibodies for mucin 2 (Cosmobio, Tokyo, Japan), followed by incubation with a secondary antibody for rabbit IgG labeled with horse radish peroxidase. Detection was performed using the ECL system (GE Healthcare, Little Chalfont, UK).

2.5. DNA Microarray and RT-QPCR

Colon tissues were ground in liquid nitrogen, followed by RNA extraction using the SV Total RNA isolation system (Promega, Madison, WI, USA) according to the manufacturer's protocol. The extracted RNA was checked by spectrophotometer analysis and electrophoresis and then stored at −80 °C for further use.

RNA samples from each treatment replicate were pooled, and 500 ng of RNA from each treatment condition were used for transcriptome analysis of the mouse colon using a GeneChip 430A 2.0 mouse array (Affymetrix, Santa Clara, CA, USA). Experiments were performed according to the manufacturer's technical manual (Affymetrix). Data obtained from scanning the arrays were normalized (robust multi-array analysis; RMA) and analyzed using Arraystar software (DNAStar, Madison, WI, USA). Genes that were up- or down-regulated by ≥2-fold in comparison with the DSS treatment were categorized using the Database for Annotation, Visualization and Integrated Database (DAVID) version 6.7 [25,26] and the annotation data of "biological processes" in Gene Ontology [27].

To validate the changes in the expression of the selected genes, we performed quantitative real time RT-PCR (QPCR). We selected guanylate binding protein 1 (Gbp1:NM_010259), Gbp2 (NM_010260) and interferon gamma receptor 2 (IFNgr2: NM_008338) for QPCR analysis. cDNA was synthesized using Superscript II reverse transcriptase (Life Technologies, Rockville, MD, USA) and oligo-dT primers. Primers for the genes were selected from PrimerBank [28,29] as described in Table 2. QPCR was performed with the specific primers and SYBR premix Ex Taq II (Takara, Kyoto, Japan) on an ABI 7300 Real-time PCR system (Applied Biosystems, Foster City, CA, USA), and the results were normalized using expression levels of the glyceraldehyde-3 phosphate dehydrogenase gene (GAPGH: NM_008084). All data are presented as the mean ± standard deviation (SD). Statistical analysis was performed by using the Student's *t*-test.

Table 2. Primer sequences used for QPCR analysis.

Gene symbol	GenBank accession	PrimerBank ID	Primer sequence (5′–3′) Forward primer/reverse primer	Amplicon size
Gbp1	NM_010259	6753948a1	ACAACTCAGCTAACTTTGTGGG TGATACACAGGCGAGGCATATTA	183
Gbp2	NM_010260	6753950a1	CTGCACTATGTGACGGAGCTA GAGTCCACACAAAGGTTGGAAA	115
IFNgr2	NM_008338	6680373a1	TCCTCGCCAGACTCGTTTTC GTCTTGGGTCATTGCTGGAAG	115
GAPDH	NM_008084	6679937a1	AGGTCGGTGTGAACGGATTTG GGGGTCGTTGATGGCAACA	123

3. Results and Discussion

3.1. Heat Treatment Affects the Characteristics of Whey Proteins

The protein profiles of LWPC and HWPC were compared by SDS-PAGE (Figure 1). LWPC showed clear peptide bands of high intensity at approximately 20, 29 and within 60–80 kD. These molecular weights represent β-lactoglobulin (20 kD), light-chain immunoglobins (29 kD), and bovine serum albumin (BSA) (66 kD) and lactoferrin (77 kD), as confirmed by Jimenez *et al.* [30] and Morin *et al.* [31]. In the HWPC lanes, proteins within the range of 60–80 kD were the most prominent, but presented a lower intensity than that in the LWPC lanes. The clearest band coincided with BSA. Both lactoferrin and light-chain immunoglobins have degraded in the HWPC lane. These proteins contain immunomodulatory, anti-inflammatory and antibacterial functions. Further, they show rapid degradation at temperatures above 65 °C. Similar results were shown in [30]. It is clear that whey protein degradation increased under the higher-temperature treatment. Jovanovic *et al.* [32] obtained the same finding, where a gradual decrease in all whey proteins was observed with treatments at increasing temperatures. Lin *et al.* [33] also obtained similar results for protein degradation with increasing temperature; however, BSA was stable at temperatures lower than 80 °C, which may explain the presence of BSA in HWPC treated at 70 °C in the present study. β-Lactoglobulin, BSA, lactoferrins and immunoglobins are known to have bioactive properties important for treating many diseases [34].

Figure 1. SDS-PAGE profiles of whey protein concentrates. Lanes 1 and 5: marker; lanes 2–4: low-temperature-processed whey protein concentrate (LWPC); lanes 6–8: high-temperature-processed whey protein concentrate (HWPC).

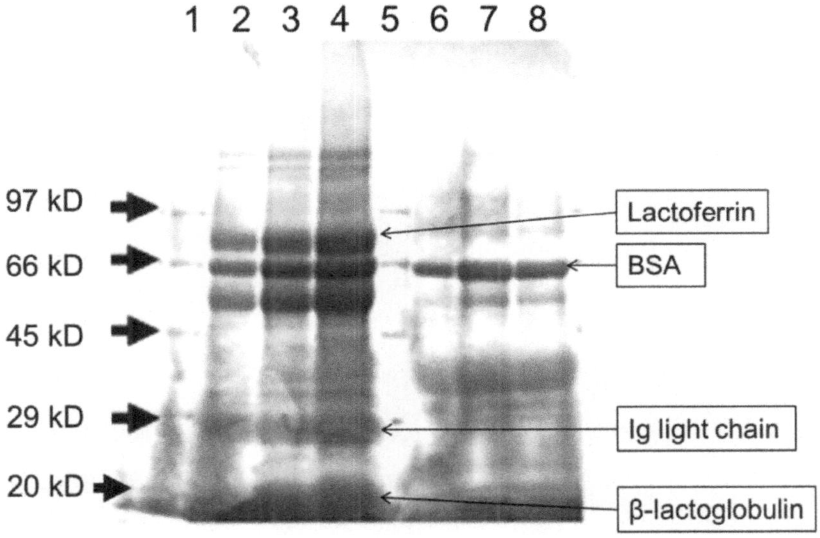

3.2. Oral Intake of LWPC Protects against DSS-Induced Colitis in Mice and Enhances Recovery from Colitis

DSS-induced intestinal injury serves as an experimental model for human ulcerative colitis and is a well-established method for the chemical induction of intestinal injury [35]. The change in body weight in the mice during the experimental period is shown in Figure 2D. The body weight of mice without DSS treatment was constant throughout the experiment period. DSS induction started to show effects by Day 8 after treatment in DSS-treated mice; these mice continued to lose weight even after DSS administration was stopped (Days 14–16). Reduction of body weight is a common effect seen in DSS colitis and is caused by reduced food intake and impaired intestinal nutrient absorption [36]. The body weight of DSS + LWPC- and DSS + HWPC-treated mice showed a reduction until Day 14. DSS + LWPC-treated mice lost less weight than DSS- and DSS + HWPC-treated mice. After the DSS treatment was stopped, the body weight of the mice steadily increased. This recovery of weight was most prominent in LWPC-treated mice. These findings indicate that LWPC treatment resulted in amelioration of DSS-induced colitis and facilitated rapid recovery from colitis after the administration of DSS was stopped. The role of LWPC in suppressing DSS-induced colitis has not been clearly determined; however, Sprong *et al.* [37] found that DSS-induced colitis in mice was ameliorated by feeding the mice cheese whey proteins, which stimulated gastric mucin expression and intestinal microflora modulation.

Figure 2. The effect of the oral intake of whey protein concentrates on body weight of dextran sodium sulfate (DSS)-treated mice. Opened circles, normal mice without DSS treatment; filled circles, DSS-treated mice; filled triangles, mice treated with DSS + HWPC feed; filled squares, mice treated with DSS + LWPC feed.

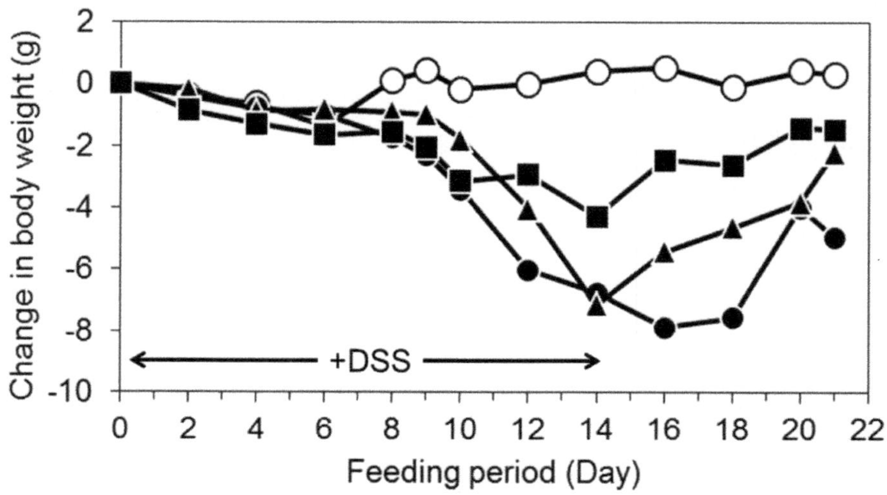

3.3. LWPC Protects the Intestinal Epithelium against DSS-Induced Inflammation

Under inflammatory conditions, excess infiltration of neutrophils can be observed clearly. Microscopic images of colon sections from mice subjected to different treatments are shown in Figure 3. The normal mucosal structure of the colon (Figure 3A) demonstrated the clear separation of crypts and the lamina propria, and infiltration of leucocytes was not observed. Epithelial cells and goblet cells were prominent within the lining of each crypt. Histopathological features of inflammation include abnormal crypt formation, goblet cell depletion and marked intestinal wall thickening, with leukocyte infiltration [24]. DSS treatment induced severe inflammation of the colon with complete loss of the mucosal structure, severe infiltration of the lamina propria and thickening of the intestinal walls (Figure 3B). The colon of DSS + HWPC-treated mice showed a slight improvement in the structure of the mucosa; however, only a few goblet cells remained in crypts with an abnormal size and shape, and severe leukocyte infiltration was present (Figure 3C). DSS + LWPC treatment (Figure 3D) produced a distinct reduction in the severity of inflammation compared to that observed for DSS + HWPC and also improved the mucosal structure. The crypt structure and shape were well maintained, although the number of goblet cells decreased, and leukocyte infiltration was moderate. To evaluate the extent of damage in the colon sections, we scored the sections based on visual criteria (Table 1). The score of the colon tissue in DSS-treated mice was 15.0, whereas the score in colon tissue from mice without

DSS treatment was 1.1 (Figure 3E). DSS + LWPC-treated mice (9.4) showed a significant decrease in histological damage to the colon compared to that in DSS-treated mice. However, a significant reduction of damage scores was not observed in the DSS + HWPC group (13.3). These results suggest that oral intake of LWPC is effective for protection from DSS-induced colitis or for regeneration of damaged colon tissue. Moreover, pasteurization at a higher temperature may cause whey protein to lose these favorable properties.

Figure 3. Histological effects of whey protein intake on DSS-induced colitis. (**A–D**) Transverse colon sections of mice on Day 21, stained with HE. (**A**) Normal mice without DSS treatment; (**B**) DSS-treated mice; (**C**) DSS mice treated with DSS + HWPC feed; (**D**) mice treated with DSS + LWPC feed (lp, lamina propria; gc, goblet cells; i, infiltration of leukocytes; ca, crypt atrophy; cab, crypt abscess; bars = 10 μm); (**E**) histological damage scores for the mouse colon on Day 21, evaluated according to Table 1. The final scores allocated are the average values of 25 slide spots of each different treatment group. * Statistically significant difference from the DSS group ($p < 0.05$). (**F**) Expression analysis of myeloperoxidase (MPO) and mucin in colon by western blot. Lane 1: normal mice without DSS treatment; lane 2: DSS-treated mice; lane 3: mice treated with DSS + HWPC feed; lane 4: mice treated with DSS + LWPC feed.

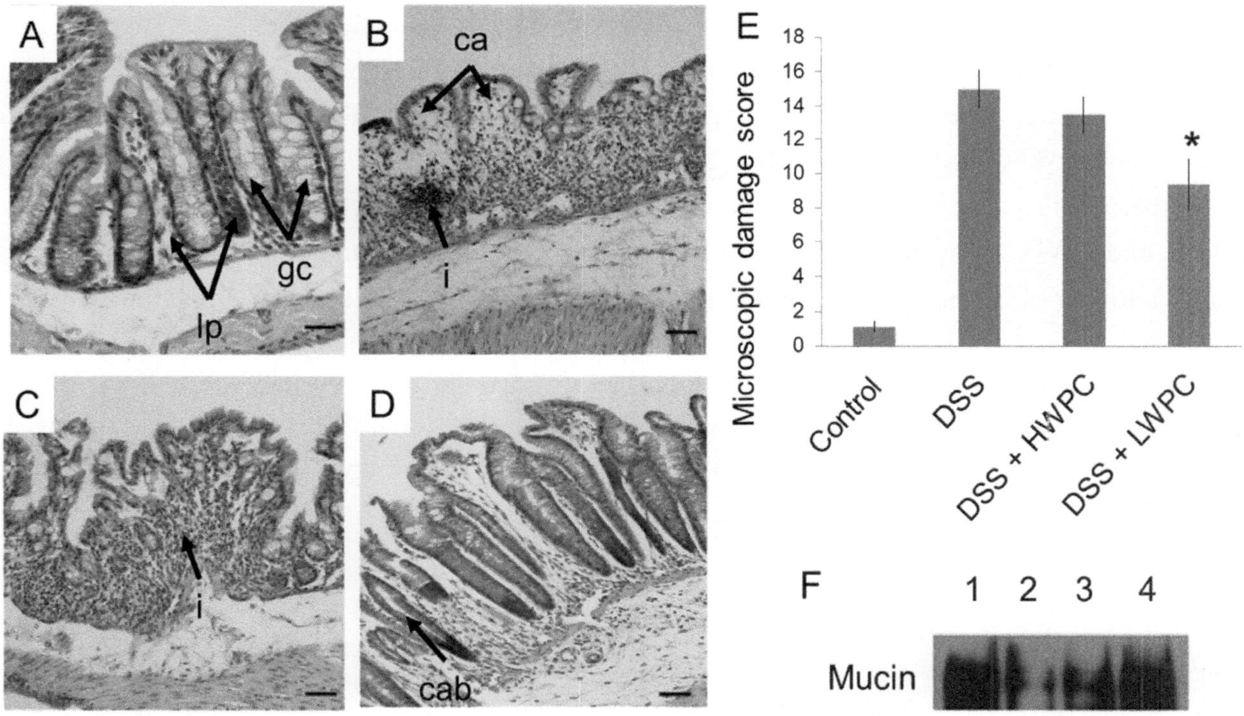

In the colon, mucins are produced by goblet cells for lubrication [38]. The availability of mucins in the colon epithelium is an indicator showing that goblet cells are healthy and functioning well. Mucin was highly detected in mice colons without DSS treatment, and mucin levels were reduced by DSS treatment, indicating that goblet cells were destroyed by DSS-induced colitis in the colon (Figure 3F). Oral intake of both HWPC and LWPC increased the mucin level in the presence of DSS. Compared to the findings for the HWPC group, the expression level of myeloperoxidase (MPO) for the LWPC treatment was lower and was similar to that in the mice without DSS treatment (data not shown).

The reduced MPO levels in the LWPC-treated mouse colons compared to the levels in the colons of control and HWPC-treated mice show that LWPC treatment reduces neutrophil recruitment into the colon, further suggesting that inflammation was suppressed. Similarly, the increase of mucin levels in the gastric epithelium indicates the recovery of the DSS-induced colon by regenerated epithelial cells and goblet cells. Compared with LWPC, HWPC had lower effects on reducing MPO activity and increasing mucin content in the colon. Sprong *et al.* [37] reported that cheese whey protein feed increased the mRNA levels of mucin 2 (MUC2) in the colon of rats induced with DSS inflammation, possibly because whey protein contains higher amounts of amino acids required for mucin synthesis. The quantity and composition of amino acids are almost identical between LWPC and HWPC. However, in our study, the protein profile was different according to the results of SDS-PAGE (Figure 1). LWPC is shown to retain more proteins than the HWPC fractions. These proteins possibly contain more bioactive peptides in a higher solubility form, which is more accessible in digestion. Bounous *et al.* [39] and Hongsprabhas *et al.* [40] also found similar results where mild heat treatment on whey proteins preserved the biological activity and showed an increased function of antioxidant and immune response, both *in vivo* and *in vitro*.

3.4. The Effect of LWPC on Gene Expression under DSS-Induced Inflammation in the Colon

To investigate the mechanism underlying the effect of LWPC, we evaluated the transcriptome of the mouse colon by DNA microarray analysis. Compared with the control group, 677 out of 22,626 probed genes were up- or down-regulated by over two-fold in the DSS treatment group (Figure 4, Table 3). Compared with the DSS group, the number of genes differentially expressed (≥2-fold change) in the LWPC group was higher (351) than that in the HWPC group (144). This finding suggests that LWPC has more bioactive peptides than HWPC. In total, 110 genes were regulated by both DSS and LWPC, whereas 51 genes were regulated by both DSS and HWPC. Therefore, LWPC may have a higher effect on the expression of genes involved in DSS-induced inflammation.

Figure 4. Venn diagram of differentially-regulated genes. Numbers show genes up- or down-regulated (≥2-fold), compared between the control (no treatment) and DSS, between DSS and DSS + LWPC and between DSS and DSS + HWPC groups.

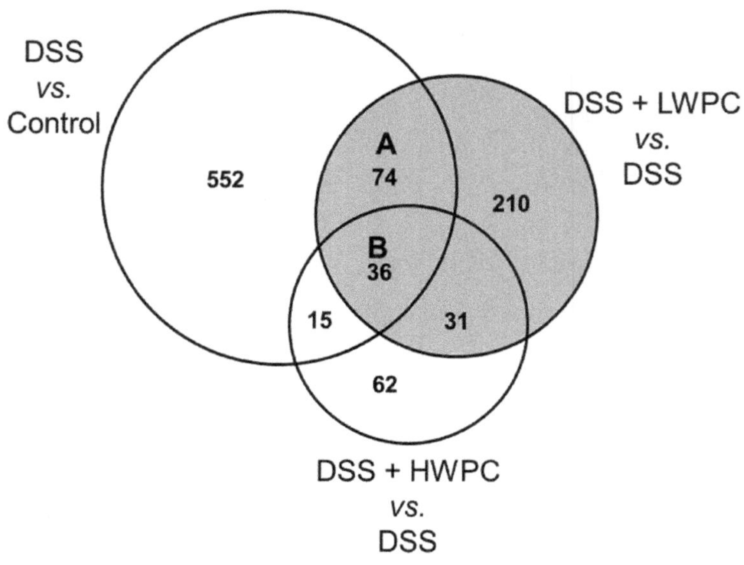

Table 3. The number of genes differentially regulated by DSS and whey protein concentrates.

Treatment group	Number of genes differentially regulated (fold change *vs.* control >2)		
	Upregulated	Downregulated	Total
DSS	348	328	677
DSS + LWPC	213	138	351
DSS + HWPC	113	31	144

The 110 genes regulated by both DSS and LWPC were further classified into functional groups using the DAVID functional annotation program according to the GO biological process in the GO ontology, as shown in Table 4. Among the genes, 12 genes were involved in "immune response" and sox genes were involved in "inflammatory response", indicating possible effects on inflammation. The expression levels of the 12 immune responsive genes (Table 5) are shown in Figure 5. All of the genes were upregulated by DSS treatment and downregulated by LWPC treatment, except for Cfd (complement factor D) and Il1rn (interleukin 1 receptor antagonist). The genes encoding guanylate-binding proteins (GBP), such as Gbp1, Gbp2 and Gbp6, were highly expressed in the colon of DSS-treated mice, but distinctly downregulated by LWPC ingestion, whereas the expression levels did not decrease upon oral administration of HWPC. Our results revealed relative expression increases of 3.5-, 2.5- and 1.7-fold in LWPC-treated mice and 8.2-, 5.4- and 3.3-fold in DSS-treated mice for Gbp1, Gbp2 and Gbp6, respectively, relative to the blank treatment (Figure 6), which clearly indicates the reduced expression of Gbps in LWPC. Furthermore, Cxcl9 (chemokine (c-x-c motif) ligand 9), Ccl8 (chemokine (c-cl motif) ligand 8) and Itgb6 (integrin beta 6) showed a pattern similar to that of Gbps. The remaining genes were downregulated in both the HWPC and LWPC groups. However, the level of decrease of the expression of these genes was lower in the HWPC group than in the LWPC group.

Table 4. Functional categories of genes differentially regulated by both DSS and LWPC

GO biological process	Number of genes (fold change > 2)	% *	*p*-Value **
GO:0006955 Immune response	12	12.2	3.1×10^{-4}
GO:0006869 Lipid transport	5	5.1	4.6×10^{-3}
GO:0010876 Lipid localization	5	5.1	6.0×10^{-3}
GO:0015918 Sterol transport	3	3.1	7.4×10^{-3}
GO:0030301 Cholesterol transport	3	3.1	7.4×10^{-3}
GO:0006954 Inflammatory response	6	6.1	9.0×10^{-3}
GO:0048593 Camera-type eye morphogenesis	3	3.1	3.0×10^{-2}
GO:0042159 Lipoprotein catabolic process	2	2.0	3.4×10^{-2}
GO:0034754 Cellular hormone metabolic process	3	3.1	3.5×10^{-2}
GO:0010817 Regulation of hormone levels	4	4.1	3.6×10^{-2}

* Differentially regulated genes/total genes in category; ** modified Fisher exact *p*-value (Expression Analysis Systematic Explorer score) adopted in the DAVID annotation chart.

Table 5. The genes in "immune response" (GO:0006955) whose expression was regulated by both DSS and LWPC.

GenBank Accession No.	Gene symbol	Gene description	Regulation by DSS and LWPC	
NM_013459	Cfd	Complement factor D (adipsin)	↓	↑
NM_001159564	Itgb6	Integrin beta 6	↑	↓
NM_001276445	Tlr1	Toll like receptor 1	↑	↓
NM_012012	Exo1	Exonuclease 1	↑	↓
NM_133829	Mfsd6	Major facilitator S domain containing 6	↑	↓
NM_008599	Cxcl9	Chemokine (c-x-c motif) ligand 9	↑	↓
NM_021443	Ccl8	Chemokine (c-cl motif) ligand 8	↑	↓
NM_011315	Saa3	Serum amyloid A3	↑	↓
NM_010259	Gbp1	Guanylate binding protein 1	↑	↓
NM_010260	Gbp2	Guanylate binding protein 2	↑	↓
NM_145545	Gbp6	Guanylate binding protein 6	↑	↓
NM_001039701	Il1rn	Interleukin 1 receptor antagonist	↑	↑

Figure 5. Expression of genes involved in the immune response that were differentially regulated by whey protein concentrate. The graph shows the relative expression level (fold change) determined by microarray analysis (white: normal mice without DSS treatment; black: DSS-treated mice; grey: mice treated with DSS + HWPC feed; hatched: mice treated with DSS + LWPC feed). The proper names of the genes are shown in Table 5.

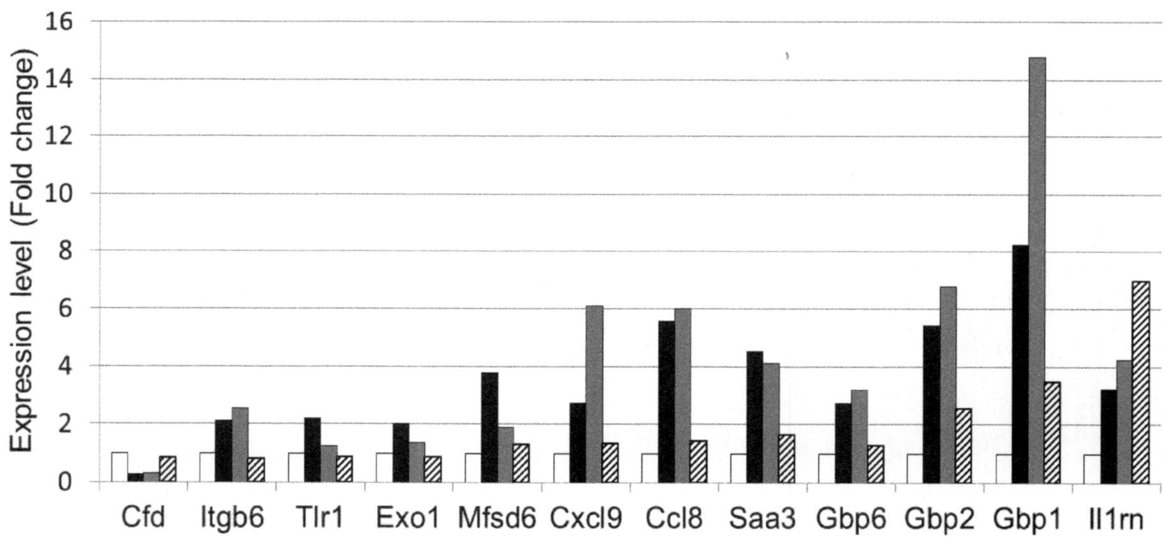

Figure 6. Gene expression analysis by real-time QPCR. The relative expression levels of Gbp1, Gbp2 and IFNgr2 are shown as ratios compared with GAPDH. * Statistically significant difference from the DSS group ($p < 0.05$).

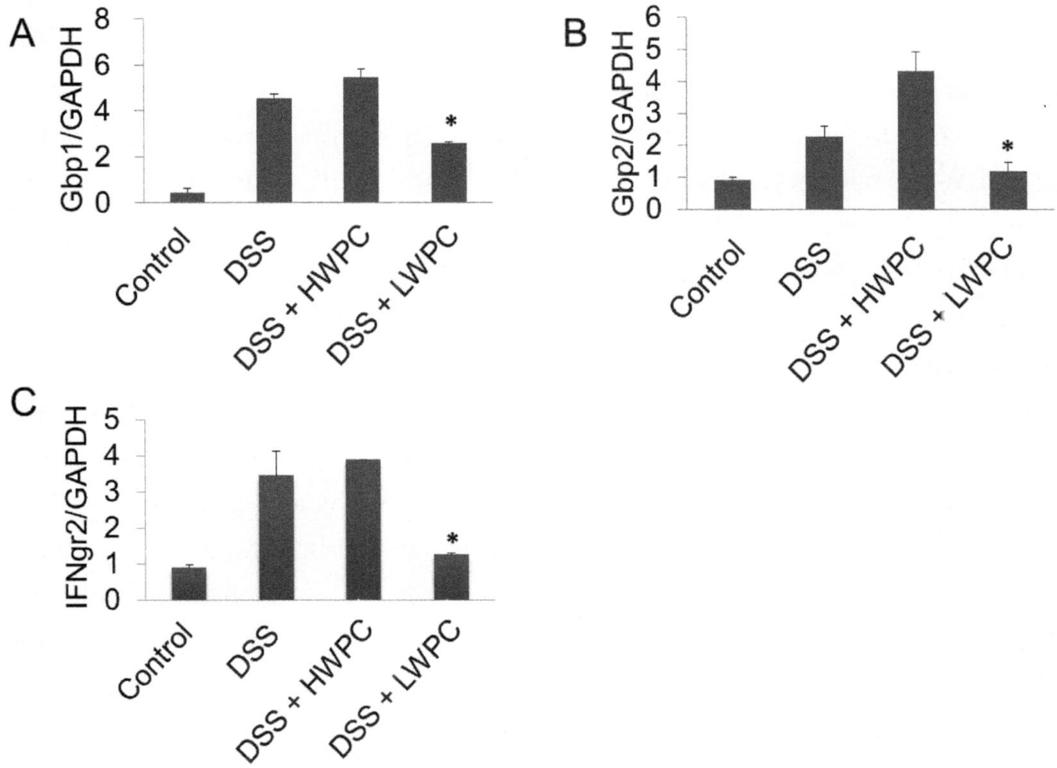

Quantitative real-time RT-PCR was performed to confirm the expression levels of Gbp1 and Gbp2 (Figure 6A,B). Both Gbp1 and Gbp2 were upregulated by DSS treatment by 4.54 ± 0.19- and 2.27 ± 0.32-fold, respectively. Oral intake of LWPC significantly decreased the expression level of both Gbp1 and Gbp2 in the colon of DSS-treated mice, with relative expression levels of 1.26 ± 0.35- and 1.18 ± 0.59- fold, respectively, relative to the blank treatment. However, HWPC did not suppress the expression of these Gbp genes. The GBP family is a group of proteins that bind to guanine nucleotides, but that are different from the G protein family. GBPs are highly expressed in endothelial cells and are strongly induced by inflammatory cytokines, such as interferon (IFN) gamma, tumor necrosis factor alpha, interleukin 1 (IL-1) alpha and IL-1beta [41]. Gbp1 also contributes to cell survival through the inhibition of apoptosis, thereby maintaining the integrity of the barrier function of the epithelium [42]. Gbp1 has also been proposed to regulate vital homeostatic functions, such as apoptosis and cell growth [43]. Britzen-Laurent *et al.* [44] found that Gbp1 mediates the anti-tumorigenic effects of IFN gamma, resulting in the inhibition of tumor growth in colorectal cancer (CRC). Thus, higher expressions of GBPs are expected in inflammatory CRC sites. Gbp1 and Gbp2 expression may be reduced upon LWPC treatment through the suppression of the IFN gamma-mediated inflammatory response. Our microarray findings also showed the expression of the chemokine, Cxcl9, which was also reported to be induced via IFN gamma signaling [45]. Thus, the decreased expression of Gbp1, Gbp2, Gbp6 and Cxcl9 may result from the suppression of IFN gamma signaling through oral intake of LWPC. IFN gamma is a critical mediator of the immune system and inflammatory responses mediated by activated T-cells. IFN gamma acts to enhance inflammation by recruiting the infiltration of leukocytes. Excess IFN gamma results in inflammatory disorders and is well established to regulate cell proliferation and apoptosis through the disassembly of tight junctions and the reduction of the rate of cell migration in the epithelium in IBD [46]. According to our microarray results, the expression level of IFN gamma was not changed by LWPC treatment. However, we noticed that the receptor for IFN gamma, namely IFNgr2, was downregulated in LWPC by 1.7-fold in our microarray results. The gene expression level of IFNgr2 was validated by QPCR, as shown in Figure 6C. In particular, IFNgr2 was upregulated by DSS treatment and downregulated by oral intake of LWPC, but not by that of HWPC. We also found that further downstream of the IFNG pathway, the signal transducers and activators of transcription1 (STAT1) was downregulated by 2.1 by LWPC, while HWPC showed no changes. Interleukin 10 (IL-10) is a unique cytokine having anti-inflammatory properties [47]. According to recent research findings [48], both Il-10 and interleukin 10 receptor 1 (IL-10R1) increased in response to higher levels of IFNg in tissues. Interleukin 27 (IL-27) is characterized as a cytokine with intestinal barrier protection through transcription of antibacterial and anti-inflammatory genes [49]. LWPC did not show any change on the IL-10, IL-27 or their receptor expression levels in our results. Therefore, we suggest that suppression of inflammation by LWPC was not induced through the anti-inflammatory cytokines above. Our histological study described above confirmed the reduced inflammation and damage in the DSS-induced colon when treated with LWPC. This result is in line with the reduced expression of IFNgr2, GBPs and Cxcl9. Therefore, we suggest that LWPC acts by downregulating IFN gamma receptor, resulting in decreased inflammation in DSS-induced colitis.

Whey protein isolates subjected to high hydrostatic pressure have been shown to affect the suppression of tumor necrosis factor-alpha, interleukin 8 and interleukin 18 in Caco-2 colon cancer cells [50]. Rusu *et al.* [51] showed that whey protein increases innate immunity by priming neutrophils. Attaallah *et al.* [52] reported that oral intake of whey protein hydrolysate protects rats against azoxymethane- and DSS-induced colon cancer. These studies suggest that whey protein modulates the immune system. In this study, we showed that LWPC suppresses DSS-induced colitis by suppressing the IFN gamma pathway. This anti-inflammatory activity was lost upon heat treatment of the whey protein, which suggests that the main active peptide of whey protein is heat labile. Some proteins in whey and peptides derived from these proteins have been reported to have bioactive properties, as reviewed by Smithers [53]. Lactoferrin, lactoperoxidase (LPO) and immunoglobulins have antimicrobial activity. Shin *et al.* [54] reported that DSS-induced inflammation was suppressed by oral administration of LPO in mice and also showed that LPO reduces IFN gamma expression. However, LPO may not be a candidate active peptide for anti-inflammation in LWPC, because LWPC did not downregulate IFN gamma expression, but rather reduced Infgr2 expression in our study. Transforming growth factor beta (TGF-beta) is a cytokine that controls proliferation, cellular differentiation and other functions in most cells. TGF-beta is also involved in the effects of whey protein and has been shown to ameliorate IFN gamma-induced intestinal barrier disturbance by upregulating claudin-4 (Cldn4) in human HT-29/B6 cells [55]. However, LWPC did not affect the expression level of Cldn4 in our microarray experiment (data not shown). Thus, we expect that the anti-inflammatory peptide in LWPC is not TGF-beta or LPO. The anti-inflammatory effect of LWPC was lost by heat treatment. However, there are no reports regarding the heat stability of these bioactive peptides in whey to date. Exploration of the active peptides is now ongoing.

4. Conclusions

DSS-induced colitis was prevented by oral intake of LWPC, which also enhanced the recovery from colitis. Comprehensive analysis of gene expression in the colon by DNA microarray analysis showed that GBPs were downregulated by LWPC intake. Expression of GBPs is known to be regulated by interferon gamma. IFNgr2, a receptor for interferon gamma, was also downregulated by LWPC, which suggests that oral intake of LWPC results in the suppression of IFN gamma-mediated pathways and leads to the suppression of inflammation.

Acknowledgments

This study was supported in part by a Grant-in-Aid for the Global Centers of Excellence Program "Frontier Program for Animal Global Health and Hygiene" from the Ministry of Education, Culture, Sports, Science and Technology of Japan.

Author Contributions

Sharmila Jayatilake and Yoshihiko Tokuji contributed to histological and gene expression analyses and manuscript writing. Katsuhito Arai, Nanami Kumada, Yoshiko Ishida and Mikio Kinoshita contributed to animal experiments and western blotting. Satoru Iwatsuki and Ichiro Tanaka prepared WPCs. Takuji Ohwada and Masao Ohnishi contributed to the interpretation of the data.

Conflicts of Interest

Satoru Iwatsuki is employed by Asama Chemical Co. Ltd. The whey protein concentrates used in this study were provided by Asama Chemical Co. Ltd.

References

1. Hugot, J.P. Genetic origin of IBD. *Inflamm. Bowel Dis.* **2004**, *10*, S11–S15.
2. MacDermott, R.P. Alterations of the mucosal immune system in inflammatory bowel disease. *J. Gastroenterol.* **1996**, *31*, 907–916.
3. Bernstein, C.N.; Fried, M.; Krabshuis, J.H.; Cohen, H.; Eliakim, R.; Fedail, S.; Gearry, R.; Goh, K.L.; Hamid, S.; Khan, A.G.; *et al.* World gastroenterology organization practice guidelines for the diagnosis and management of IBD in 2010. *Inflamm. Bowel Dis.* **2010**, *16*, 112–124.
4. Nguyen, G.C.; Tuskey, A.; Dassopoulos, T.; Harris, M.L.; Brant, S.R. Rising hospitalization rates in inflammatory bowel disease in the Unites States between 1998–2004. *Inflam. Bowel Dis.* **2007**, *13*, 1529–1535.
5. Rogler, G. Gastrointestinal and liver adverse effects of drugs used for IBD. *Best Pract. Res. Clin. Gastroentrol.* **2010**, *24*, 157–165.
6. Young, D.; Ibuki, M.; Nakamori, T.; Fan, M.; Mine, Y. Soy-derived di & tripeptides alleviates colon and ileum inflammation in pigs with dextran sulfate sodium-induced colitis. *J. Nutr.* **2012**, *142*, 363–368.

7. Hwang, J.W.; Lee, S.J.; Kim, Y.S.; Kim, E.K.; Ahn, C.B.; Jeon, Y.J.; Moon, S.H.; Jeon, B.T.; Park, P.J. Purification and characterization of a novel peptide with inhibitory effects on colitis induced mice by dextran sulfate sodium from enzymatic hydrolysis of *Crassotea gigas*. *Fish Shellfish Immonol.* **2012**, *33*, 993–999.
8. Fox, P.F.; McSweeney, P.L.H. *Dairy Chemistry and Biochemistry*; Springer: Berlin, Germany, 1998; pp. 186–187.
9. McCarthy, R.; Mills, S.; Ross, R.P.; Fitzgerald, G.F.; Stanton, C. Bioactive peptides from casein and whey proteins. In *Milk and Dairy Products as Functional Foods*; Kanekanian, A., Ed.; John Wiley & Sons: Oxford, UK, 2014; pp. 23–32.
10. Korhonen, H.; Pihlanto, A. Technological options for the production of health-promoting proteins and peptides derived from milk and colostrum. *Curr. Pharm. Des.* **2007**, *13*, 829–843.
11. Gill, H.S.; Cross, M.L. Anticancer properties of bovine milk. *Br. J. Nutr.* **2000**, *84*, S161–S166.
12. Cornish, J.; Callon, K.E.; Naot, D.; Palmano, K.P.; Banovic, T.; Bava, U. Lactoferrin is a potent regulator of bone cell activity and increases bone formation *in vivo*. *Endocrinology* **2004**, *145*, 4366–4374.
13. Krissansen, G.W. Emerging health properties of whey proteins and their clinical implications. *J. Am. Col. Nutr.* **2007**, *26*, 713s–723s.
14. McIntosh, G.H.; Royle, P.J.; Leu, R.K.L.; Regester, G.O.; Johnson, M.A.; Grinsted, R.L.; Kenward, R.S.; Smithers, G.W. Whey proteins as functional food ingredients? *Int. Dairy J.* **1998**, *8*, 425–434.
15. Chatterton, D.E.W.; Smithers, G.; Roupas, P.; Brodkorb, A. Bioactivity of β-lactoglobulin and α- lactalbumin—Technological implications for processing. *Int. Dairy J.* **2006**, *16*, 1229–1240.
16. Maqueda, D.M.; Miralles, B.; Teresa, S.D.P.; Reveron, I.; Munoz, R.; Recio, I. Food derived peptide stimulate mucin sectretion and gene expression in intestinal cells. *J. Agric. Food Chem.* **2012**, *60*, 8600–8605.
17. Posadas, R.L.; Requena, P.; Gonzalez, R.; Suarez, M.D.; Zarsuelo, A.; Medina, F.S.; Augustin, O.M. Bovine glycomacropeptide has intestinal anti-inflammatory effect in rats with dextran sulfate-induced colitis. *J. Nutr.* **2010**, *140*, 2014–2019
18. Li, Y.; Ostergaard, M.V.; Jiang, P.; Chatterton, D.E.W.; Thymann, T.; Kvistgaard, A.S.; Sangild, P.T. Whey protein processing influences formula induced gut maturation in preterm pigs. *J. Nutr.* **2013**, *143*, 1934–1942.
19. Elfstrand, L.; Mansson, H.C.; Paulsson, M.; Nyberg, L.; Akesson, B. Immunoglobins, growth factors and growth hormones in bovine colostrum and the effects of processing. *Int. Dairy J.* **2002**, *12*, 879–887.
20. Laemmli, U.K. Cleavage of structural proteins during assembly of the head of bacteriophage T4. *Nature* **1970**, *227*, 680–685.
21. Reeves, P.G.; Nielsen, F.H.; Fahey, G.C. AIN-93 purified diets for laboratory rodents: Final report of the American Institute of Nutrition Ad Hoc Writing Committee on the reformulation of the AIN-76A rodent diet. *J. Nutr.* **1993**, *123*, 1939–1951.
22. Lien, E.L.; Boyle, F.G.; Wrenn, J.M.; Perry, R.W.; Thompson, C.A.; Borzelleca, J.F. Comparison of AIN-76A and AIN-93G diets: A 13-week study in rats. *Food Chem. Toxicol.* **2001**, *30*, 385–392.

23. Okayasu, I.; Hatakeyama, S.; Yamada, M.; Ohkusa, T.; Inagaki, Y.; Nakaya, R. A novel method in the induction of reliable experimental acute and chronic ulcerative colitis in mice. *Gastroenterology* **1990**, *98*, 694–702.
24. McCafferty, D.M.; Sihota, E.; Muscara, M.; Wallace, J.L.; Sharkey, K.A.; Kubes, P. Spontaneously developing chronic colitis in IL-10/iNOS double-deficient mice. *Am. J. Phys. Gast. Liver Phys.* **2000**, *279*, G90–G99.
25. Huang, D.W.; Sherman, B.T.; Lempicki, R.A. Systematic and integrative analysis of large gene lists using DAVID bioinformatics resources. *Nat. Protocols* **2009**, *4*, 44–57.
26. DAVID Bioinformatics Resources 6.7. Available online: http://david.abcc.ncifcrf.gov/ (accessed on 15 June 2012).
27. The Gene Ontology. Available online: http://www.geneontology.org (accessed on 15 June 2012).
28. Spandidos, A.; Wang, X.; Wang, H.; Seed, B. PrimerBank: A resource for human and mouse PCR primer pairs for gene expression detection and quantification. *Nuc. Acids Res.* **2009**, *38*, D792–D799.
29. PrimerBank. Available online: http://pga.mgh.harvard.edu/primerbank/ (accessed on 2 May 2013).
30. Jimenez, X.T.; Cuenca, A.A.; Jurado, A.T.; Corona, A.A.; Urista, C.R.M. Traditional methods for whey protein isolation and concentration: Effects on nutritional properties and biological activities. *J. Mex. Chem. Soc.* **2012**, *56*, 369–377.
31. Morin, P.; Pouliot, Y.; Florez, R.J. A comparative study of the fractionation of regular buttermilk and whey buttermilk by microfiltration. *J. Food. Eng*. **2006**, *77*, 521–528.
32. Jovanovic, S.; Barac, M.; Macej, O.; Vucic, T.; Lacnjevac, C. SDS-PAGE analysis of soluble proteins in reconstituted milk exposed to different heat treatments. *Sensors* **2007**, *7*, 371–383.
33. Lin, S.; Sun, J.; Cao, D.; Cao, J.; Jiang, W. Distinction of differently heat treated bovine milks by native-PAGE fingerprinting of their whey proteins. *Food Chem.* **2010**, *121*, 803–808.
34. Tavares, T.G.; Malcata, X.F. Whey proteins as source of bioactive peptides against hypertension. In *Bioactive Food Peptides in Health and Disease*; Hernandez-Ledesma, B., Hsieh, C., Eds.; Intech: Rijeka, Croatia, 2013.
35. Yanaba, K.; Yoshizaki, A.; Asano, Y.; Kadono, T.; Tedder, T.F.; Sato, S. IL-10-producing B10 cells inhibit intestinal injury in a mouse model. *Am. J. Pathol.* **2011**, *178*, 735–743.
36. Cook, M.D.; Martin, S.A.; Williams, C.; Whitlock, K.; Wallig, M.A.; Pence, B.D.; Woods, J.A. Forced treadmill exercise training exacerbates inflammation and cause mortality while voluntary wheel training is protective in a mouse model of colitis. *Brain Behav. Immun.* **2013**, *33*, 46–56.
37. Sprong, R.C.; Schonewille, A.J.; van der Meer, R. Dietary cheese whey protein protects rats against mild dextran sulfate sodium-induced colitis: Role of mucin and microbiota. *J. Dairy Sci.* **2010**, *93*, 1364–1371
38. Gersemann, M.; Becker, S.; Nuding, S.; Antoni, L.; Ott, G.; Fritz, P.; Oue, N.; Yasui, W.; Wehkamp, J.; Stange, E.F. Olfactomedin-4 is a glycoprotein secreted into mucus in active IBD. *J. Crohns Colitis* **2012**, *6*, 425–434
39. Bounous, G.; Gold, P. The biological activity of undenatured dietary whey proteins: Role of glutathione. *Ciln. Invest. Med.* **1991**, *14*, 296–309.

40. Hongsprabhas, P.; Kerdchouay, P.; Sakulsom, P. Loweirng the maillard reaction products (MRPs) in heated whey protein products and their cytotoxicity in human cell models by whey protein hydrolysates. *Food. Res. Int.* **2011**, *44*, 748–754.
41. Lubeseder-Martellato, C.; Guenzi, E.; Jörg, A.; Töpolt, K.; Naschberger, E.; Kremmer, E.; Zietz, C.; Tschachler, E.; Hutzler, P.; Schwemmle, M.; *et al.* Guanylate-binding protein-1 expression is selectively induced by inflammatory cytokines and is an activation marker of endothelial cells during inflammatory diseases. *Am. J. Pathol.* **2002**, *161*, 1749–1759.
42. Mirpuri, J.; Brazil, J.C.; Berardinelli, A.J.; Nasr, T.R.; Cooper, K.; Schooner, M.; Lin, P.W.; Parkos, C.A.; Louis, N.A. Commensal *Escherichia coli* reduces epithelial apoptosis through IFN-αA mediated induction of GBP-1 in human and murine models of developing intestine. *J. Immunol.* **2010**, *184*, 7186–7195.
43. Guenzi, E.; Töpolt, K.; Cornali, E.; Lubeseder-Martellato, C.; Jörg, A.; Matzen, K.; Zietz, C.; Kremmer, E.; Nappi, F.; Schwemmle, M.; *et al.* The helical domain of GBP-1 mediates the inhibition of endothelial cell proliferation by inflammatory cytokines. *EMBO J.* **2001**, *20*, 5568–5577.
44. Britzen-Laurent, N.; Lipnik, K.; Ocker, M.; Naschberger, E.; Schellerer, V.S.; Croner, R.S.; Vieth, M.; Waldner, M.; Steinberg, P.; Hohenadl, C.; Stürzl, M. GBP-1 acts as a tumor suppressor in colorectal cancer cells. *Carcinogenesis* **2013**, *34*, 153–162.
45. Gorbachev, A.V.; Kobayashi, H.; Kudo, D.; Tannenbaum, C.S.; Finke, J.H.; Shu, S.; Farber, J.M.; Fairchild, R.L. CXC chemokine ligand 9/monokine induced by IFN-gamma production by tumor cells is critical for T cell-mediated suppression of cutaneous tumors. *J. Immunol.* **2007**, *178*, 2278–2286.
46. Wang, F.; Graham, W.V.; Wang, Y.; Witkowski, E.D.; Schwarz, B.T.; Turner, J.R. Interferon-gamma and tumor necrosis factor-alpha synergize to induce intestinal epithelial barrier dysfunction by up-regulating myosin light chain kinase expression. *Am. J. Pathol.* **2005**, *166*, 409–419.
47. Kominsky, D.J.; Campbell, E.L.; Ehrentraut, S.F.; Wilson, K.E.; Kelly, C.J.; Glover, L.E.; Collins, C.B.; Bayless, A.J.; Saeedi, B.; Colgan, S.P.; Bruder, D. IFN-γ mediated of an apical IL-10 receptor on polarized intestinal epithelia. *J. Immunol.* **2014**, *193*, 1267–1276.
48. Riley, J.K.; Taked, K.; Akira, S.; Schreiber, R.D. Interleukin-10 receptor signaling through the JAK- STAT pathway; requirement for two distinct receptor-derived signals for anti-inflammatory action. *J. Biol. Chem.* **1999**, *274*, 16513–16521.
49. Diegelmann, I.; Olzsal, T.; Goke, B.; Blumberg, R.S.; Brand, S. A novel role for interleukin-27(IL-27) as mediator of intestinal epithelial barrier protection mediated via differential signal transducer activator of transcription (STAT) protein signaling and induction of antibacterial and anti-inflammatory proteins. *J. Biol. Chem.* **2012**, *287*, 286–295.
50. Piccolomini, A.F.; Iskandar, M.M.; Lands, L.C.; Kubow, S. High hydrostatic pressure pre-treatment of whey proteins enhances whey protein hydrolysate inhibition of oxidative stress and IL-8 secretion in intestinal epithelial cells. *Food Nutr. Res.* **2012**, *56*, doi:10.3402/fnr.v56i0.17549.
51. Rusu, D.; Drouin, R.; Pouliot, Y.; Gauthier, S.; Poubelle, P.E. A bovine whey protein extract can enhance innate immunity by priming normal human blood neutrophils. *J. Nutr.* **2008**, *139*, 386–393.

52. Attaallah, W.; Yilmaz, A.M.; Erdoğan, N.; Yalçin, A.S.; Aktan, A.O. Whey protein *versus* whey protein hydrolyzate for the protection of azoxymethane and dextran sodium sulfate induced colonic tumors in rats. *Path. Oncol. Res.* **2012**, *18*, 817–822.
53. Smithers, G.W. Whey and whey proteins—From 'gutter-to-gold'. *Int. Dairy J.* **2008**, *18*, 695–704.
54. Shin, K.; Horigome, A.; Yamauchi, K.; Yaeshima, T.; Iwatsuki, K. Orally administered lactoperoxidase ameliorates dextran sulfate sodium-induced colitis in mice by up-regulating colonic interleukin-10 and maintaining peripheral regulatory T cells. *Int. Immunopharmacol.* **2009**, *9*, 1387–1393.
55. Hering, N.A.; Andres, A.; Fromm, A.; Tol, E.A.; Amasheh, M.; Mankertz, J.; Fromm, M.; Schulzke, D. Transforming growth factor-β, a whey protein component, strengthens the intestinal barrier by upregulating claudin-4 in HT-29/B6 cells. *J. Nutr.* **2011**, *141*, 783–789.

3

Total Environmental Impact of Three Main Dietary Patterns in Relation to the Content of Animal and Plant Food

Luciana Baroni [1,†], Marina Berati [2,†], Maurizio Candilera [3,†] and Massimo Tettamanti [2,†,*]

[1] Primary Care Unit, District No. 4, U.L.S.S. No. 9, Treviso 31100, Italy; E-Mail: lbaroni@ulss.tv.it
[2] Nutrition Ecology International Center (NEIC), Torino 10100, Italy;
E-Mail: marina.berati@mclink.it
[3] Department of Mathematics, University of Padua, 35121 Padova, Italy;
E-Mail: candiler@math.unipd.it

† These authors contributed equally to this work.

* Author to whom correspondence should be addressed;
E-Mail: massimo.tettamanti@nutritionecology.org

Abstract: Based on a review of the most recent available scientific evidence, the new Dietary Guidelines for Americans 2010 (USDA DG) provide information and advice for choosing a healthy diet. To compare the environmental impacts of, respectively, omnivorous (OMN), lacto-ovo-vegetarian (LOV) and vegan (VEG) dietary patterns as suggested in the USDA DG, we analyzed the three patterns by Life Cycle Assessment (LCA) methodology. The presence of animal food in the diet was the main determinant of environmental impact. The major impact always stemmed from land and water use. The second largest impact came from energy use. Emission of toxic inorganic compounds into the atmosphere was the third cause of impact. Climate change and acidification/eutrophication represented other substantial impacts.

Keywords: Life Cycle Assessment (LCA); Dietary Guidelines; environmental impact

---

1. Introduction

The evidence for a link between high consumption of meat and other animal foods and poor health outcomes is growing. The main contributing factors are likely to be the high saturated fat content in animal products, the high salt content in processed foods, and the fact that the consumption of animal products limits the consumption of health-promoting foods such as fruit, vegetables, nuts, beans and grains [1–13].

According to the World Health Organization (WHO), malnutrition affects one in every three people worldwide; it is present in all age groups and populations, and plays a major role in half of the 10.4 million annual child deaths in the developing world; malnutrition also plays a role in causing disease and disability in the children who survive [14,15].

A European study tackled the problem of sustainable food production and assessed the environmental impact of human food consumption and its related processes [16]. The study assumed as the basis for calculations a complete diet (defined as the total amount of food that one single person eats in one week), whereas previous research had been mainly limited to single foods or to specific comparisons [17–19].

Its results confirmed the need to educate people living in developed countries to shift their eating habits towards an increase in the direct consumption of plant foods.

The 2010 Dietary Guidelines for Americans, USDA DG [20], released in 2011, provide information and advice for choosing a healthy diet; namely, one that focuses on nutrient-dense foods, and that contributes to achieving and maintaining a healthy weight. They are meant to be used in developing educational materials for the general public, and to aid policymakers in designing and carrying out nutrition-related programs, including federal nutrition assistance and educational programs. To get the full benefit, it is recommended that individuals abide by the USDA DG recommendations in their entirety, as part of an overall approach to a healthy lifestyle.

Although the USDA DG 2010 declare that the recommendations are based on a review of the most recent available scientific evidence, it must be noticed that there is not unanimous agreement among the nutrition science community that the recommendations do in fact reflect the *best* available scientific evidence. In particular, Harvard School of Public Health judges that the USDA recommendations do not reflect "the best eating choices" and proposes its Healthy Eating Plate, an eating guide for omnivorous people that is more selective on the quality of diet and more limited on the amount of dairy food [21].

The aim of the present study is to further explore and compare the environmental impacts of different but "homogeneous" food patterns. We chose to use the omnivorous USDA food patterns (OMN) and their vegetarian adaptations, respectively, lacto-ovo-vegetarian (LOV) and vegan (VEG) patterns, as suggested in Appendices 7, 8 and 9 of the USDA DG [20].

The "omnivorous" (OMN) food pattern includes animal flesh and animal products, and any plant food. The "lacto-ovo-vegetarian" (LOV) food pattern includes any plant food and milk, dairy products and eggs, while excluding any type of animal flesh (meat, poultry or fish); the "vegan" (VEG) food pattern is a plant-only diet which excludes any food of animal origin: not only meat, poultry and fish but also milk, dairy and eggs. The three dietary patterns suggested in the USDA DG are recommended as healthy and nutritionally adequate [20].

The assessment of the environmental impact of the dietary patterns was conducted using the Life Cycle Assessment (LCA) method, an internationally standardized procedure (ISO 14040) [22].

2. Materials and Methods

2.1. Life Cycle Assessment

The analysis was performed using the Life Cycle Assessment (LCA), an objective procedure for the evaluation of the energy and environmental impacts of a process or activity. More relevant results stem from the application of the LCA methodology to the total environmental impact of a complete dietary pattern, rather than from its partial application to single steps or single impact subcategories of a production process, or to specific food items. The LCA approach allows to:

1. systematically estimate the complete environmental consequences and analyze all the energetic and material exchanges occurring in the environment, and
2. quantify the various emissions into air, water and land in every life cycle phase, and
3. detect any significant change in the environmental effects in an objective way, and
4. estimate the effects of material consumptions and environmental emissions on the health of human beings and on the ecosystem as related to food production.

Usually, it is carried out through the identification of the energy and raw material consumption and the release of waste into the environment: the assessment includes the whole life cycle of a real process or a real activity, from the extraction and processing of raw materials to the production, transportation, distribution, use, reuse and recycling, and final disposal.

Since the aim of our study was to evaluate the pure food-related impacts, focusing on theoretical diets and keeping other variables fixed (*i.e.*, not to compare locally-produced, low-impact foods *versus* imported, high-impact foods), we did not consider any difference related to geographical zone or transportation; the import-export food fluxes have not been considered, nor related emissions during cooking and storing in the household/in restaurants have been taken into account. The system boundaries included the following steps in the process chain: agricultural production, processing and packaging.

According to ISO 14040 standards for LCA [22], four phases have been performed: 1, Goal and Scoping; 2, Life Cycle Inventory; 3, Life Cycle Impact Assessment, LCIA; 4, Life Cycle Interpretation.

2.1.1. Goal and Scoping

The goal of the study was to compare the environmental impact of the OMN, LOV and VEG dietary patterns proposed in the USDA DG for Americans [20]. We took into account only food produced by intensive, non-organic farming, both because this is, and is likely to remain, the most widespread method, and because previous research had already shown that the production method (non-organic or organic) [16] and transportation [23] have much less influence on the overall environmental impact compared to the source (animal or plant) of the food. The software we selected to carry out the Inventory Analysis and the Impact Assessment was SimaPro 7.3.3 [24].

2.1.2. Life Cycle Inventory

In this phase, which is the core of any LCA, all data were collected, and a model representing the whole life cycle of the products, processes and activities was prepared. In some cases, as stated in the USDA DG, it was necessary to subsume individual foods into overall categories, in order to compare new results with existing databases or previous literature which, sometimes, presented simplified data for "fruit", "vegetables" and "cheese".

Input/output data on processes in the food sector have been collected from a variety of sources. Data on production in agriculture and fishery have been determined by a "top-down" approach, where statistical data on a national level have been broken down to represent specific processes. Specific data collection was performed from textbooks/scientific papers describing specific case-studies [16,18,19,25–37].

2.1.3. Life Cycle Impact Assessment (LCIA)

In the LCIA phase, the collected data were used to evaluate the various environmental impacts, and to quantify the impact of each single process on the overall damage.

The elements necessary to this assessment are:

1. selection of impact categories (environmental effects) and of the environmental indicators representing them;
2. attribution of the results of inventory analysis to the selected impact categories (classification), according to the effects they exert or may exert on the environment.

The software assigns to each component of the diet a "weight", *i.e.*, an a-dimensional value, which represents the intensity of the effect that each component exerts on the environment. The "total weight" of each diet, called single score, is expressed in points (*Pt*), the unit of measure used by the software to assign a numeric value to the overall environmental impact of the diet. The higher the "score" in *Pt*, the higher the damage to the environment.

In order to obtain a complete analysis, to facilitate comparison with data from other studies, and to minimize bias, the assessment phase has been conducted using all the indicators made available by the SimaPro software.

Ecoindicator99

A damage-oriented indicator that analyses the following impacts, which can be further categorized according to three large damage categories:

1. damages to human health (substances which have a negative impact on respiration, organic and inorganic compounds, carcinogenesis, climate change and ozone, ionizing radiations);
2. damages to ecosystems quality (ecotoxicity, acidification and eutrophication);
3. damages to resources (use of primary resources—land and water—and of fuel).

Ecopoint

An indicator designed to evaluate the impacts due to the release of chemicals into the environment (NO_x, SO_x, NH_3, CO_2, Metals, COD, DUST PM_{10}, *etc.*). The Swiss Ecopoints 1997 (environmental scarcity) is an update of the 1990 method, based on actual pollution and critical targets, derived from the Swiss policy for Environment. It comes in 3 versions, with identical evaluation and indicator values but different in the normalization factor; the version number 2 is used in SimaPro [38,39].

EDIP

A method adapted for LCA food database projects, representing the most used and widespread indicator to evaluate different forms of toxicity (global warming, acidification, eutrophication, ecotoxicity, human toxicity).

The LCIA was carried out three times, once for each indicator. LCA experts assume a general uncertainty of 10% to 20% in the results [40].

2.1.4. Life Cycle Interpretation

In this phase, all the results of the Inventory and/or of the Impact Assessment were processed, according to the objective and purpose of the research, in order to formulate conclusions and recommendations. It is the final phase of an LCA and its purpose is to propose the necessary changes to reduce environmental impacts.

2.2. Diets

Beside omnivorous (OMN) dietary patterns (Appendix 7), USDA DG suggest vegetarian adaptations, respectively lacto-ovo-vegetarian (LOV) and vegan (VEG) patterns (Appendices 8 and 9), from 1000 to 3200 kcal, that meet the nutritional needs of a well-balanced diet for healthy individuals [20].

The three food patterns share a main overlapping component, which is the same for each type of pattern for kind and amounts of suggested foods, and which is composed by "fruits", "vegetables" and "grains" groups. The three patterns differ in the "oils" group (for amount of servings and the presence of fish oil in the OMN pattern), "dairy" group (which is substituted with non-dairy milk and derivatives for VEG pattern), and "protein foods" group. The latter, in each of three patterns, includes nuts, seeds, and soy products, in different amounts. In the OMN pattern, the "protein foods" group includes also seafood, meat and poultry, while in LOV and VEG patterns it includes again the "beans and peas" subgroup (already present in the "vegetable" group, but in addition to the amounts recommended in it). Moreover, it is important to note that "nuts and seeds" and "soy products" represent a single protein food subgroup in the LOV and VEG patterns, while in the OMN patterns they are aggregated in the "nuts, seeds and soy products" protein subgroup; similarly, "eggs" represent a single protein food subgroup in the LOV patterns, while they are aggregated in the "meat, poultry, eggs" protein food subgroup in the OMN patterns. The relative contribution of each food of these aggregated categories in the OMN patterns has been calculated on the basis of food consumption reported in the FAO database [28].

2.2.1. "*Whole Diet*" Study

To compare the environmental impact of the OMN, LOV and VEG dietary patterns proposed in the USDA DG for Americans, we choose for each pattern three plans with calorie intakes of 1600,

2400 and 3200 kilocalories respectively: 1600 OMN, 2400 OMN, 3200 OMN; 1600 LOV, 2400 LOV, 3200 LOV; 1600 VEG, 2400 VEG, 3200 VEG [20]. We called this part of the analysis the "*whole diet*" study, given that it analyzed the whole composition of the diet.

2.2.2. "*Delta*" Study

In order to better evaluate the differences among the corresponding impacts of the three patterns, we focused only on the difference in food components, limiting the analysis to the 2400 kcal patterns. We called this part of the analysis the "*delta*" study, given that it was based only on the delta, the difference, among the three patterns: it allowed us to focus on the consequences on the overall environmental impact of apparently small modifications in the composition of a diet.

As mentioned above, USDA DG propose eating patterns which, even in the OMN and LOV patterns, are largely based on plant foods. Therefore, each of the three patterns has a major overlapping area of plant food content (about 81% in mass). As a consequence, the differences in the environmental impacts of the three patterns stem only from the 19% in mass in which the three recommended eating patterns differ, mainly in the different protein sources. To obtain the food component of the *delta* study, for each of the three patterns the overlapping component have been subtracted, *i.e.*:

1. the "fruits", "vegetables" and "grains" groups;
2. the total amounts of nuts, seeds, soy products listed in the OMN patterns;
3. the total amount of oils listed in the VEG patterns.

Therefore, for OMN patterns, the delta component included seafood, meat, poultry, eggs, dairy, oils; for LOV patterns, the delta component included eggs, beans and peas (in addition to the amount recommended in the "vegetable" group), soy products, nuts and seeds, dairy, oils; for VEG patterns, the delta component included beans and peas (in addition to the amount recommended in the "vegetable" group), soy products, nuts and seeds, non-dairy substitutes.

2.3. Sensitivity

In a previous paper we performed an accurate sensitivity analysis, showing that, for this kind of evaluation, any variation of the examined data elicited a variation in the results that came to be perceptually acceptable and unable to modify the interpretation of the results [16].

More recent studies from other authors confirmed the significant stability of LCA methodology when applied to the analysis of dietary patterns [41].

3. Results and Discussion

3.1. "Whole Diet" Study

The results obtained for the three food patterns (whole diet study), according to each method of analysis (indicator), are reported in Table 1.

The analysis results are expressed in points for week (*Pt/week*): the higher the "score" in *Pt*, the higher the damage to the environment.

Table 1. Whole diet study: total and partial environmental impacts (indicated as single score, *Pt* per week) of the three complete food patterns, according to each indicator.

Pattern/kcal	VEG 1600	VEG 2400	VEG 3200	LOV 1600	LOV 2400	LOV 3200	OMN 1600	OMN 2400	OMN 3200
Ecoindicator99									
Total	0.64800481	0.94997163	1.3071604	2.3859409	2.6923735	3.0492296	3.7122608	4.40574	4.9341334
Carcinogens	0.00019843	0.000286384	0.00036383	0.00133281	0.00142583	0.00150253	0.001630978	0.00180755	0.001925124
Respiratory organics	0.00018305	0.000251572	0.0003258	0.00030875	0.00037811	0.00045206	0.000417358	0.000518085	0.000606523
Respiratory inorganics	0.10049695	0.14412463	0.18855662	0.17767523	0.22237083	0.26670832	0.28722378	0.36469605	0.42409415
Climate change	0.03919139	0.055214859	0.0709742	0.13229725	0.14852923	0.16416739	0.17863998	0.20869219	0.23059325
Radiation	0.00002247	3.42×10^{-5}	4.34×10^{-5}	0.00038047	0.00039248	0.00040169	0.000398937	0.000416055	0.000427788
Ozone layer	0.00001066	1.47×10^{-5}	1.90×10^{-5}	3.40×10^{-5}	3.85×10^{-5}	4.27×10^{-5}	5.57×10^{-5}	6.64×10^{-5}	7.36×10^{-5}
Ecotoxicity	0.00106054	0.001434384	0.00171095	0.0026162	0.00302211	0.00329787	0.003002629	0.003519975	0.003860926
Acidification/Eutrophication	0.02526500	0.036874031	0.04925412	0.04261296	0.05475953	0.0671279	0.090794324	0.11735622	0.13629427
Land and water use	0.29042511	0.43349405	0.63636379	1.0764882	1.2228196	1.4256888	2.117066	2.5657137	2.9009677
Minerals	0.00006435	9.65×10^{-5}	0.00012433	0.00053008	0.00056272	0.00059049	0.000579085	0.00062538	0.000659985
Fossil fuels	0.19108686	0.27814636	0.35942432	0.95166495	1.0380746	1.1192498	1.0324519	1.1423284	1.2346302
Ecopoint									
Total	6962.153	9977.3835	13418.176	13599.654	16700.341	20134.632	26697.481	33726.743	38929.907
NOx	1190.3383	1687.0413	2174.7931	2741.075	3239.3288	3725.3904	3351.1462	4029.0262	4599.7719
SOx	769.90348	1087.0439	1383.1478	1016.3942	1333.0982	1628.8733	1078.1917	1416.3569	1723.7377
NMVOC	139.32288	192.98861	253.43336	273.88218	328.37026	388.55252	350.76288	427.2963	497.94655
NH3	842.99251	1250.5947	1703.2695	1192.126	1627.3129	2079.9873	3490.5891	4614.2057	5379.5064
Dust PM10	9.0847823	12.923818	15.191499	65.534052	69.802741	72.053644	72.707679	79.03103	82.418179
CO2	1377.9895	1943.5007	2498.664	4725.9375	5298.3987	5849.4224	6340.5217	7394.1754	8163.2418
Ozone layer	0.6610465	0.90966159	1.1767778	2.1511266	2.4272521	2.689679	3.4930708	4.1519191	4.5977102
Pb (air)	0.32147787	0.47727671	0.60871325	1.5374657	1.702998	1.833917	2.0107655	2.3081678	2.5067236
Cd (air)	1.5009171	2.1788895	2.7819723	9.748633	10.451046	11.049965	11.405062	12.571727	13.398671
Zn (air)	0.34491482	0.51571004	0.66441941	1.0479235	1.2308866	1.3790125	1.6420776	1.9914576	2.2250812
Hg (air)	1.1709015	1.7566871	2.2428252	11.655909	12.243954	12.729547	12.321511	13.092723	13.669511
COD	4.5777467	6.6868027	8.6133306	25.25851	27.433316	29.349171	30.326905	33.929372	36.545081
P	466.47307	688.61262	932.40825	605.62586	831.28441	1075.0789	1216.5565	1613.6763	1938.9971
N	2080.5034	2992.2043	4299.9698	2621.3589	3578.1404	4885.9003	10,403.361	13,712.094	16,063.676
Cr (water)	0.18439044	0.27280425	0.34372097	1.9087695	2.0000171	2.0706068	2.1334244	2.2856304	2.3864404
Zn (water)	0.11026105	0.16221025	0.20594354	0.94313822	0.99673879	1.0402484	1.0679982	1.1560351	1.2165352
Cu (water)	0.15052732	0.22462302	0.28297967	1.7211955	1.7973628	1.855504	1.9056096	2.0314536	2.1143221
Cd (water)	0.18614116	0.27169122	0.35127599	0.55325881	0.64589623	0.72481966	0.91269133	1.1064944	1.2360639
Hg (water)	0.7057345	1.0552604	1.3499184	5.9398762	6.286672	6.5811908	6.2440261	6.6747297	7.0105413
Pb (water)	0.0305093	0.045206483	0.05642818	0.30565965	0.32090765	0.33209206	0.33619951	0.35977884	0.37519415
Ni (water)	0.02068235	0.030679134	0.03847235	0.23622184	0.24660089	0.25435586	0.26447911	0.28248352	0.29403315
AOX (water)	0.00530489	0.007279036	0.00937398	0.01604704	0.01823847	0.02029386	0.027071622	0.032408888	0.03596629
Metals (soil)	0.21573554	0.31068211	0.39921194	1.1959458	1.2942133	1.382073	1.433325	1.5988108	1.7190422
Pesticide (soil)	15.523208	20.69761	25.872013	15.523208	20.69761	25.872013	15.523208	20.69761	25.872013
Energy	59.835651	86.870388	112.30158	277.97726	304.81086	330.2091	302.59665	336.61044	365.40786

Table 1. *Cont.*

Pattern/kcal	VEG 1600	VEG 2400	VEG 3200	LOV 1600	LOV 2400	LOV 3200	OMN 1600	OMN 2400	OMN 3200
					EDIP				
Total	0.013394846	0.018757951	0.02442873	0.03742483	0.0429144	0.04856047	0.049088143	0.057968728	0.065202436
Global warming (GWP 100)	0.001047775	0.001477754	0.00189988	0.00359432	0.0040296	0.00444857	0.004823125	0.005624618	0.006209534
Acidification	0.000544453	0.000788231	0.00104088	0.00087095	0.00112338	0.00137577	0.001664129	0.002154371	0.002515469
Eutrophication	0.003267925	0.004453371	0.00574918	0.00361988	0.00482479	0.00611978	0.006304814	0.008318734	0.00997571
Photochemical smog	0.000170838	0.000235238	0.0003047	0.0003023	0.0003675	0.0004367	0.000415924	0.00051399	0.000598279
Ecotoxicity water chronic	0.001493251	0.002078688	0.00284514	0.00456145	0.00519663	0.00595413	0.007208578	0.008596149	0.009714592
Ecotoxicity water acute	0.001243991	0.001714344	0.0022113	0.00420647	0.0047279	0.00521615	0.006717897	0.00795496	0.008786304
Ecotoxicity soil chronic	0.002577418	0.003584406	0.00468446	0.00595102	0.00695597	0.0080556	0.006466586	0.00761594	0.00878632
Human toxicity air	0.000509268	0.000720343	0.00092095	0.00152606	0.00173505	0.00193547	0.001620564	0.001857016	0.002070554
Human toxicity water	7.44×10^{-5}	0.000110764	0.00014493	0.00061465	0.00065113	0.00068526	0.000657524	0.000705894	0.00074591
Human toxicity soil	0.002465551	0.003594811	0.00462731	0.01217773	0.01330246	0.01433303	0.013209003	0.014627056	0.015799764

For all indicators, the results indicated that VEG patterns always had the lowest single score: LOV patterns had single scores of 3 ± 0.7 times higher than VEG patterns, and OMN patterns had single scores of 4.7 ± 1 times higher than VEG patterns, depending to the calorie intake.

3.2. "Delta" Study

The results obtained in the delta study, according to each of the above-mentioned indicators, are reported in Table 2.

Table 2. Delta study: total and partial environmental impacts (indicated as single score, *Pt* per week) of the 19% component in which the three 2400 kcal patterns differ (delta), according to each indicator.

Pattern	VEG Delta	LOV Delta	OMN Delta
	Ecoindicator99		
Total	0.21163256	1.9540345	3.6643576
Carcinogens	8.05×10^{-5}	0.001219979	0.001596435
Respiratory Organics	4.01×10^{-5}	0.000166618	0.000305573
Respiratory Inorganics	0.021479271	0.09972547	0.24028832
Climate change	0.011307043	0.10462142	0.16357382
Radiation	4.92×10^{-5}	0.000363216	0.000386427
Ozone layer	7.65×10^{-5}	3.15×10^{-5}	5.91×10^{-5}
Ecotoxicity	0.00015814	0.001745867	0.002227464
Acidification/Eutrophication	0.005841891	0.023727391	0.08568815
Land and water use	0.14422255	0.93354814	2.2793129
Minerals	1.37×10^{-5}	0.000479934	0.000541406
Fossil fuels	0.028476768	0.78840499	0.89037801

Table 2. *Cont.*

Pattern	VEG Delta	LOV Delta	OMN Delta
	Ecopoint		
Total	1966.647	8736.5711	25,386.999
NO*x*	244.3271	1792.9067	2564.5648
SO*x*	100.58787	345.87779	424.46498
NMVOC	40.724424	175.43345	272.92316
NH3	210.51048	584.99251	3527.0838
Dust PM10	2.9942366	59.304326	68.263183
CO2	398.16059	3813.7786	5905.1117
Ozone layer	0.46604541	1.9578271	3.640162
Pb (air)	0.08821719	1.3213139	1.9105317
Cd (air)	0.53593394	8.5764778	10.615507
Zn (air)	0.092202355	0.8064707	1.5367188
Hg (air)	0.17861332	10.371796	11.187239
COD	1.3626347	22.224598	28.637096
P	107.57917	249.26705	1025.8589
N	849.72417	1431.9097	11270.471
Cr (water)	0.045675248	1.7736143	2.0564881
Zn (water)	0.029915182	0.85359468	1.0089572
Cu (water)	0.032329539	1.5634239	1.7889283
Cd (water)	0.07970724	0.46663514	0.92955853
Hg (water)	0.081921167	5.3177828	5.7018097
Pb (water)	0.006726089	0.28355506	0.32204421
Ni (water)	0.00528008	0.22105271	0.25655257
AOX (water)	0.003837047	0.014917741	0.029135601
Metals (soil)	0.078326533	1.0618578	1.3636445
Energy	8.9515701	226.28606	257.2731
	EDIP		
Total	0.003713746	0.027870192	0.042734028
Global warming (GWP 100)	0.000297634	0.002849477	0.004412835
Acidification	0.000112545	0.000447691	0.001467579
Eutrophication	0.000333923	0.000705344	0.004129094
Photochemical smog	3.79×10^{-5}	0.000170155	0.000315601
Ecotoxicity water chronic	0.001190023	0.004307961	0.00768123
Ecotoxicity water acute	0.000872404	0.003885964	0.007087624
Ecotoxicity soil chronic	0.000396519	0.003768079	0.004424933
Human toxicity air	4.81×10^{-5}	0.001062836	0.00118261
Human toxicity water	1.70×10^{-5}	0.000557353	0.000611296
Human toxicity soil	0.000407688	0.010115332	0.011421227

For all indicators, the results of the delta study showed that, compared with the single score of the VEG pattern, the single score of the LOV pattern was up to 9.2 times higher, and the single score of the OMN pattern was up to 17.3 times higher. The single score of the delta study for the OMN pattern (2400 kcal), for each indicator, was: 3.66 (Ecoindicator99), 25,387 (Ecopoint) and 0.043 (EDIP). It was higher than (or equal to, for LOV-EDIP) the single score calculated in the whole diet study for both the other diets, that was, respectively: 0.95 (VEG) and 2.69 (LOV), calculated with the

Ecoindicator99; 9,977 (VEG) and 16,700 (LOV), calculated with the Ecopoint; 0.019 (VEG) and 0.043 (LOV), calculated with the EDIP.

3.3. "Whole Diet" Study vs. *"Delta" Study*

It can be noticed that the animal food component in the OMN pattern, while making up only 19% of the total weight of the diet, accounted for about 73%–83% of its total environmental impact: for Ecoindicator99 the single score was 3.66 (delta) out of 4.41 (whole); for Ecopoint, 25,387 (delta) out of 33,726 (whole); for EDIP, 0.043 (delta) out of 0.058 (whole).

Other interesting results were the ratios of the single score of the delta study to the single score of the whole diet study: 0.83 (OMN), 0.73 (LOV), and 0.22 (VEG). Moreover, the ratios between the single scores of the delta study over the single score of the plant food common component of the diet (81% in mass) was shown to be 1.21 (OMN), 0.98 (LOV), and 0.30 (VEG).

3.4. Distribution of the Sources of Relative Impact within the Dietary Patterns

The different components of the overall environmental impacts, accordingly to each indicator, within the same dietary pattern, can be summarized as follows:

3.4.1. Ecoindicator99

The results obtained using Ecoindicator99 showed that the major impact, from 45% to 60% of the overall impact, always stemmed from land and water use. The second largest impact, from 25% to 50% of the total, came from energy use. The third cause of impact, from 7% to 16% of the total, was due to the emission of toxic inorganic compounds into the environment. Effects on climate change (5%–6%) and on acidification/eutrophication (2%–4%) represented another substantial impact.

3.4.2. Ecopoint

The results of the Ecopoint analysis showed that the most impacting factor, from 35% to 55% of the total, was the contamination from inorganic nitrogen and phosphorous compounds. The second major cause of impact, from 21% to 37% of the total, was the emissions into the atmosphere, while the third source of impact, from 16% to 27% of the total, was due to the emission of oxides into the atmosphere. The inclusion of N-oxides among the Greenhouse Gas (GHG) subcategories put the GHG emissions at the same level of impact of inorganic nitrogen and phosphorous compounds (34%–57%).

3.4.3. EDIP

The impacts detected by the above two indicators translated into a considerable number and variety of toxic impacts, that could be further evaluated by the EDIP indicator. Results of EDIP analysis showed that the highest impact, from 25% to 33% of the total, was due to human toxicity caused by soil contamination. The second-highest impact, from 20% to 29% of the total, was due to acute and chronic eco-toxicity of the water. The third-highest impact, from 13% to 19%, was due to chronic eco-toxicity of the soil. The fourth-highest impact, from 10% to 18%, was due to water eutrophication, while the fifth impact, from 8% to 10%, was given by various contributions to global warming.

3.5. Absolute Values of the Impacts in the Different Dietary Patterns

The above mentioned percentages represented the proportions of the various kinds of impact within the same dietary pattern, and they were very similar for all the examined dietary patterns, that is, for each dietary pattern its main impact was land and water use, then energy and so on.

But when taking into consideration the absolute values of those impacts, they varied dramatically among the various patterns. In fact, the total impact (single score) for the VEG pattern was 35% and 22%, respectively, than the one for the LOV and OMN patterns (these data apply specifically to the 2400 kcal scenario for Ecoindicator99). Therefore, even if, for example, the land and water use accounted for 50% of the impact both in VEG and in OMN diets, the absolute value in the VEG diet was 78% lower than the absolute value for the OMN diet for the same impact.

The overall results of our study showed that OMN diets had the highest impact, while VEG diets had the lowest environmental impact, independently of the calorie intake.

In most cases, the differences were just as significant also between a LOV and an OMN diet, so much so that the overall impact of a 3200 kcal LOV was always lower than that a 1600 kcal OMN one. The presence of animal food in the diet resulted to be the main impacting factor.

3.6. Subcategories of Impact

The study of the total impacts (represented by the single score, *Pt/week*) included not only the analysis of some well-known, commonly used impact subcategories, *i.e.*, GHG, land and water use, but also other less commonly studied impact subcategories, all listed in Tables 1 and 2: the various subscores of impact subcategories contributed to the final value of the single score of the respective indicator. Although more complex, the single score represents an index of the total environmental impact of food production, more accurate and comprehensive than the score of each single impact subcategory. For example, for the 2400 kcal OMN diet, the subscore for land and water use, a subcategory of Ecoindicator99, was respectively 5.92 and 2.10 times the score of the VEG and LOV patterns for the whole study (Table 3), and 15.80 and 2.44 for the delta study (Table 4). For comparison, the single score of the total impact category analyzed by Ecoindicator99 was 4.64 and 1.64 times for the whole study and 17.31 and 1.88 times for the delta study. It is worth to underline that in the whole diet study, land and water use subscore contributed for 45%–46% in the vegetarian patterns and 58% in the OMN patterns, to the total impact, as represented by the value of the single score (Table 3). Similarly, a recent study conducted in Germany by Meier *et al.* [42], which analyzed consumption data derived from a National Nutrition Survey, showed a land-saving potential effect of plant-based diets, which was maximum for VEG diets.

Table 3. Whole diet study: land and water use, and GHG subscores (indicated as single score, *Pt* per week) of the three complete food patterns, according to each indicator.

Pattern	**VEG**			**LOV**			**OM**		
kcal	1600	2400	3200	1600	2400	3200	1600	2400	3200
					Ecoindicator99				
TOTAL single score	0.64800	0.94997	1.30716	2.38594	2.69237	3.04923	3.71226	4.40574	4.93413
versus VEG	-	-	-	-	-	-	573%	464%	377%
versus LOV	-	-	-	-	-	-	156%	164%	162%
versus OMN	17%	22%	26%	64%	61%	62%	-	-	-
Land and water use subscore	0.29043	0.43349	0.63636	1.07649	1.22282	1.42569	2.11707	2.56571	2.90097
versus TOTAL	45%	46%	49%	45%	45%	47%	57%	58%	59%
versus VEG	-	-	-	-	-	-	729%	592%	456%
versus LOV	-	-	-	-	-	-	197%	210%	203%
versus OMN	14%	17%	22%	51%	48%	49%	-	-	-
Climate change subscore	0.03919	0.05521	0.07097	0.13230	0.14853	0.16417	0.17864	0.20869	0.23059
versus TOTAL	6%	6%	5%	6%	6%	5%	5%	5%	5%
versus VEG	-	-	-	-	-	-	456%	378%	325%
versus LOV	-	-	-	-	-	-	135%	141%	140%
versus OMN	22%	26%	31%	74%	71%	71%	-	-	-
					Ecopoint				
TOTAL single score	6962.1	9977.3	13,418.1	13,599.6	16,700.3	20,134.6	26,697.4	33,726.7	38,929.9
versus VEG	-	-	-	-	-	-	383%	338%	290%
versus LOV	-	-	-	-	-	-	196%	202%	193%
versus OMN	26%	30%	34%	51%	50%	52%	-	-	-
Greenhouse Gas (GHG) subscore	2707.6	3823.5	4926.8	7740.8	8866.0	9963.3	10,042.4	11,850.4	13,260.9
versus TOTAL	39%	38%	37%	57%	53%	49%	38%	35%	34%
versus VEG	-	-	-	-	-	-	371%	310%	269%
versus LOV	-	-	-	-	-	-	130%	134%	133%
versus OMN	27%	32%	37%	77%	75%	75%	-	-	-
					EDIP				
TOTAL single score	0.01339	0.01876	0.02443	0.03742	0.04291	0.04856	0.04909	0.05797	0.06520
versus VEG	-	-	-	-	-	-	366%	309%	267%
versus LOV	-	-	-	-	-	-	131%	135%	134%
versus OMN	27%	32%	37%	76%	74%	74%	-	-	-
Global warming (GWP 100) subscore	0.00105	0.00148	0.00190	0.00359	0.00403	0.00445	0.00482	0.00562	0.00621
versus TOTAL	8%	8%	8%	10%	9%	9%	10%	10%	10%
versus VEG	-	-	-	-	-	-	460%	381%	327%
versus LOV	-	-	-	-	-	-	134%	140%	140%
versus OMN	22%	26%	31%	75%	72%	72%	-	-	-

Table 4. Delta study: land & water use and GHG subscores (indicated as single score, *Pt* per week) of the 19% component in which the three 2400 kcal patterns differ (delta), according to each indicator.

Pattern	VEG	LOV	OMN
Ecoindicator99			
TOTAL single score	0.21163	1.95403	3.66436
versus VEG	-	-	1731%
versus LOV	-	-	188%
versus OMN	6%	53%	-
Land and water use subscore	0.14422	0.93355	2.27931
versus TOTAL	68%	48%	62%
versus VEG	-	-	1580%
versus LOV	-	-	244%
versus OMN	6%	41%	-
Climate change subscore	0.01131	0.10462	0.16357
versus TOTAL	5%	5%	4%
versus VEG	-	-	1447%
versus LOV	-	-	156%
versus OMN	7%	64%	-
Ecopoint			
TOTAL single score	1966.6	8736.5	25,386.9
versus VEG	-	-	1291%
versus LOV	-	-	291%
versus OMN	8%	34%	-
Greenhouse Gas (GHG) subscore	683.2	5782.1	8742.5
versus TOTAL	35%	66%	34%
versus VEG	-	-	1280%
versus LOV	-	-	151%
versus OMN	8%	66%	-
EDIP			
TOTAL single score	0.00371	0.02787	0.04273
versus VEG	-	-	1151%
versus LOV	-	-	153%
versus OMN	9%	65%	-
Global warming (GWP 100) subscore	0.00030	0.00285	0.00441
versus TOTAL	8%	10%	10%
versus VEG	-	-	1483%
versus LOV	-	-	155%
versus OMN	7%	65%	-

GHG emissions (calculated as GHG emissions in kilograms of carbon dioxide equivalents, $kgCO_2e$), have been assessed in two recent studies in UK and Northern USA [43,44]. In the UK study, the real 2000 kcal diet of 55,504 subjects was analyzed [43], and the average production of $kgCO_2e$/day resulted to be, in medium-meat-eaters, 1.95 and 1.48 times the amounts produced by VEG and LOV subjects, respectively. Soret [44] reported similar results for the average emissions of CO_2e

per year in 73,308 American nonvegetarians, respectively 1.41 and 1.28 times the amounts produced by vegetarians and semivegetarians, for an average calorie intake of about 1700 kcal.

In our LCA study, GHG emissions were analyzed by the indicators Ecoindicator99 (climate change subcategory), Ecopoint (NO*x*, NMVOC-Non Methane Volatile Organic Compounds, CO_2 subcategories) and EDIP (global warming subcategory), and contributed to the single score of the respective indicator for 5%–6% (Ecoindicator99), 34%–57% (Ecopoint) and 8%–10% (EDIP). For the lower calorie patterns, the 1600 kcal diets, the subscore for GHG emissions in OMN pattern was respectively: 4.56 and 1.35 times the score of the VEG and LOV patterns in Ecoindicator99; 3.71 and 1.30 times the score of the VEG and LOV patterns in Ecopoint; 4.60 and 1.34 times the score of the VEG and LOV patterns in EDIP. Again, it is worthwhile to underline that in the 1600 kcal OMN pattern, the single score of the total impact categories was 5.73 and 1.56 times (Ecoindicator99), 3.83 and 1.96 times (Ecopoint) and 3.66 and 1.31 times (EDIP) the single score of the VEG and LOV patterns, respectively. The data referred to the above mentioned subcategories, for all the dietary patterns and all the indicators, were summarized in Tables 3 and 4.

Although the importance of the above mentioned studies relies on the analysis of real diets, they have been performed with an approach different from the present study. In fact, our study evaluated theoretical diets, so we did not consider any difference in geographical zone or transportation, import-export food fluxes and related emissions during cooking and storing in the household/in restaurants. For these reasons, we think that a comparison among the results of the different studies is not possible, even if these studies on real diets confirm the lowest environmental impact of plant based diets: future studies, performed in other countries and evaluating the total impacts, are warranted.

4. Conclusions

The results of our study confirmed the findings reached over the last few years by the research in this field, and showed that the environmental impact of a diet is mainly related to the consumption of animal products. This is the main reason why the total environmental impact of various dietary patterns can differ so much, as we found in our study. This is true from every perspective: climate change, energy consumption, water requirements, waste disposal, soil usage, deforestation, chemical use, and impacts both environmental and social aspects—namely the possibility of feeding all the world's citizens.

The United Nations' UNEP report, "Assessing the Environmental Impacts of Consumption and Production", urges a global move towards an animal-product-free diet, identifying animal product consumption as one of the primary sources of environmental impact, pollution, greenhouse effect and resource waste. Factory farming is among the first four sectors labeled in the report as "First Priority", and we find meat and dairy processing in the first positions of the "Second Priority" sectors. The report conclusions recommend a "substantial worldwide diet change, away from animal products" in order to reach a sustainable food production and to be able to feed an increasing human population [45].

From the viewpoint of human health, the USDA DG implicitly lead in the same direction: the environmental impacts of the various diets proposed by the DG are relatively low, exactly because they contain a high proportion of plant food and a very limited amount of animal food; basically, they recommend a dietary pattern much more slanted towards a direct consumption of plant foods than the average dietary pattern followed by people in industrialized countries all over the world, whose animal food consumption is much higher than the USDA DG recommends [16,20,28].

In relation to this, it is important to notice that the dietary OMN and LOV USDA patterns are quite different from the most widespread and common dietary patterns followed by people in industrialized countries all over the world: the average person's consumption of animal food is much higher than that recommended by the DG [16]. As a consequence, there is likely to be a considerable difference between the estimated environmental impact of the "ideal" OMN and LOV diets recommended by the USDA DG and the impact of the "real" diets followed by most people in industrialized countries. This difference should be carefully assessed in future studies performed in different countries, as recently done in Germany [42], UK [43], and USA [44].

The composition of the healthy OMN and LOV diets recommended by the USDA DG is evidence of the desirability of a shift towards a much higher consumption of plant food, and a correspondingly much lower consumption of animal food, not only to reduce environmental impacts but also for health reasons. The consequences of a radical shift towards a plant-based diet would be many, all of them positive: a substantial influence on climate change, a profitable decrease in energy use and water waste, a lessening of the impact of deforestation, a much more rational use of soil (also leading to a dramatic decrease of chemicals use in agriculture).

The 2010 USDA DG should stimulate not only the scientific community but also national governments, international and scientific institutions, and the media, to promote a cultural shift: there is much that national and worldwide institutions, and the scientific community itself, can do in order to speed up the transition towards more environmentally sustainable, and healthier, dietary habits.

Acknowledgments

The authors thank the anonymous reviewers who made many useful suggestions to improve the manuscript. The authors wish to thank Carmen Dell'Aversano, of Pisa University, for her consulting role and helpful discussions. The authors are indebted to the owner of *Simapro*, Società Scientifica di Nutrizione Vegetariana-SSNV, Italy, who supplied the software.

Author Contributions

All authors contributed equally to this work.

Conflicts of Interest

The authors declare no conflict of interest.

References

1. Sinha, R.; Cross, A.; Graubard, B.; Leitzmann, M.; Schatzkin, A. Meat intake and mortality. *Arch. Intern. Med.* **2009**, *169*, 562–571.
2. American Institute for Cancer Research (AICR) & World Cancer Research Fund (WRF). *Food, Nutrition, Physical Activity and the Prevention of Cancer: A Global Perspective*; AICR: Washington, DC, USA, 2007.

3. Yokoyama, Y.; Nishimura, K.; Barnard, N.D.; Takegami, M.; Watanabe, M.; Sekikawa, A.; Okamura, T.; Miyamoto, Y. Vegetarian diets and blood pressure: A meta-analysis. *JAMA Intern. Med.* **2014**, *174*, 577–587.
4. Crowe, F.L.; Appleby, P.N.; Travis, R.C.; Key, T.J. Risk of hospitalization or death from ischemic heart disease among British vegetarians and nonvegetarians: Results from the EPIC-Oxford cohort study. *Am. J. Clin. Nutr.* **2013**, *97*, 597–603.
5. Key, T.; Fraser, G.; Thorogood, M.; Appleby P.; Beral, V.; Reeves, G.; Burr, M.L.; Chang-Claude, J.; Frentzel-Beyme, R.; Kuzma, J.W.; *et al.* Mortality in vegetarians and nonvegetarians: Detailed findings from a collaborative analysis of 5 prospective studies. *Am. J. Clin. Nutr.* **1999**, *70S*, S516–S524.
6. Larsson, S.; Virtamo, J.; Wolk, A. Red meat consumption and risk of stroke in Swedish men. *Am. J. Clin. Nutr.* **2011**, *94*, 417–421.
7. Pan, A.; Sun, Q.; Bernstein, A.; Schulze, M.; Manson, J.; Willett, W.; Hu, F.B. Red meat consumption and risk of type 2 diabetes: 3 cohorts of US adults and an updated meta-analysis. *Am. J. Clin. Nutr.* **2011**, *94*, 1088–1096.
8. Rizzo, N.S.; Sabaté, J.; Jaceldo-Siegl, K.; Fraser, G.E. Vegetarian dietary patterns are associated with a lower risk of metabolic syndrome: The adventist health study 2. *Diabetes Care* **2011**, *34*, 1225–1227.
9. Bradbury, K.E.; Crowe, F.L.; Appleby, P.N.; Schmidt, J.A.; Travis, R.C.; Key, T.J. Serum concentrations of cholesterol, apolipoprotein A–I and apolipoprotein B in a total of 1694 meat-eaters, fish-eaters, vegetarians and vegans. *Eur. J. Clin. Nutr.* **2014**, *68*, 178–183.
10. Tantamango-Bartley, Y.; Jaceldo-Siegl, K.; Fan, J.; Fraser, G. Vegetarian diets and the incidence of cancer in a low-risk population. *Cancer Epidemiol. Biomarkers Prev.* **2013**, *22*, 286–294.
11. Micha, R.; Wallace, S.; Mozaffarian, D. Red and processed meat consumption and risk of incident coronary heart disease, stroke and diabetes mellitus. A systematic review and meta-analysis. *Circulation* **2010**, *121*, 2271–2283.
12. Scarborough, P.; Allender, S.; Clarke, D.; Wickramasinghe, K.; Rayner, M. Modelling the health impact of environmentally sustainable dietary scenarios in the UK. *Eur. J. Clin. Nutr.* **2012**, *66*, 710–715.
13. Scarborough, P.; Nnoaham, K.E.; Clarke, D.; Capewell, S.; Rayner, M. Modelling the impact of a healthy diet on cardiovascular disease and cancer mortality. *J. Epidemiol. Community Health* **2012**, *66*, 420–426.
14. World Health Organization (WHO). *Micronutrient Malnutrition—Half of the World's Population Affected*; WHO: Geneva, Switzerland, 1996; Volume 78, pp. 1–4.
15. World Health Organization (WHO). *Turning the Tide of Malnutrition: Responding to the Challenge of the 21st Century*; WHO: Geneva, Switzerland, 2000.
16. Baroni, L.; Cenci, L.; Tettamanti, M.; Berati, M. Evaluating the environmental impact of various dietary patterns combined with different food production systems. *Eur. J. Clin. Nutr.* **2007**, *61*, 279–286.
17. Pimentel, D.; Pimentel, M. Sustainability of meat-based and plant-based diets and the environment. *Am. J. Clin. Nutr.* **2003**, *78*, S660–S663.

18. Beeton, R. Sustainably managing food production resources to maximise human nutritional benefit. *Asia Pac. J. Clin. Nutr.* **2003**, *12*, S50.
19. Imhoff, M.L.; Bounoua, L.; Ricketts, T.; Loucks, C.; Harriss, R.; Lawrence, W.T. Global patterns in human consumption of net primary production. *Nature* **2004**, *429*, 870–873.
20. U.S. Department of Agriculture. USDA Dietary Guidelines for Americans, 2010. Available online: http://www.cnpp.usda.gov/DietaryGuidelines.htm (accessed on 7 April 2014).
21. Harvard School of Public Health. Healthy Eating Plate & Healthy Eating Pyramid. Available online: http://www.hsph.harvard.edu/nutritionsource/healthy-eating-plate/ (accessed on 15 June 2014).
22. International Organization for Standardization (ISO). *ISO 14040. Environmental Management: Life Cycle Assessment, Principles and Guidelines*; ISO: Geneva, Switzerland, 2006.
23. Weber, C.L.; Matthews, H.S. Food-miles and the relative climate impacts of food choices in the United States. *Environ. Sci. Technol.* **2008**, *42*, 3508–3513.
24. Pré—Product Ecology Consultants. *SimaPro 7.3.3 Software*; Product Ecology Consultants: Amersfoort, The Netherlands, 2012.
25. Cassidy, E.S.; West, P.W.; Gerber, J.S.; Foley, J.A. Redefining agricultural yields: From tonnes to people nourished per hectare. *Environ. Res. Lett.* **2013**, *8*, 034015, doi:10.1088/1748-9326/8/3/034015.
26. Cederberg, C. *Life Cycle Assessment of Milk Production. SIK Report No. 643*; SIK, The Swedish Institute for Food and Biotechnology: Göteborg, Sweden, 1998.
27. CORINAIR. *The EMEP/CORINAIR Atmospheric Emission Inventory Guidebook*; European Environment Agency: Copenhagen, Denmark, 1996.
28. FAO. Food Balance Sheet; 2012. Available online: http://data.fao.org/dataset?entryId=48dc9161-53e2-4883-93c0-8f099e5e67ab (accessed on 7 April 2013).
29. Høgass Eide, M.; Ohlsson, T. A comparison of two different approaches to inventory analysis of diaries. *Int. J. Life Cycle Assess.* **1998**, *3*, 209–215.
30. Joyce, A.; Dixon, S.; Comfort, J.; Hallett, J. Reducing the environmental impact of dietary choice: Perspectives from a behavioural and social change approach. *J. Environ. Public Health.* **2012**, *2012*, 978672, doi:10.1155/2012/978672.
31. Leitzmann, C. Nutrition ecology: The contribution of vegetarian diets. *Am. J. Clin. Nutr.* **2003**, *78*, S657–S659.
32. Mekonnen, M.M.; Hoekstra, A.Y. A global assessment of the water footprint of farm animal products. *Ecosystems* **2012**, *15*, 401–415.
33. Mekonnen, M.M.; Hoekstra, A.Y. The green, blue and grey water footprint of crops and derived crop products. *Hydrol. Earth Syst. Sci.* **2011**, *15*, 1577–1600.
34. Reijnders, L.; Soret, S. Quantification of the environmental impact of different dietary protein choices. *Am. J. Clin. Nutr.* **2003**, *78*, S664–S668.
35. Renault, D.; Wallender, W.W. Nutritional water productivity and diets. *Agric. Water Manag.* **2000**, *45*, 275–296.
36. Rojstaczer. S.; Sterling, S.M.; Moore, N.J. Human appropriation of photosynthesis products. *Science* **2001**, *294*, 2549–2552.

37. Vanhama, D.; Mekonnen, M.M.; Hoekstrab, A.Y. The water footprint of the EU for different diets. *Ecol. Indic.* **2013**, *32*, 1–8.
38. Braunschweig, A. *Bewertung in Ökobilanzen mit der Methode der ökologischen Knappheit. Ökofaktoren 1997*; Methodik Für Oekobilanzen: Buwal Schriftenreihe Umwelt, 297, 1998 (in German). Available online: http://www.bafu.admin.ch/publikationen/publikation/00436/index.html?lang=de/ (accessed on 7 April 2013).
39. Frischknecht, R.; Büsser Knöpfel, S. Swiss Eco-Factors 2013 according to the Ecological Scarcity Method. Methodological Fundamentals and Their Application in Switzerland. Environmental Studies No. 1330; Federal Office for the Environment: Bern, 2013. Available online: http://www.bafu.admin.ch/publikationen/publikation/01750/index.html?lang=en&download=NHzLpZig7t,lnp6I0NTU042l2Z6ln1ad1IZn4Z2qZpnO2Yuq2Z6gpJCHdnx9e2ym162dpYbUzd,Gpd6emK2Oz9aGodetmqaN19XI2IdvoaCVZ,s-.pdf (accessed on 13 May 2014).
40. Blonk, H.; Ponsioen, T.; Kool, A.M.; Marinussen, M. *The Agri-Footprint Method; Methodological LCA Framework, Assumptions and Applied Data, Version 1.0*; Blonk Milieu Advies: Gouda, The Netherlands, 2011.
41. Vieux, F.; Darmon, N.; Touazi, D.; Soler. L.G. Changing the diet structure or consuming less? *Ecol. Econ.* **2012**, *75*, 91–101.
42. Meier, T.; Christen, O. Environmental impacts of dietary recommendations and dietary styles: Germany as an example. *Environ. Sci. Technol.* **2013**, *47*, 877–888.
43. Scarborough, P.; Appleby, P.N.; Mizdrak, A.; Briggs, A.D.M.; Travis, R.C.; Bradbury, K.E.; Key, T.J. Dietary greenhouse gas emissions of meat-eaters, fish-eaters, vegetarians and vegans in the UK. *Clim. Change* **2014**, doi:10.1007/s10584-014-1169-1 (Available online: http://link.springer.com/article/10.1007/s10584-014-1169-1 (accessed on 19 June 2014).
44. Soret, S.; Mejia, A.; Batech, M.; Jaceldo-Siegl, K.; Harwatt, H.; Sabaté, J. Climate change mitigation and health effects of varied dietary patterns in real-life settings throughout North America. *Am. J. Clin. Nutr.* **2014**, *100*, S490–S495.
45. United Nations Environment Programme (UNEP). *Assessing the Environmental Impacts of Consumption and Production: Priority Products and Materials, A Report of the Working Group on the Environmental Impacts of Products and Materials to the International Panel for Sustainable Resource Management*; UNEP: Nairobi, Kenya, 2010.

4

High Pressure Processing of Bivalve Shellfish and HPP's Use as a Virus Intervention †

David H. Kingsley

Food Safety and Intervention Technologies Research Unit, Agricultural Research Service, U.S. Department of Agriculture, James W.W. Baker Center, Delaware State University, Dover, DE 19901, USA; E-Mail: David.Kingsley@ars.usda.gov

† USDA is an equal opportunity provider and employer. Mention of trade names or commercial products in this article is solely for the purpose of providing specific information and does not imply recommendation or endorsement by the U.S. Department of Agriculture.

Abstract: Bivalve shellfish readily bioconcentrate pathogenic microbes and substance, such as algal and dinoflagulate toxins, fecal viruses and bacteria, and naturally present vibrio bacteria. High pressure processing (HPP) is currently used as an intervention for *Vibrio vulnificus* bacteria within molluscan shellfish and its potential to inactivate food-borne viruses and bacteria are discussed. Mechanisms of action of high pressure against bacteria and viruses, as well as how time of pressure application, pressure levels, and pre-pressurization temperature influence inactivation are described. Matrix influences such as ionic strength are noted as important additional considerations. The potential of HPP to influence spoilage and enhance shelf-life of shucked shellfish is also discussed.

Keywords: seafood safety; high pressure processing; bivalve shellfish

---

1. Introduction

High pressure processing (HPP) is an increasingly popular method of separating shellfish meat from the shell for molluscan shellfish (*i.e.*, oysters and clams) as well as crustacea (*i.e.*, crabs and lobsters). HPP shucking of shellfish results in completely intact meat and can be a considerable labor saving method [1]. Also HPP-treated shellfish can be opened easily by restaurant staff and does not require a dedicated onsite shucker (*i.e.*, raw bar). Furthermore a consumer survey has demonstrated that while consumers are not particularly knowledgeable about the concept of high pressure processing, they are apparently not resistant to eating foods treated by high pressure [2].

The pressures ordinarily used for food processing range from 200 to 600 megaPascals (MPa). One MPa is approximately equal to 9.87 atms of pressure and approximately 145 pounds per square inch (PSI). By comparison, the pressure at the bottom of the Marianas Trench in the Pacific is approximately 100 MPa. Essentially, HPP is a non-thermal process in that foods are not specifically heated. Bivalve shellfish have a number of potential food safety issues associated with their ability to bioconcentrate toxic chemicals and pathogens from water. These include human intestinal viruses and some fecal bacteria from wastewater and sewage, bacteria that are naturally present in estuarine waters, and toxins derived from plankton and dinoflagulates, which can have serious neurologic consequences for shellfish consumers.

1.1. Toxins

Harmful algae blooms and ocean dinoflagulates can produce a number of chemical toxins that can have rather nasty neurological properties. Examples include saxotin, paralytic shellfish poisoning, and amnesiac shellfish poisoning [3–11], the net result of which can be brain or motor nerve damage or even death. Testing for these agents is performed using GC-mass spectrophotometry [12] or bioassay using mice [10]. These toxins are thermally resistant, so even thorough cooking will not inactivate them. Unfortunately HPP is not believed to have any discernable effect on these toxins since HPP does not alter chemical bonds [13].

1.2. Vibrio

These bacteria are naturally found in abundance within warm tropical waters year round and in temperate waters during the summer season [14,15]. The most serious of these is *Vibrio vulnificus* (*Vv*) which is a substantial problem for the oyster industry. For ordinary persons with no underlying health issues, this bacterium does not pose a serious threat. However for the approximately 7% of the US population that have liver problems due to excessive alcohol consumption, hepatitis, or have diabetes, are immunocompromised, or have elevated blood iron levels, contracting a *Vv* infection carries a 50% risk of mortality [16–18]. While perhaps less than one hundred people contract this annually in the US by various routes of infection, there are as many as 10–20 mortalities associated with consumption of shellfish contaminated with *Vv* [19]. A second bacterium, *Vibrio parahaemolyticus* (*Vp*) is associated with oyster consumption, typically causes mild to moderate gastrointestinal illness which ordinarily persists for a few days but can last for several weeks [20]. While quite unpleasant, *Vp* is not associated with a high mortality rate.

Risk of contracting *Vv* and *Vp* by consumption of raw shellfish becomes substantially greater as increasing amounts of these of bacteria are consumed. Vibrio levels are known to increase dramatically after harvest in response to the shellfish being out of the water and in a warm environment [21–25]. As a result, many regulatory authorities require that loads of harvested shellfish be shaded from direct sunlight, harvested at cooler times of the day or even after dark, and mandate cooling protocols that cool shellfish to less than 10 °C within a few hours of harvest to reduce outgrowth of Vibrio bacteria [26,27]. These efforts have reduced the incidence of oyster-associated illness but these measures have not completely eliminated them. As an additional precaution, post-harvest processing is now required for *Vv* in some jurisdictions. [28]. Post-harvest *Vv* interventions used currently include irradiation [29], quick-freeze [30,31], flash heat [32] and HPP [33,34].

2. Fecal Bacteria and Sanitary Standards for Shellfish

Oysters were first recognized as vectors for typhoid fever in New York around the turn of the 20th century when market shellfish were being exposed to sewage effluent from the city [35]. Today, typhoid fever associated with raw shellfish consumption is virtually unheard of due to sewage treatment, institution of fecal coliform (FC) standards, and its limited circulation among the general public. Currently two separate, but conceptually similar FC standards are used in the US and EU [36,37]. The US classifies its shellfish growing waters based on regular testing to determine the levels of FC bacteria in waters, essentially measuring hygienic water quality. Other localities, such as the EU, measure the amount of fecal coliforms found directly within shellfish from different harvest locations, essentially measuring the quantity of FC bacteria found directly within oysters. With some caveats, these standards do a reasonable job of measuring human and animal fecal impact on growing areas and identifying areas where shellfish can be and should not be harvested [38]. That said, the potential to contract *Salmonella enterica* infections from raw oysters remains a qualified concern. In the US, one report indicated salmonella bacteria in US market shellfish had an incidence of 7.8% [39], but a subsequent report put this number at about 1% [40]. In addition, a recent report seems to indicate that salmonella is able to persist within live shellfish more efficiently than other coliform bacteria [41].

3. HPP as a Microbiologic Intervention

3.1. Bacteria and Spoilage

Generally Gram-negative spoilage bacteria are more sensitive to HPP than Gram-positive bacteria. For example, the Gram-positive bacterial population fraction within oysters were reported to increase from 56% to 84% Gram-positive after a 500 MPa treatment and storage at 2 °C for 28 days [42,43]. Because much of the typical off-odor associated with spoilage is due to the growth of Gram-negative bacteria, HPP is capable of extending the refrigerated shelf-life of oysters [44]. After HPP, a prominent reduction of bacterial diversity also occurs. For example after a 500 MPA treatment of oysters (*Crassostrea gigas*) and 28 days of storage at 7 °C Linton *et al.* [42] reported that 96% of bacterial isolates were limited to *Bacillus*, *Acintobacter*/*Moraxella* and lactic acid bacteria.

3.2. Spores and Fungi

He *et al.* [44] report that many fungi are inactivated by pressures ranging from 300 to 600 MPa. Bacterial spores are a challenge for HPP, although spore reduction has been noted by double-cycle HPP treatment in which pressure is applied, the pressure is released for a brief period to permit spore germination, and then reapplied to kill germinated spores [45]. Unfortunately some viable spores are able to survive these double-cycle treatments.

3.3. *Vibrio*

HPP is highly effective in reducing *Vv* in oyster meat to "non-detectable levels" when approximately 275–300 MPa are applied for 3 min at ambient temperatures [33,34]. While processing interventions are not currently mandated for *Vp* in the US, indications are that pressures slightly above 300 MPa can be effective as an intervention for raw shellfish [33,46,47].

3.4. *Fecal Bacteria*

Current commercial HPP treatments for oysters are not thought to be high enough to substantially reduce salmonella, but treatments >350 MPa in culture media did generally show 3-$\log_{10}$ reductions of *Salmonella enterica* [48], suggesting HPP has potential for inactivating this bacteria in shellfish. Other bacteria such as campylobacter [49], and shigella [50] are occasionally implicated in oyster-vectored food outbreaks. HPP is capable of inactivating these bacteria, but like salmonella, exact conditions required for inactivation within live shellfish have yet to be defined. Many of the food-borne bacteria found in shellfish are Gram-negative, which as stated earlier, are more susceptible to pressure inactivation than Gram-positive bacteria [51]

3.5. *Viruses*

Viruses remain a vexing problem for the shellfish industry. There are a large number of pathogenic viruses that can be shed from the human gastro-intestinal tract [52]. These viruses can enter shellfish growing areas as a result of sewage overflows and floods, defective septic systems, overboard waste discharge, or even as a result of a vomiting event [53]. Enteric viruses are shed at high levels, perhaps billions of particles per illness, and are typically highly infectious, with only a handful of particles required to establish an infection [54]. Viruses are very stable in the environment, and can persist in shellfish growing waters and even within shellfish for extended periods [55,56]. Because these viruses require a human gut to replicate, they do not grow, or amplify, within shellfish as a result of temperature abuse as can occur for bacteria. Rather they simply contaminate shellfish as a result of filter feeding activities, typically sequestering themselves at relatively low levels. Because direct testing for viruses is currently impractical, classification of shellfish harvest areas are regulated based on the levels of fecal bacteria in growing waters or directly within the shellfish meats, as described previously. While this classification system prevents a great deal of unsanitary shellfish from potentially reaching the dinner table, it is now recognized that low FC levels in growing waters, or within shellfish meats, do not necessarily indicate that there are no viruses present [57]. This is principally due to the fact that viruses can persist within shellfish tissues for longer periods than FCs [55,56,58]. Presumably this is due to virus' resistance of the acidic digestive processes of shellfish [56]. Thus, a suitable intervention for viruses potentially sequestered within raw shellfish would be of key significance.

Cooking and depuration, two traditional means of sanitizing shellfish, are of limited effectiveness against viruses. Depuration is a method in which live shellfish are placed in clean water for 2–3 days to permit the bivalves to pump and purge pathogens. This method is generally effective against fecal bacteria which can be reduced by several orders of magnitude, but it is accepted that pathogenic human viruses do not purge efficiently enough to make the process a viable intervention for virus contamination. Cooking is thought to inactivate viruses to a substantial degree, but it is unclear what temperatures and cooking times are completely effective against these viruses within shellfish. In fact, several documented outbreaks have been associated with "properly cooked shellfish" [59–61]. Furthermore, consumers often prefer uncooked or lightly-cooked shellfish.

Although there are many different types of fecal viruses that can be potentially transmitted by shellfish, the two principal shellfish-borne virus threats are recognized to be human norovirus (HuNoV) and hepatitis A virus (HAV). Human norovirus is now arguably considered to cause the majority of food-borne incidents worldwide [19,62]. Approximately half of the food-borne incidents associated with shellfish are due to norovirus and the overall fraction of food-borne noroviruses attributed to molluscan shellfish is approximately 13% [63,64].

Initial research on the potential of HPP to inactivate norovirus focused on genetically-related surrogate research viruses, such as feline calicivirus [65–68] and murine norovirus [69], because human norovirus strains have not been reproducibly propagated in the laboratory [70,71] and there are no suitable small animal research models for the virus. Surrogate work pointed to reasonable prospects for inactivation of norovirus. Feline calicivirus was found to be highly sensitive to HPP with 5 min room temperature treatments of 275 MPa being sufficient to inactivate 7-$\log_{10}$ of the virus [68,72]. The subsequent isolation and discovery of murine norovirus, which was propagable, made it possible to evaluate a closer genetic relative of human norovirus [73]. Results indicated that higher pressures, on the order of 400 MPa, were needed to inactivate substantial quantities of this virus [69]. However successful inactivation of murine norovirus within oysters was demonstrated at this pressure level [69,74]. Also work with immunocompromised mice confirmed that inactivation by HPP *in vitro*, as assessed by tissue culture and *in vivo*, as assessed using mice were essentially equivalent [75].

More recent work has looked at the potential of HPP to inactivate human norovirus. A human volunteer study evaluated conditions required to inactivate the prototype norovirus strain (GI.1 Norwalk). Four-$\log_{10}$ PFU of Norwalk virus was injected into oysters and three 5-min pressure treatments were performed at 400 MPa, 22 °C; 400 MPa, 6 °C; and 600 MPa, 6 °C. Unfortunately only the 600 MPa treatment was sufficient to protect all volunteers [76]. Based on a reduced illness frequency, it was postulated that the 400 MPa, 6 °C treatment may have inactivated some norovirus virus. Subsequent investigations using the newly developed porcine gastric mucin binding assay (PGM-MB) binding assay, which can assess norovirus inactivation, have confirmed that norovirus is sensitive to HPP at about 400 MPa [77–79].

HAV contamination of shellfish is now uncommon in most parts of the developed world due to improved hygienic standards and vaccination campaigns [80] but it remains a problem in the developing world and the Mediterranean region [81]. HAV illness can be quite serious, often resulting in hospitalization and occasionally in mortality. Morbidity and mortality due to HAV is often age related, with persons over the age of 50 being more prone to mortality and with young children often having only unapparent infections [82,83]. Considering that exposure induces immunity that is generally thought lifelong, HAV vaccination campaigns are often not given much emphasis in endemic regions since the local populations largely become immune at young ages. Unfortunately shellfish and uncooked fruits and vegetable products grown in endemic regions and sold in developed countries can become vectors for HAV outbreaks [84,85]. A tissue culture-adapted strain of HAV has been evaluated for sensitivity to HPP. Results indicate that a 5 min room temperature treatment at 450 MP is sufficient to inactivate 7-$\log_{10}$ of HAV virus stock [68]. Evaluation of HAV-contaminated oysters demonstrated a 3-$\log_{10}$ reduction after a 1 min-400 MPa treatment at 9 °C [86].

It was hoped that HPP would also inactivate other pernicious viruses which could potentially contaminate shellfish such as Aichi virus [87], hepatitis E virus (HEV; [88]), coxsackie viruses, *etc.* HEV has yet to be evaluated but Aichi and a number of other members of the picornavirus family have proven more tolerant to high pressure, requiring either pressures well above 400 MPa, or even being completely as resistant to 600 MPa treatments [89,90].

3.6. Parameters

Research has shown that there are a number of considerations for inactivating bacteria and viruses with HPP. Of course the primary determinant for pathogen inactivation is the pressure level applied, which generally follows first-order kinetics since plotting $\log_{10}$ pathogen reduction *versus* increasing pressure applied gives a straight line. Beyond pressure levels applied, time under pressure and pre-pressurization temperature can have a considerable influence on inactivation levels. For all pressure-sensitive viruses tested to date, increased application time does increase the amount of virus inactivation observed but the amount of increase observed asymptotically decreases, matching log-logistic or weibull kinetics [43,52,91]. Solutes, such as salt and sugar generally decrease the effectiveness of HPP inactivation for viruses and bacteria [43,51,67,68,92]. Formally speaking the reason for this is unknown, but presumably, the presence of solutes may tend to prevent compression and addition of more water molecules into the solvation cage surrounding the protein. Although currently undefined, this may be an important consideration for shellfish grown in different salinities, since the salt content of bivalves mimics the waters from which they have been harvested. Also, generally speaking, bacteria that are actively growing in exponential phase are more sensitive to pressure than bacteria in stationary phase [51].

Perhaps more intriguing is the concept that temperature has a substantial influence on inactivation. For vegetative bacteria, pressure applied above and below room temperature generally appears to enhance inactivation [51]. For viruses, the temperature effect is variable since different viruses react differently for HPP at different temperatures. For noroviruses and all caliciviruses tested to date, refrigeration temperatures dramatically enhance inactivation, often by several logs [52,69]. Curiously, HAV is the reverse. Room temperature and above dramatically enhance inactivation by HPP as compared to refrigeration temperatures [92,93]. Unlike the *Caliciviridae*, other picornaviruses have shown variable inactivation with respect to temperature [90]. Why viruses behave differently under pressure at different temperatures is currently unknown.

Presumably, HPP-treated shellfish would have a neutral pH, but for other food matrices, pH is another important consideration. Low pH generally enhances HPP inactivation of vegetative bacteria, but viruses respond differently to low pH under pressure. For human norovirus, low pH appears to be inhibitory for HPP inactivation [78]. For HAV, a virus that is known to be tolerant of pH 1, lower pH actually enhances HPP inactivation [92,94]. It is also important to note that weak organic acids, such as acetic acid, become stronger acids under pressure. Why HAV and HuNoV behave differently under pressure in acidic pH is currently unknown.

There can be a substantial temperature increase associated with the application of pressure. As anyone who has ever filled a scuba tank knows, when air inside the tank is compressed, heat is generated, making the tank warm. Likewise, anyone who has fully opened the valve of a pressured tank knows that as the pressure is released, the tank becomes quite cold. These effects are due to a principle called adiabatic heating and cooling [95]. Although the incremental amount of heating/cooling does vary somewhat with the initial temperature at which pressure is applied, the temperature increase is approximately 3–3.5 °C per 100 MPa for water-based commercial units. Also, the degree to which this heat dissipates to the environment varies with the size of HPP units, with smaller units dissipating heat to the environment more rapidly due to increased surface area to volume ratio.

3.7. Inactivation Mechanisms

HPP is not known to directly damage nucleic acids. However high pressure is thought to either distort or destroy lipid membranes in bacteria by causing phase inversion. Leakage of cytosolic contents as a result of membrane integrity loss accounts for much of vegetative bacterial inactivation. A second mechanism of inactivation is via protein denaturation. On a molecular level, water forms a solvation cage around proteins. When high pressure is applied, more water is forced into this solvation cage resulting in changes in the tertiary and quaternary structure of proteins disrupting function. Non-enveloped viruses, such as norovirus and HAV, by definition do not have lipids associated with them. Therefore inactivation of food-borne viruses is via protein denaturation.

3.8. Organoleptic Considerations

The color, taste, texture, appearance and smell of raw oysters are of paramount concern to the industry and consumers alike. Currently there is a perception by industry that pressures above 300 MPa result in undesirable changes to oyster quality. It is true that the degree to which HPP changes a raw oyster's characteristics is a function of the pressure applied and the temperature at which pressure is applied. Some whitening or blanching occurs when treating oysters at 600 MPa at room temperature but this is minimized when 600 MPa is performed at 5 °C. Overall appearance of oysters and clams on-the-half-shell is much better when shucked by HPP [96], since the bivalve meat is completely intact (see Figure 1). This differs from a hand-shucked oyster which can often be sliced by the shucking knife. There are reports that HPP can induce some firmness or chewiness in seafoods, but this change is relatively subtle and may be considered desirable, since firmness can be considered an attribute of freshness [13]. Juiciness and flavor are enhanced by HPP, since shellfish take up liquid from the surrounding liquor within the shell [44,97,98]. This attribute results in a "yield" increase since the shucked oyster becomes more voluminous due to absorption of liquor fluid. One drawback is that liquid taken up does not remain within the bivalve tissues over time. Thus an originally full jar of HPP-shucked oysters will be reduced in volume a week later. Also depending on what the oysters were feeding on when harvested, sometimes this liquid in the jar can be an unappealing yellow or greenish color.

Figure 1. A hand-shucked clam (left; *Mercenaria mercenaria*) is compared to a high pressure processing (HPP)-treated clam (right). Note: this figure is reprinted with permission from [9]. Copyright Elsevier B.V. 2014.

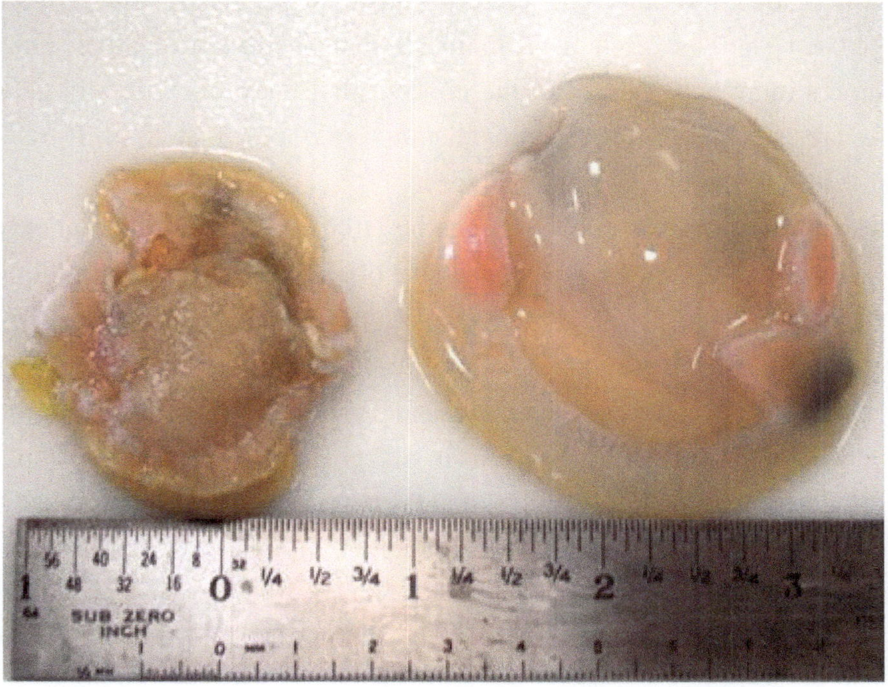

Studies clearly indicate that HPP-treated oysters are well received after treatments at pressures that are higher than are currently used commercially to shuck oysters and inactivate vibrio [99]. For example, recent results (shown in Table 1) from an organoleptic study evaluating the taste of oysters has shown that oysters treated at 400–600 MPa at 6 °C and 300–500 MPa at 22 °C are actually preferred to manually-shucked oysters [100]. Another challenge for HPP-treated oysters is time between processing and consumption may be several days which may compromise their perceived quality as compared to a fresh-shucked oyster.

Table 1. Organoleptic analysis of HPP-treated [#] oysters * using untrained volunteers.

	Control	300 MPa 22 °C	400 MPa 22 °C	500 MPa 22 ° C	400 MPa 6 °C	500 MPa 6 °C	600 MPa 6 °C	*F* value (Sig *p* value)
Appearance	4.11 ± 1.6	5.46 ± 1.5	5.39 ± 1.4	5.20 ± 1.6	5.39 ± 1.5	5.45 ± 1.4	5.22 ± 1.7	5.78 (0.000)
Texture	4.54 ± 1.9	5.23 ± 1.7	5.36 ± 1.8	5.55 ± 1.6	5.20 ± 1.6	5.47 ± 1.5	5.43 ± 1.7	2.46 (0.024)
Flavor	4.64 ± 1.7	5.04 ± 1.8	5.05 ± 1.7	5.13 ± 1.7	4.86 ± 1.6	5.35 ± 1.6	5.27 ± 1.6	1.24 (0.287)
Aroma	4.90 ± 1.4	5.27 ± 1.3	5.04 ± 1.2	5.30 ± 1.3	5.27 ± 1.4	5.33 ± 1.3	5.33 ± 1.4	0.94 (0.469)
Acceptability	4.64 ± 1.6	5.14 ± 1.6	5.13 ± 1.6	5.28 ± 1.6	5.02 ± 1.5	5.53 ± 1.4	5.38 ± 1.6	2.05 (0.058)

[#] All HPP treatments were for 5 min; * analysis performed using triploid oysters (*Crassostrea virginica*) obtained from Cape May NJ August 2013; ± represents standard deviation.

3.9. Challenges and Future Directions

There are a number of challenges to the widespread application of HPP to the shellfish industry. First HPP is relatively expensive with the minimal cost for a commercial scale unit being several hundred thousand US dollars. Thus to be economically viable, shellfish harvesters and processors must be relatively large in scale. Currently most shell aquaculture and fishing operations are too small to successfully amortize this expense. Development of a low cost pressure unit for limited shellfish quantities that is suitable for use near or at the point of consumption would be a boon for the shellfish industry and consumers alike. Another challenge is consumer resistance and regulations regarding in shell-shellfish. A closed shell is often used to judge that shellfish are alive and fresh. HPP kills shellfish, so consumers must be educated to the concept that these are safe and good to eat. In most jurisdictions, there are no longer prohibitions against sale of in-shell HPP-treated shellfish but regulations against theses may remain in some places.

Another potential challenge is that shellfish must be HPP-treated relatively quickly after harvest. If the liquor inside the shell dries out or is reduced in volume permitting air within the shell, as can happen after a day or two of cold storage, the shells may be cracked, or even crushed when pressure is applied due to air pockets under the shell (personal observation). Lastly, should HPP treatment become mainstream as a pathogen intervention technique for shellfish, it will be important that current hygienic quality standards for safe harvest remain in place. HPP should only be applied in addition to current hygienic standards, not in lieu of these standards or as a way to utilize shellfish grow under non-hygienic conditions.

4. Conclusions

High pressure processing is a viable nonthermal intervention for prominent food-borne pathogens associated with raw bivalve shellfish.

Conflicts of Interest

The author declares no conflict of interest.

References

1. Martin, D.E.; Hall, S.G. Oyster shucking technologies: Past and present. *Int. J. Food Sci. Technol.* **2006**, *41*, 223–232.
2. Hicks, D.T.; Pivarnik, L.F.; McDermott, R.; Richard, N.; Hoover, D.G.; Kniel, K.E. Consumer awareness and willingness to pay for high pressure processing of ready-to eat food. *J. Food Sci. Ed. Res.* **2009**, *8*, 32–38.
3. Andjelkovic, M.; Vandevijvere, S.; van Klaveren, J.; van Oyen, H.; van Loco, J. Exposure to domoic acid through shellfish consumption in Belgium. *Environ. Int.* **2012**, *49*, 115–119.
4. Ciminello, P.; Fattorusso, E. *Bivalve molluscs* as vectors of marine biotoxins involved in seafood poisoning. *Prog. Mol. Subcell. Biol.* **2006**, *43*, 53–82.
5. Gerssen, A.; Pol-Hofstad, I.E.; Poelman, M.; Mulder, P.P.; van den Top, H.J.; de Boer, J. Marine toxins: Chemistry, toxicity, occurrence and detection, with special reference to the Dutch situation. *Toxins* **2010**, *2*, 878–904.
6. Hinder, S.L.; Hays, G.C.; Brooks, C.J.; Davies, A.P.; Edwards, M.; Walne, A.W.; Gravenor, M.B. Toxic marine microalgae and shellfish poisoning in the British Isles: History, review of epidemiology and future implications. *Environ. Health* **2011**, *10*, doi:10.1186/1476-069X-10-54.

7. Jeffery, B.; Barlow, T.; Moizer, K.; Paul, S.; Boyle, C. Amnesic shellfish poison. *Food Chem. Toxicol.* **2004**, *42*, 545–557.
8. Lee, K.J.; Mok, J.S.; Song, K.C.; Yu, H.; Jung, J.H.; Kim, J.H. Geographic and annual variation in lipophilic shellfish toxins from oysters and mussels along the south coast of Korea. *J. Food Prot.* **2011**, *74*, 2127–2133.
9. Lefebvre, K.A.; Robertson, A. Domoic acid and human exposure risks: A review. *Toxicon* **2010**, *56*, 218–230.
10. Suzuki, H. Susceptibility of different mice strains to okadaic acid, a diarrhetic shellfish poisoning toxin. *Food Addit. Contam. A Chem. Anal. Control Expo. Risk Assess.* **2012**, *29*, 1307–1310.
11. Watkins, S.M.; Reich, A.; Fleming, L.E.; Hammond, R. Neurotoxic shellfish poisoning. *Mar. Drugs* **2008**, *6*, 431–455.
12. McNabb, P.S.; Selwood, A.I.; van Ginkel, R.; Boundy, M.; Holland, P.T. Determination of brevetoxins in shellfish by LC/MS/MS: Single-laboratory validation. *J. AOAC Int.* **2012**, *95*, 1097–1105.
13. Murchie, L.W.; Cruz-Promero, M.; Kerry, J.P.; Linton, M.; Patterson, M.F.; Smiddy, M.; Kelly, A.L. High pressure processing of shellfish: A review of microbiological and other quality aspects. *Innov. Food Sci. Emerg. Technol.* **2005**, *6*, 257–270.
14. Turner, J.W.; Malayil, L.; Guadagnoli, D.; Cole, D.; Lipp, E.K. Detection of *Vibrio parahaemolyticus*, *Vibrio vulnificus* and *Vibrio cholerae* with respect to seasonal fluctuations in temperature and plankton abundance. *Environ. Microbiol.* **2014**, *4*, 1019–1028.
15. Motes, M.L.; DePaola A.; Cook, D.W.; Veazey, J.E.; Hunsucker, J.C.; Garthright, W.E.; Blodgett, R.J.; Chirtel, S.J. Influence of water temperature and salinity on *Vibrio vulnificus* in northern Gulf and Atlantic Coast oysters (*Crassostrea virginica*). *Appl. Environ. Microbiol.* **1998**, *64*, 1459–1465.
16. Desenclos, J.A.; Klontz, K.C.; Wolfe, L.E.; Hoecheri, S. The risk of Vibrio illness in the Florida raw oyster eating population, 1981–1988. *Am. J. Epidemiol.* **1991**, *134*, 290–297.
17. Drake, S.L.; DePaola, A.; Jaykus, L. An overview of *Vibrio vulnificus* and *Vibrio parahaemolyticus*. *Comp. Rev. Food Sci. Saf.* **2007**, *6*, 120–144.
18. Thacket, C.O.; Brenner, F.; Blake, P.A. Clinical features and epidemiological study of *Vibrio vulnificus* infections. *J. Infect. Dis.* **1984**, *149*, 558–561.
19. Mead, P.S.; Slutsker, L.; Dietz, V.; McCaig, L.F.; Bressee, J.S.; Shapiro, C.; Griffin, P.M.; Tauxe, R.V. Food-related illness and death in the United States. *Emerg. Infect. Dis.* **1999**, *5*, 607–625.
20. Altekruse, S.F.; Bishop, R.D.; Baldy, L.M.; Thompson, S.G.; Wilson, S.A.; Ray, B.J.; Griffin, P.M. Vibrio gastroenteritis in the US Gulf of Mexico region: The role of raw oysters. *Epidemiol. Infect.* **2000**, *124*, 489–495.
21. Cook, D.W. Effect of time and temperature on multiplication of *Vibrio vulnificus* in postharvest Gulf coast shellstock oysters. *Appl. Environ. Microbiol.* **1994**, *60*, 3483–3484.
22. Cook, D.W.; Ruple, A.D. Indicator bacteria and *Vibrionaceae* multiplication in post-harvest shellstock oysters. *J. Food Prot.* **1989**, *52*, 343–349.

23. Gooch, J.A.; DePaola, A.; Bowers, J.; Marshall, D.L. Growth and survival of *Vibrio parahaemolyticus* in postharvest American oysters. *J. Food Prot.* **2002**, *65*, 970–974.
24. Johnson, W.G.; Salinger, A.C.; King, W.C. Survival of *Vibrio parahaemolyticus* in oyster shellstock at two different storage temperatures. *Appl. Microbiol.* **1973**, *26*, 122–123.
25. Lorca, T.A.; Pierson, M.D.; Flick, G.J.; Hackney, C.R. Levels of *Vibrio vulnificus* and organoleptic quality of raw shellstock oysters (*Crassostrea virginica*) maintained at different storage temperatures. *J. Food Prot.* **2001**, *64*, 1716–1721.
26. Burnham, V.E.; Janes, M.E.; Jaykus, L.A.; Supan, J.; DePaola, A.; Bell, J. Growth and survival differences of *Vibrio vulnificus* and *Vibrio parahaemolyticus* strains during cold storage. *J. Food Sci.* **2009**, *74*, 314–318.
27. Cook, D.W. Refrigeration of oyster shellstock: Conditions which minimize the outgrowth of *Vibrio vulnificus*. *J. Food Prot.* **1997**, *60*, 349–352.
28. Muth, M.K.; Karns, S.A.; Anderson, D.W.; Murray, B.C. Effects of post-harvest treatment requirements for the markets for oysters. *Agric. Resour. Econ. Rev.* **2002**, *31*, 171–186.
29. Jakabi, M.; Gelli, D.S.; Torre, J.C.; Rodas, M.A.; Franco, B.D.; Destro, M.T.; Landgrafi, M. Inactivation by ionizing radiation of *Salmonella enteritidis*, *Salmonella infantis*, and *Vibrio parahaemolyticus* in oysters (*Crassostrea brasiliana*). *J. Food Prot.* **2003**, *66*, 1025–1029.
30. Liu, C.; Lu, J.; Su, Y. Effects of flash freezing, followed by frozen storage, on reducing *Vibrio parahaemolyticus* in Pacific raw oysters (*Crassostrea gigas*). *J. Food Prot.* **2009**, *72*, 174–177.
31. Parker, R.W.; Maurer, E.M.; Childers, A.B.; Lewis, D.H. Effect of frozen storage and vacuum packaging on survival of *Vibrio vulnificus* in Gulf Coast oysters (*Crassostrea virginica*) *J. Food Prot.* **1994**, *57*, 604–606.
32. Andrews, L.S.; Park, D.L.; Chen, Y.P. Low temperature pasteurization to reduce the risk of vibrio infections from raw shell-stock oysters. *Food. Add. Cont.* **2000**, *17*, 787–791.
33. Berlin, D.L.; Herson, D.S.; Hicks, D.T.; Hoover, D.G. Response of pathogenic *Vibrio* species to high hydrostatic pressure. *Appl. Environ. Microbiol.* **1999**, *65*, 2776–2780.
34. Ye, M.; Huang, Y.; Gurtler, J.B.; Niemera, B.A.; Sites, J.E.; Chen, H. Effects of pre- and post-processing storage conditions on high hydrostatic pressure inactivation of *Vibrio parahaemolyticus* and *V. vulnificus* in oysters. *Int. J. Food Microbiol.* **2013**, *163*, 146–152.
35. Rippey, S.R. Infectious diseases associated with molluscan shellfish consumption. *Clin. Microbiol. Rev.* **1994**, *7*, 419–425.
36. Calci, K.R.; DePaola, A.; Burkhardt, W., III. Chapter 49: Molluscan shellfish; oysters, mussels, clams. In *Compendium of Methods for the Microbiological Examination of Foods*; Doores, S., Salfinger, Y., Tortorello, M.L., Wilcke, B.W., Eds.; American Public Health Association: Washington, DC, USA, 2013.
37. Olivera, J.; Cunha, A.; Catilho, F.; Romalde, J.L.; Pereira, M.J. Microbial contamination and purification of bivalve shellfish: Crucial aspects in monitoring and future perspectives—A mini-review. *Food Cont.* **2011**, *22*, 805–816.
38. Hood, M.A.; Ness, G.E.; Blake, N.J. Relationship among fecal coliforms, *Escherichia coli* and *Salmonella* spp. in shellfish. *Appl. Environ. Microbiol.* **1983**, *45*, 122–126.

39. Brands, D.A.; Inman, A.E.; Gerba, C.P.; Maré, C.J.; Billington, S.J.; Saif, L.A.; Levine, J.F.; Joens, L.A. Prevalence of *Salmonella* ssp. in oysters in the United States. *Appl. Environ. Microbiol.* **2005**, *71*, 893–897.
40. DePaola, A.; Jones; J.L.; Woods, J.; Burkhardt, W., III; Calci, K.R.; Krantz, J.A.; Bowers, J.C.; Kasturi, K.; Byars, R.H.; Jacobs, E.; *et al*. Bacterial and viral pathogens in live oysters: 2007 United States market survey. *Appl. Environ. Microbiol.* **2010**, *76*, 2754–2768.
41. Morrison, C.M.; Armstrong, A.E.; Evans, S.; Mild, R.M.; Langdon, C.J.; Joens, L.A. Survival of *Salmonella newport* in oysters. *Int. J. Food Microbiol.* **2011**, *148*, 93–98.
42. Linton, M.; McClements, J.M.J.; Patterson, M.F. Changes in the microbial quality of shellfish brought about by treatment with high hydrostatic pressure. *Int. J. Food Sci. Technol.* **2003**, *38*, 713–727.
43. Smelt, J.P.P.M. Recent advances in the microbiology of high pressure processing. *Trends Food Sci. Technol.* **1998**, *9*, 152–158.
44. He, H.; Adams, R.M.; Farkas, D.F.; Morrissey, M.T. Use of high pressure processing for oyster shucking and shelf-life extension. *J. Food Prot.* **2002**, *67*, 640–645.
45. Shearer, A.E.; Dunne, C.P.; Sikes, A.; Hoover, D.G. Bacterial spore inhibition and inactivation in foods by pressure, chemical preservatives, and mild heat. *J. Food Prot.* **2000**, *63*, 1503–1510.
46. Kural, A.G.; Shearer, A.E.H.; Kingsley, D.H.; Chen, H. Conditions for high pressure inactivation of *Vibrio parahaemolyticus* in oysters. *Int. J. Food Microbiol.* **2008**, *127*, 1–5.
47. Ma, L.; Su, Y.C. Validation of high pressure processing for inactivating *Vibrio parahaemolyticus* in Pacific oysters (*Crassostrea gigas*). *Int. J. Food Microbiol.* **2011**, *144*, 469–474.
48. Sherry, A.E.; Patterson, M.F.; Madden, R.H. Comparison of 40 *Salmonella enterica* servars injured by thermal, high pressure and irradiation stress. *J. Appl. Microbiol.* **2004**, *96*, 887–893.
49. Griffin, M.R.; Dalley, E.; Fitzpatrick, M.; Austin, S.H. Campylobacter gastroenteritis associated with raw clams. *J. Med. Soc. N. J.* **1983**, *80*, 607–609.
50. Reeve. G.; Martin, D.L.; Pappas, J.; Thompson, R.E.; Greene, K.D. An outbreak of shigellosis associated with the consumption of raw oysters. *N. Engl. J. Med.* **1989**, *321*, 224–227.
51. Considine, K.M.; Kelly, A.L.; Fitzgerald, G.F.; Hill, C.; Sleator, R.D. High pressure processing—Effects on microbial food safety and food quality. *FEMS Microbiol. Lett.* **2008**, *281*, 1–9.
52. Kingsley, D.H. High pressure processing and its application to the challenge of virus-contaminated Foods. *Food Environ. Virol.* **2013**, *5*, 1–12.
53. Kingsley, D.H. Food-borne noroviruses. In *Genomes of Food- and Water-borne Pathogens*; Fratamico, P., Kathariou, S., Liu, Y., Eds.; ASM Press: Washington, DC, USA, 2011; pp. 237–245.
54. Teunis, P.F.; Moe, C.L.; Liu, P.; Miller, S.E.; Lindesmith, L.; Baric, R.S.; le Pendu, J.; Calderon, R.L. Norwalk virus: How infectious is it? *J. Med. Virol.* **2008**, *80*, 1468–1476.
55. Kingsley, D.H.; Richards, G.P. Persistence of hepatitis A virus within oysters. *J. Food Prot.* **2003**, *66*, 331–334.
56. Provost, K.; Dancho, B.A.; Ozbay, G.; Anderson, R.; Richards, G.; Kingsley, D.H. Hemocytes are sites of persistence for enteric viruses within oysters. *Appl. Environ. Microbiol.* **2011**, *77*, 8360–8369.

57. Love, D.C.; Lovelace, G.L.; Sobsey, M.D. Removal of *Escherichia coli*, *Enterococcus fecalis*, coliphage MS2, poliovirus, and hepatitis A virus from oysters (*Crassostrea virginica*) and hard shell clams (*Mercinaria mercinaria*) by depuration. *Int. J. Food Microbiol.* **2010**, *143*, 211–217.
58. Grohmann, G.S.; Murphy, A.M.; Christopher, P.J.; Auty, E.; Greenberg, H.B. Norwalk gastroenteritis in volunteers consuming depurated oysters. *Aust. J. Exp. Biol. Med. Sci.* **1981**, *59*, 219–228.
59. Chalmers, J.W.; McMillan, J.H. An outbreak of viral gastroenteritis associated with adequately prepared oysters. *Epidemiol. Infect.* **1995**, *115*, 163–167.
60. Kirkland, K.B.; Meriwether, R.A.; Leiss, J.K.; MacKenzie, W.R. Steaming oysters does not prevent Norwalk-like gastroenteritis. *Public Health Rep.* **1996**, *111*, 527–530.
61. McDonnell, S.; Kirkland, K.B.; Hlady, W.G.; Aristeguieta, C.; Hopkins, R.S.; Monroe, S.S.; Glass, R.I. Failure of cooking to prevent shellfish associated gastroenteritis. *Arch. Int. Med.* **1997**, *157*, 111–116.
62. Scallan, E.; Hoekstra, R.M.; Angulo, F.J.; Tauxe, R.V.; Widdowson, M.A.; Roy, S.L.; Jones, J.L.; Griffin, P.M. Foodborne illness acquired in the United States—Major pathogens. *Emerg. Infect. Dis.* **2011**, *17*, 7–15.
63. Butt, A.A.; Aldridge, K.E.; Sanders, C.V. Infections related to the ingestion of seafood part I: Viral and bacterial infections. *Lancet Infect. Dis.* **2004**, *4*, 201–212.
64. Hall, A.J.; Eisenbart, V.G.; Etingüe, A.L.; Gould, L.H.; Lopman, B.A.; Parashar, U.D. Epidemiology of foodborne norovirus outbreaks, United States, 2001–2008. *Emerg. Infect. Dis.* **2012**, *18*, 1566–1573.
65. Buckow, R.; Isbarn, S.; Knorr, D.; Hienz, V.; Lehmacher, A. Predictive model for inactivation of feline calcivirus, a norovirus surrogate, by heat and high hydrostatic pressure. *Appl. Environ. Microbiol.* **2008**, *74*, 1030–1038.
66. Chen, H.; Hoover, D.G.; Kingsley, D.H. Temperature and treatment time influence high hydrostatic pressure inactivation of feline calicivirus, a norovirus surrogate. *J. Food Prot.* **2005**, *68*, 2389–2394.
67. Kingsley, D.H.; Chen, H. Aqueous matrix composition influences feline calicivirus inactivation by high pressure processing. *J. Food Prot.* **2008**, *71*, 1598–1603.
68. Kingsley, D.H.; Hoover, D.; Papafragkou, E.; Richards, G.P. Inactivation of hepatitis A virus and a calicivirus by high hydrostatic pressure. *J. Food Prot.* **2002**, *65*, 1605–1609.
69. Kingsley, D.H.; Holliman, D.R.; Calci, K.R.; Chen, H.; Flick, G.J. Inactivation of a norovirus by high pressure processing. *Appl. Environ. Microbiol.* **2007**, *73*, 581–585.
70. Duizer, E.; Schwab, K.J.; Neill, F.H.; Atmar, R.L.; Koopmans M.P.G.; Estes, M.K. Laboratory efforts to cultivate noroviruses. *J. Gen. Virol.* **2004**, *85*, 79–87.
71. Herbst-Kralovetz, M.M.; Radtke, A.L.; Lay, M.K.; Hjelm, B.E.; Bolick, A.N.; Sarker, S.S.; Atmar, R.L.; Kingsley, D.H.; Arntzen, C.J.; Estes, M.K.; *et al.* Lack of norovirus replication and histo-blood group antigen expression in 3-D intestinal epithelial cell cultures. *Emerg. Infect. Dis.* **2013**, *19*, 431–438.
72. Grove, S.F.; Forsyth, S.; Wan, J.; Coventry, J.; Cole, M.; Stewart, C.M.; Lewis, T.; Ross, T.; Lee, A. Inactivation of hepatitis A virus, poliovirus, and a norovirus surrogate by high pressure processing. *Innov. Food Sci. Emerg. Technol.* **2008**, *9*, 206–210.

73. Wobus, C.E.; Thackray, L.B.; Virgin, H.W., IV. Murine norovirus: A model system to study norovirus biology and pathogenesis. *J. Virol.* **2006**, *80*, 5104–5112.
74. Arcangeli G.; Terregino, C.; de Benedictis, P.; Zecchin, B.; Manfrin, A.; Rossetti, E.; Magnabosco, C.; Mancin, M.; Brutti, A. Effect of high hydrostatic pressure on murine norovirus in Manila clams. *Lett. Appl. Microbiol.* **2012**, *54*, 325–329.
75. Gogal, R.M., Jr.; Kerr, R.; Kingsley, D.H.; Granata, L.A.; LeRoith, T.; Holliman, S.D.; Dancho, B.A.; Flick, G.J., Jr. High hydrostatic pressure processing of murine norovirus-1 contaminated oysters inhibits oral infection in STAT-1$^{-/-}$ deficient female mice. *J. Food Prot.* **2011**, *74*, 209–214.
76. Leon, J.S.; Kingsley, D.H.; Montes, J.S.; Richards, G.P.; Lyon, G.M.; Abdulhafid, G.M.; Seitz, S.R.; Fernandez, M.L.; Teunis, P.F.; Flick, G.J.; *et al.* Randomized, double-blinded clinical trial for human norovirus inactivation in oysters by high hydrostatic pressure processing. *Appl. Environ. Microbiol.* **2011**, *77*, 5476–5482.
77. Dancho, B.A.; Chen, H.; Kingsley, D.H. A method to discriminate between infectious and inactive human noroviruses. *Int. J. Food Microbiol.* **2012**, *155*, 222–226.
78. Li, X.; Chen, H.; Kingsley, D.H. The influence of temperature, pH, and water immersion on the high hydrostatic pressure inactivation of GI.1 and GII.4 human noroviruses. *Int. J. Food Microbiol.* **2013**, *167*, 138–143.
79. Ye, M.; Li, X.; Kingsley, D.H.; Jiang, X.; Chen, H. Inactivation of human norovirus in contaminated oysters and clams by high-hydrostatic pressure. *Appl. Environ. Microbiol.* **2014**, *80*, 2248–2253.
80. Jacobsen, K.H.; Koopman, J.S. Declining hepatitis A virus seroprevalence: A global review and analysis. *Epidemiol. Infect.* **2004**, *132*, 1005–1022.
81. Franco, E.; Meleleo, C.; Serino, L.; Sorbara, D.; Zaratti, L. Hepatitis A: Epidemiology and prevention in developing countries. *World J. Hepatol.* **2012**, *4*, 68–73.
82. Fiore, A.E. Hepatitis A transmitted by food. *Clin. Infect. Dis.* **2004**, *38*, 705–715.
83. FitzSimons, D.; Hendrickx, G.; Vorsters, A.; van Damme, P. Hepatitis A and E: Update on prevention and epidemiology. *Vaccine* **2010**, *28*, 583–588.
84. Hepatitis A outbreak associated with green onions at a restaurant—Monaca, Pennsylvania, 2003. *Morb. Mortal. Wkly. Rep.* **2003**, *52*, 1155–1157.
85. Sánchez, G.; Pintó, R.M.; Vanaclocha, H.; Bosch, A. Molecular characterization of hepatitis A virus isolates from a transcontinental shellfish-borne outbreak. *J. Clin. Microbiol.* **2002**, *40*, 4148–4155.
86. Calci, K.R.; Meade, G.K.; Tetzloff, R.C.; Kingsley, D.H. High pressure inactivation of hepatitis A virus within oysters. *Appl. Environ. Microbiol.* **2005**, *71*, 339–343.
87. Le Guyader, F.S.; le Saux, J.C.; Ambert-Balay, K.; Krol, J.; Serais, O.; Parnaudeau, S.; Giraudon, H.; Delmas, G.; Pommepuy, M.; Pothier, P.; *et al.* Aichi virus, norovirus, astrovirus enterovirus and rotavirus involved in clinical cases from a French oyster-related gastroenteritis outbreak. *J. Clin. Microbiol.* **2008**, *46*, 4011–4017.
88. Crossan, C.; Baker, P.J.; Craft, J.; Takeuchi, Y.; Dalton, H.R.; Scobie, L. Hepatitis E virus genotype 3 in shellfish United Kingdom. *Emerg. Infect. Dis.* **2012**, *18*, 2085–2087.

89. Kingsley, D.H.; Chen, H.; Hoover, D. Inactivation of selected picornaviruses by high hydrostatic pressure. *Virus Res.* **2004**, *102*, 221–224.
90. Kingsley, D.H.; Li, X.; Chen, H. Temperature effects for high pressure processing of picornaviruses. *Food Environ. Virol.* **2014**, *6*, 58–61.
91. Chen, H.; Hoover, D.G. Use of Weibull model to describe and predict pressure inactivation of *Listeria monocytogenes* Scott A in whole milk. *Innov. Food Sci. Emerg. Technol.* **2004**, *5*, 269–276.
92. Kingsley, D.H.; Chen, H. Influence of pH, salt, and temperature on pressure inactivation of hepatitis A virus. *Int. J. Food Microbiol.* **2009**, *130*, 61–64.
93. Kingsley, D.H.; Guan, D.; Hoover, D.G.; Chen, H. Inactivation of hepatitis A virus by high pressure processing: The role of temperature and pressure oscillation. *J. Food Prot.* **2006**, *69*, 2454–2459.
94. Kingsley, D.H.; Guan, D.; Hoover, D.G. Pressure inactivation of hepatitis A virus in strawberry puree and sliced green onions. *J. Food Prot.* **2005**, *68*, 1748–1751.
95. Balasubramanian, S.; Balasubramaniam, V.M. Compression heating influence of pressure transmitting fluids on bacteria inactivation during high pressure processing. *Food Res. Int.* **2003**, *36*, 661–668.
96. Mootian, G.K.; Flimlin, G.E.; Karwe, M.; Schaffner, D.W. Inactivation of *Vibrio parahaemolyticus* in hard clams (*Mercanaria mercanaria*) by high hydrostatic pressure (HHP) and the effect of HHP on the physical characteristics of hard clam meat. *J. Food Sci.* **2013**, *78*, E251–E257.
97. Cruz-Romero, M.; Kelly, A.L.; Kerry, J.P. Effects of high-pressure and heat treatments on physical and biochemical characteristics of oysters (*Crassostrea gigas*). *Innov. Food Sci. Emerg. Technol.* **2007**, *8*, 30–38.
98. Narwanker, S.P.; Flimlin, G.E.; Schaffner, D.W.; Tepper, B.J.; Karwe, M.V. Microbial safety and consumer acceptability of high-pressure processed hard clams (*Mercenaria mercenaria*). *J. Food Sci.* **2011**, *76*, M375–M380.
99. Lopez-Caballero, M.E.; Perez-Mateos, M.; Montero, P.; Borderias, A.J. Oyster preservation by high-pressure treatment. *J. Food Prot.* **2000**, *63*, 196–201.
100. Kingsley, D.H.; Kuhn, D.D.; Flick, G.J.; Oh, J.; Lawson, L.S.; Meade, G.K.; Giesecke, C.C. Desirability of oysters treated by high pressure processing at different temperatures and elevated pressures. *Am. J. Food Technol.* **2014**, in press.

5

Presence of *Listeria monocytogenes* in Mediterranean-Style Dry Fermented Sausages

Domenico Meloni

Department of Veterinary Medicine, University of Sassari, Via Vienna 2, 07100, Sassari, Italy; E-Mail: dmeloni@uniss.it

Academic Editor: Christopher J. Smith

Abstract: The morphological, physiological and epidemiological features of *L. monocytogenes*, together with the severity of human listeriosis infections, make *L. monocytogenes* of particular concern for manufacturers of cold-stored "ready to eat" (RTE) foods. *L. monocytogenes* has been isolated from a wide variety of RTE foods and is responsible for several outbreaks associated with the consumption of RTE meat, poultry, dairy, fish and vegetable products. Although *L. monocytogenes* is among the most frequently-detected pathogens in dry fermented sausages, these products could be included in the category of RTE products in which the growth of *L. monocytogenes* is not favored and have rarely been implicated in listeriosis outbreaks. However, *L. monocytogenes* is highly difficult to control in fermented sausage processing environments due to its high tolerance to low pH and high salt concentration. In many Mediterranean-style dry fermented sausages, an empirical application of the hurdle technology often occurs and the frequent detection of *L. monocytogenes* in these products at the end of ripening highlights the need for food business operators to properly apply hurdle technology and to control the contamination routes of *L. monocytogenes* in the processing plants. In the following, through an up-to-date review of (personal and un-) published data, the main aspects of the presence of *L. monocytogenes* in Mediterranean-style dry fermented sausages will be discussed.

Keywords: *Listeria monocytogenes*; Mediterranean area; dry fermented sausages; hurdle technology

---

1. Introduction to the Main Features of *Listeria monocytogenes*

1.1. Taxonomy

The genus Listeria comprises fifteen species, *i.e.*, *L. monocytogenes*, *L. ivanovii*, *L. innocua*, *L. welshimeri*, *L. seeligeri*, *L. grayi*, *L. marthii*, *L. rocourtiae*, *L. leichmannii*, *L. weihenstephanensis*, *L. floridensis*, *L. aquatic*, *L. cornellensis*, *L. riparia* and *L. grandensis* (Meloni, 2014 [1]).

1.2. Morphology of Listeria monocytogenes

Members of the genus *Listeria* are non-spore-forming, facultative anaerobic and small Gram-positive rods (0.5–4 µm in diameter and 0.5–2 µm in length). Peritrichous flagella give them a typical tumbling motility, occurring at 20–25 °C. Based on somatic (O) and flagellar (H) antigens, 13 serotypes of *L. monocytogenes* have been recognized. These are identified alphanumerically: 1/2a, 1/2b, 1/2c, 3a, 3b, 3c, 4a, 4ab, 4b, 4c, 4d, 4e and 7 (Meloni, 2014 [1]). Serotypes 1/2a, 1/2b and 1/2c are the most frequently isolated from food or the food production environment.

1.3. Physiology of Listeria monocytogenes

L. monocytogenes is catalase-positive, oxidase-negative and is able to survive between 0 and 45 °C. The optimum growth temperature is around 30–37 °C. *L. monocytogenes* can grow at pH ranges between 4.5 and 9.0 (optimum pH between 6 and 8) and is able to multiply in food matrices at water activity (a_w) values of 0.92 and in NaCl concentrations of 12%, generally lethal to other microorganisms. *L. monocytogenes* is a ubiquitous organism, widely distributed in the environment: the principal reservoirs are soil, forage and water (Sauders and Wiedmann, 2007 [2]; Todd and Notermans, 2011 [3]; European Food Safety Authority (EFSA), 2014 [4]). Other reservoirs include healthy humans and animals (International Life Sciences Insistute (ILSI), 2005 [5]) or infected domestic and wild animals (EFSA, 2014 [4]). *L. monocytogenes* is a psychrotrophic bacterium, can multiply at low temperatures, both under aerobic and anaerobic conditions, adapt to disinfectants and adhere to various surfaces (Arevalos-Sánchez *et al.*, 2012 [6]). *L. monocytogenes* is widespread in food processing facilities and has been isolated from different processing environments. Once introduced into the processing plants, it is able to survive and persist for a long time under adverse conditions (Farber and Peterkin, 1991 [7]; Gram *et al.*, 2007 [8]; Gandhi and Chikindas, 2007 [9]). The biofilm forming ability is an important cause for such persistence (Cruz and Fletcher; 2012 [10]; Fonnesbech Vogel *et al.*, 2001 [11]). In the pork meat supply chain, *L. monocytogenes* has been repeatedly isolated (Nesbakken *et al.*, 1996 [12]), with an increase of contamination along the production line (Chasseignaux *et al.*, 2002 [13]).

2. *Listeria monocytogenes* as a Foodborne Pathogen

2.1. Pathogenicity of Listeria monocytogenes

L. monocytogenes is the etiologic agent of listeriosis. Human cases of listeriosis are almost exclusively caused by *L. monocytogenes*. Very rare cases of infections are attributed to *L. ivanovii* and

L. seeligeri. The difference in the pathogenic potential of *L. monocytogenes* strains has been demonstrated by means of *in vivo* bioassay and *in vitro* cell assay (Soni *et al.*, 2014 [14]). Whereas some *L. monocytogenes* strains are naturally virulent, inflicting high morbidity and mortality, others are non-virulent and unable to infect the mammalian host (Liu *et al.*, 2003 [15]; Velge and Roche, 2010 [16]). The discrimination between pathogenic and non-pathogenic strains is imperative to assess the possible significance of this microorganism from food safety and public health aspects (Jensen *et al.*, 2008 [17]; Roberts *et al.*, 2009 [18]). Rasmussen *et al.* (1995 [19]) and Wiedmann *et al.* (1997 [20]) demonstrated that molecular typing methods can also allow *L. monocytogenes* to divide into three evolutionary lineages characterized by different pathogenic potentials: Lineage I, strains associated with epidemic outbreaks of listeriosis (serotypes 1/2b, 3b, 4b, 4d and 4e); Lineage II, strains isolated from sporadic cases of listeriosis (serotypes 1/2a, 1/2c, 3a and 3c); Lineage III, strains rarely associated with cases of listeriosis (serotypes 4a and 4c) (Wiedmann, 2002 [21]). The lineage status of serotypes 4ab and 7 still remains unclear due to limited availability of such strains (World Organization for Animal Health (OIE), 2014 [22]). The majority of the infections caused by *L. monocytogenes* are thought to be food-borne, and infections most often affect the central nervous system, the bloodstream and the pregnant uterus. Two forms of listeriosis have been described in humans, and symptoms vary, ranging from febrile gastroenteritis in healthy people (Piana *et al.*, 2005 [23]), to life-threatening invasive infections characterized by septicemia and meningoencephalitis in risk groups, such as young, old, pregnant and immune-compromised (YOPI) people (De Cesare *et al.*, 2007 [24]).

2.2. Epidemiology of Listeriosis

Listeriosis is an important disease in Europe: it is the fourth most common zoonotic disease, and it has an annual incidence of 0.41 cases per 100,000 population, with the highest notification rates in Finland, Spain and Denmark (EFSA, 2014 [4]). In 2012, 1642 confirmed human cases were reported, mostly domestically acquired. A statistically significant increasing trend was observed over the period 2008–2012, with a +10.5% increase compared with 2011 (EFSA, 2014 [4]). As reported by previous authors, the highest notification rates were reported in persons aged 65 years and above (Denny and McLauchlin, 2008 [25]) and in persons aged below one year. Transmission during pregnancy was highlighted in 79% of the cases reported in newborns, with the spread of the infection to the fetus which is born severely ill. Listeriosis has the highest hospitalization rate cases of all zoonoses under EU surveillance: 91.6% of the cases with supplementary long-term sequelae. Listeriosis is the third leading cause of death in the EU after West Nile fever and trichinellosis and the first among food-borne pathogens, with an estimated case fatality rate of 17.8% (EFSA, 2014 [4]). A total of 198 deaths were reported in 2012, the highest number of fatal cases reported since 2006 (EFSA, 2014 [4]). The morphological, physiological and epidemiological features of *L. monocytogenes*, together with the severity of human listeriosis infections, make *L. monocytogenes* of particular concern for manufacturers of cold-stored "ready to eat" (RTE) foods (Romanova *et al.*, 2002 [26]; Van Coillie *et al.*, 2004 [27]; Shen *et al.*, 2006 [28]). *L. monocytogenes* has been isolated from a wide variety of RTE foods and is responsible for several outbreaks associated with the consumption of RTE meat, poultry, dairy, fish and vegetable products (Aureli *et al.*, 2000 [29]; Gillespie *et al.*, 2006 [30]; Public

Health Agency Canada (PHAC), 2009 [31]; U.S. Department of Agriculture's Food Safety and Inspection Service (USDA/FSIS), 2010 [32]; Todd and Notermans, 2011 [3]). The EU legislation (Regulation (EC) No. 2073/2005 [33]) lays down food safety criteria for *L. monocytogenes* in RTE foods, and it is generally considered that concentrations of *L. monocytogenes* greater than 100 CFU/g are required to cause human disease in healthy populations. Qualitative results alone are not necessarily an indicator of public health risk. In RTE products intended for infants and for special medical purposes, *L. monocytogenes* must be absent in 25 g, while in RTE products in which the growth of *L. monocytogenes* is not favored (pH $\leq$ 4.4 or $a_w \leq 0.92$ or pH $\leq$ 5.0 and $a_w \leq 0.94$), *L. monocytogenes* must not be present at levels exceeding 100 CFU/g during the shelf-life. In RTE foods able to support its growth, *L. monocytogenes* must be absent in 25 g at the time of leaving the production plant. However, if the producer can demonstrate that the product will not exceed the limit of 100 CFU/g throughout its shelf-life, this criterion does not apply (European Commission, 2005 [33]). In 2012 the highest reported levels of non-compliance in RTE samples taken at processing were observed in fishery products (8.0%) and unspecified cheeses (3.4%). The highest proportions of food samples exceeding the legal safety limit, at retail, were observed in fishery products (0.5%) and fermented meat sausages (0.4%) (EFSA, 2014 [4]). Pork meat products, e.g., fermented sausages that are contaminated by *L. monocytogenes* at more than 100 CFU/g and that are to be consumed without further heat treatment, are considered to form a direct risk to human health.

3. *Listeria monocytogenes* in the Pork Meat Processing Industry

3.1. Swine Slaughterhouses

L. monocytogenes has been previously found in every stage along the pork processing industry (Thévenot *et al.*, 2006 [34]; López *et al.*, 2008 [35]), including swine slaughterhouses (Sammarco *et al.*, 1997 [36]; Korsak *et al.*, 1998 [37]). The sources of *L. monocytogenes* contamination during swine slaughtering are pig-related and environmental (Bonardi *et al.*, 2002 [38]). *L. monocytogenes* is spread to the carcass mainly from the carrier animal: the pathogen has been occasionally isolated from feces and from the skin of healthy carriers with intestinal colonization (Autio *et al.*, 2000 [39]).

3.1.1. Prevalence of *Listeria monocytogenes*

The prevalence in feces is generally comprised between 0% and 50% (Belœil *et al.*, 2003 [40]). This wide range of prevalence is probably due to the emptying of the rectum before evisceration, an operation that usually helps to reduce the extent of dissemination and the consequent fecal contamination of carcasses (Kanuganti *et al.*, 2002 [41]). The role of live animals as a source of processing environment contamination and, consequently, pork carcasses has been demonstrated: the most contaminated areas are usually represented by the area of stunning/hanging (Gobat and Jemmi, 1991 [42]; Nesbakken *et al.*, 1994 [43]; Saide-Albornoz *et al.*, 1995 [44]; Borch *et al.*, 1996 [45]). Contamination may occur during the evisceration, because of the breaking of the intestine (Adesyiun *et al.*, 1995 [46]), and reaches a prevalence of around 60%–65% (Thévenot *et al.*, 2005 [47]; López *et al.*, 2008 [35]), highlighting the profound influence of good hygiene practices and equipment cleanliness on carcass contamination (Bonardi *et al.*, 2002 [38]; Meloni *et al.*, 2013 [48]). Several authors (Buncic *et al.*, 1991 [49]; Ripamonti *et al.*, 2002 [50]; Kanuganti *et al.*, 2002 [41]; Autio *et al.*, 2003 [51]; Fabbi *et al.*, 2005 [52]) correlated the contamination of equipment and consequently of the carcasses with the presence of the pathogen in other niches, such as the tongue (14%) and tonsils (7%–61%). This wide range of prevalence is probably due to differences in sampling techniques and/or methods of farm management.

3.1.2. Serotypes of *Listeria monocytogenes*

The most frequent serotypes found in carcasses and slaughterhouse environments are 1/2a, 1/2c (Hof and Rocourt, 1992 [53]; Thévenot *et al.*, 2005 [47]; Meloni *et al.*, 2013 [48]), while serotype 1/2b is generally found at low prevalence rates.

3.2. Meat Processing Plants

The level of *L. monocytogenes* contamination tends to increase along the pork supply chain (López *et al.*, 2008 [35]). Raw meat is an important source of contamination of working environments and equipment. The most contaminated zones are the areas of the receipt of raw materials, the cells of refrigeration and the processing rooms (Chasseignaux *et al.*, 2002 [13]).

3.2.1. Prevalence of *Listeria monocytogenes*

A higher prevalence is found in raw meat (45%–50%) compared to the muscles of freshly slaughtered pigs (0%–2%). Raw meat represents the primary source of contamination of final products by *L. monocytogenes* (Giovannacci *et al.*, 1999 [54]; Kathariou, 2002 [55]; Kanuganti *et al.*, 2002 [41]; Thévenot *et al.*, 2005 [47]). In turn, due to the presence of favorable conditions for growth and multiplication during the processing stages of cooling and cutting, the prevalence of *L. monocytogenes* in minced meat intended to be processed ranges between 16% and 50% (Jay, 1996 [56]; Chasseignaux *et al.*, 2002 [13]). The level of contamination increases significantly up to 70%–100% during the processing stages of grinding and bagging (Nesbakken *et al.*, 1996 [12]; Thévenot *et al.*, 2005 [47]). Pork meat can also be cross-contaminated through contact with work surfaces and equipment. The level of contamination of the surfaces in contact and without contact with meat during processing ranges between 17%–50% and 11%–25%, respectively (Thévenot *et al.*, 2005 [47]; Mureddu *et al.*, 2014 [57]).

3.2.2. Persistence of *Listeria monocytogenes*

Once introduced into the plants, *L. monocytogenes* can persist over time in the processing environment (López *et al.*, 2008 [35]), forming assemblages of surface-associated microbial cells enclosed in hydrated extracellular polymeric substances and growing in biofilms (Mafu *et al.*, 1990 [58]; Gandhi and Chikindas, 2007 [9]; Meloni *et al.*, 2012 [59]). In recent surveys carried out in fermented sausage processing plants located in Italy (Meloni *et al.*, 2012 [59]; Meloni *et al.*, 2014 [60]; Mureddu *et al.*, 2014 [57]), the evaluation of the *in vitro* biofilm production of *L. monocytogenes* strains isolated from several surfaces in contact and without contact with meat showed a low short-time persistence (3–4 months) capacity and weak or moderate ability to form biofilm after 24 to 40 h of incubation. Isolates from serotypes 1/2a, 1/2b and 4b presented higher adherence when compared to isolates from serotype 1/2c (Meloni *et al.*, 2014 [60]). Harborage sites, such as meat grinders, work tables or floor drains, can be critical sites for the processing plant environment (Tompkin, 2002 [61]). This could be due to the common presence of meat in these environmental niches: this would likely produce a common food-conditioning film, which might select for adhesion, growth and biofilm formation by isolates with a common prophage type (Verghese *et al.*, 2011 [62]). Decontaminating surfaces in contact and without contact with meat is especially challenging, because, when entrapped in a biofilm, *L. monocytogenes* is afforded unusual protection against available disinfectants and treatments (Zhao *et al.*, 2004 [63]). Without suitable sanitization procedures, the presence of *L. monocytogenes* increases the food safety risk (Samelis and Metaxopoulos, 1999 [64]; Meloni *et al.*, 2014 [60]). Previous authors have reported that the recommended concentrations of commercial sanitizers are higher than required (Cruz *et al.*, 2012 [10]). In a recent survey (Mureddu *et al.*, 2014 [57]), the *in vitro* evaluation of the resistance to disinfectants (chlorine substances and quaternary ammonium compounds, both at 37% concentration) showed a reduction of *L. monocytogenes* growth after 24, 48 and 72 h of incubation in isolates from processing environments and finished products.

3.2.3. Serotypes of *Listeria monocytogenes*

Several studies have shown that strains of *L. monocytogenes* isolated from meat-processing environments belong mainly to serotypes 1/2c and 1/2a (Chasseignaux *et al.*, 2002 [13]; Thévenot *et al.*, 2006 [34]; Mureddu *et al.*, 2014 [57]; Meloni *et al.*, 2014 [60]).

4. Production of Mediterranean-Style Dry Fermented Sausages

Mediterranean-style dry fermented sausages are characterized by their relatively longer shelf-life and the exceptional hygienic background, which is brought about by the production of lactic acid in the fermentation process (pH < 4.5–5) and low water activity (<0.90) of the final product (Ordóñez and de la Hoz, 2007 [65]). In general, in the manufacturing of fermented sausages, meat and fat involves selection, chopping and mincing and mixing with curing ingredients, spices and authorized additives. At the end of the ripening and drying process, they come out as cured meat products (Ordóñez and de la Hoz, 2007 [65]). Traditionally, fermented sausages are made using lactic acid bacteria (LAB) and Gram-positive catalase positive cocci, in particular coagulase-negative staphylococci (CNS) naturally present in the meat or with the inoculation of starter cultures at the chopping step. The mixture is then filled in natural or artificial casings, left to ferment and then dried. In the Mediterranean area, regional customs, environmental variations and family recipes have given rise to a wide range of fermented sausages, and it can be said that there are almost as many types of sausages as there are geographical regions or even manufacturers, although their production process always requires the combination of fermentation and dehydration (Ordóñez and de la Hoz, 2007 [65]). The very wide range of Mediterranean dry fermented sausages can be classified according to a range of criteria (Table 1), such as the acidity, the mincing size of meat and fat, the addition or absence of molds on the surface, the addition of ingredients and the diameter and type of casing used (Ordóñez and de la Hoz, 2007 [65]).

4.1. Preliminary Stages

The meat used depends on eating habits, customs and the preferences prevailing in the geographical region where the fermented sausage is produced (Table 1). This is usually pork, sometimes mixed with beef (Ordóñez and de la Hoz, 2007 [65]). The fat should be firm, with a high melting point and a low content of polyunsaturated fatty acids, because this causes the fermented sausage to turn rancid more quickly (Frey, 1985 [66]). Mincing of the meat and fat is done at low temperatures (between −5 and 0 °C) to achieve a clean cut and to avoid the release of intramuscular fat from fatty meats, which could cause changes in the color and the drying process during ripening (Frey, 1985 [66]). Once the meat and fat have been comminuted, the starter culture (LAB) and the nitrate reducing CNS, curing salts, additives (nitrates, nitrites, glutamate) and other ingredients (sugars, aromatic herbs and spices) are added. The mixture, after refrigerated storage overnight, is placed in a kneader and stuffed into natural or synthetic casings (Greco *et al.*, 2005 [67]; Ordóñez and de la Hoz, 2007 [65]). The sausages are then ripened.

4.2. Fermentation and Ripening

After filling and the first warming up at 20–22 °C for 4–6 h, the fermentation stage for the manufacture of a standard dry fermented sausage can be summarized as follows: one to two days at 18–24 °C and 60% relative humidity (RH) and five days at 15 °C and 70% RH (Ordóñez and de la Hoz, 2007 [65]). After fermentation, ripening is carried out for five to 15 days in store rooms at 15 °C and 70%–75% RH. These conditions are maintained until the end of the ripening period, during which many flavor compounds develop (Ordóñez and de la Hoz, 2007 [65]). The normal pH of the majority of Mediterranean-style fermented sausages is close to 4.5/5.4, which has several beneficial effects on both the manufacturing process and the shelf-life (Greco *et al.*, 2005 [67]; Ordóñez and de la Hoz, 2007 [65]). However, in some low acid fermented sausages (e.g., Soudjouk, Fuet), the final pH is close to 6.0/6.7. The suppression of the acid hurdle can compromise the safety of these products (Ordóñez and de la Hoz, 2007 [65]; Jofré *et al.*, 2009 [68]). At the end of ripening, the water activity of fermented sausages is close to 0.90, which inhibits bacterial growth. The water activity hurdle is strengthened with time and is largely responsible for the stability of fermented sausages (Ordóñez and de la Hoz, 2007 [65]).

4.3. Hurdle Technology in Mediterranean-Style Dry Fermented Sausages

L. monocytogenes is inhibited in fermented sausages by sequential steps: the "hurdle technology" concept includes several sequential hurdles, essential at different stages of the fermentation or ripening process (Barbuti and Parolari, 2002 [69]). Due to the sequence of these hurdles, pathogenic and spoilage bacteria are effectively inhibited in Mediterranean-style dry fermented sausages, and the desired competitive flora (especially LAB) is selected (Leistner, 1995 [70]). These hurdles are essential in different steps of the fermentation or ripening process and lead to stable and safe final products (Leistner and Gould, 2002 [71]). In the early steps of the fermentation process, nitrite and salt added together in the form of nitrite-curing salts inhibit many bacteria in the initial product, such as pseudomonads and other Gram-negative oxidative organisms, which rapidly multiply and spoil uncured meats in the presence of oxygen (Leistner and Gould, 2002 [71]). Other bacteria, such as CNS, are able to multiply, use up the oxygen and cause the decrease of the redox potential of the product to decrease. CNS are also important, because of other biochemical-metabolic properties, such as lipolytic activity. Together with the nitrate reduction, these affect the quality (color and flavor) and the stability of the products (Selgas *et al.*, 1994 [72]; Greco *et al.*, 1999 [73]; Mazzette *et al.*, 1999 [74]). This, in turn, favors the redox potential hurdle, which inhibits aerobic bacteria and promotes the selection of LAB (Leistner and Gould, 2002 [71]). This competitive microflora flourish by metabolizing the added sugars, producing lactic acid, bacteriocins and inhibitory metabolites, causing a decrease in pH value and an increase of the pH hurdle (Papa, *et al.*, 1993 [75]; Torriani *et al.*, 1994 [76]; Grazia *et al.*, 1998 [77]; Leroy and de Vuyst, 1999 [78]; Hebert *et al.*, 2000 [79]; Lucke, 2000 [80]). These properties explain why LAB are important as starter cultures in the manufacturing of dry fermented sausages (Greco *et al.*, 2005 [67]). This is of particular importance for the microbial stability of quick-ripened fermented sausages, which are not greatly dried. In long-ripened fermented sausages, nitrite is depleted, and lactic acid bacteria slowly die. On the contrary, the redox potential and pH increase again (Leistner and Gould, 2002 [71]). Only the water activity hurdle is strengthened with time, and this hurdle is then largely responsible for the stability of long-ripened sausages (Leistner, 1987 [81]). This sequence of hurdles inhibits pathogenic and spoilage bacteria inside Mediterranean-style dry fermented sausages, whereas undesirable mold growth on the surface of the sausages is inhibited by smoke or by the use of desirable mold starter cultures (Leistner and Gould, 2002 [71]).

Table 1. Formulation (g/100 g) of typical Mediterranean-style dry fermented sausages *.

Formulation	Salchichón/Saucisson (Spain/France) (1)	Salami (Italy) (1)	Salsiccia Sarda (Italy) (2)	Fuet (Spain) (1)	Chorizo (Spain/Portugal) (1)	Lukanka (Bulgaria) (3)
Lean pork	35/70	45–84	85–87	60–70	65–80	25
Pork fat	10–25	14–25	5–8	30–40	20–40	20
Lean beef	0–50	0–37	-	0–20	0–20	55
Sugars	0.2–0.5	0.3–0.7	0.8	0.1–0.4	0.6–0.8	-
Curing salts	2.0–2.4	1.8–2.5	3	2.0–2.4	1.8–2.1	2.24
Whole/ground black pepper	0/0.2–0.2/0.4	0/0.2–0.1/0.14	0.25	0/0.2–0/0.3	-	0.30
White pepper	-	0–0.2	0.8	-	0–0.3	-
Paprika	-	-	-	-	0/1.5–1.5/2.5	-
Red pepper	-	-	-	-	-	0.20
Cumin	-	-	-	-	-	0.20
Garlic	-	0–0.2	0.15	-	0.2–1.2	-
Sodium Glutamate	0.25	-	-	0–0.15	-	-
Powdered milk	0–0.6	0–2.5	-	-	0–2.5	-
Caseinate	0–0.6	-	-	0–1.0	-	-
Liquid smoke	-	-	-	-	-	0.20

* Adapted and modified from Ordóñez and de la Hoz, 2007 [65]. Sources: (1) Ordóñez and de la Hoz, 2007 [65]; (2) Meloni *et al.*, 2012 [59]; (3) Balev *et al.*, 2005 [82].

5. *Listeria monocytogenes* in Mediterranean-Style Dry Fermented Sausages

5.1. Prevalence of Listeria monocytogenes

Fermented meat products may be contaminated by *L. monocytogenes* at several stages. The raw materials may be contaminated from the slaughterhouse environment, during the production process or by contact with contaminated unprocessed raw materials, unclean surfaces or people (Chasseignaux *et al.*, 2002 [13]; Thévenot *et al.*, 2006 [34]) in the post-processing stages (Colak *et al.*, 2007 [83]). *L. monocytogenes* is among the most frequently-detected pathogens in dry fermented sausages, and several studies have documented the prevalence of the pathogen in fermented sausages (Table 2), reaching prevalence levels of up to 40%–45% (Cantoni *et al.*, 1989 [84]; Cordano and Rocourt, 2001 [85]; Levine *et al.*, 2001 [86]; Thévenot *et al.*, 2005 [47]; De Cesare *et al.*, 2007 [24]; Meloni *et al.*, 2009 [87]; Meloni *et al.*, 2012 [59]; Mureddu *et al.*, 2014 [57]; Meloni *et al.*, 2014 [60]; Doménech *et al.*, 2015 [88]).

5.2. Levels of Contamination by Listeria monocytogenes

Fermented sausages contaminated with *L. monocytogenes* have rarely been implicated in critical listeriosis outbreaks (EFSA, 2014 [4]). Fermented sausages have moderate rates of consumption and serving sizes in many countries. The risk per serving is low (2.1×10^{-12}), and the global number of annual cases per 100,000 people is only 0.0000055 (USDA/FSIS, 2010 [32]). Many Mediterranean-style dry fermented sausages could be included in the category of RTE products in which the growth of *L. monocytogenes* is not favored, although there is great variability depending on the local traditions that influence fermentation and ripening (Hospital *et al.*, 2012 [89]). Mostly in the manufacturing of traditional fermented sausages marketed locally or regionally, an empirical application of the hurdle technology often occurs. Some manufacturers tend to reduce the ripening period in order to increase profitability. As a matter of fact, the pH and aw of these products are often within the limits for growth of *L. monocytogenes* (Hospital *et al.*, 2012 [89]). Insufficiently dried sausages may have water activity levels close to 0.92–0.94 (Meloni *et al.*, 2014 [60]), and *L. monocytogenes* is able to survive during sausage fermentation, overcoming the hurdles encountered during the manufacturing process. In general, the contamination levels at the end of ripening are always lower than 100 CFU/g (Farber and Peterkin, 1991 [7]), because *L. monocytogenes* cannot compete with the prevailing lactic acid bacteria. Only without competitive microflora *L. monocytogenes* is able to multiply and reach high levels of contamination (higher than 1000 CFU/g), representing a major public health concern (McLauchlin *et al.*, 2004 [90]; Thévenot *et al.*, 2006 [34]).

5.3. Serotypes of Listeria monocytogenes

As already noted for raw meat and meat-processing environments, also in the Mediterranean-style sausages at the end of ripening, serotypes 1/2c, 1/2a and 1/2b are more often detected (Jay, 1996 [56]; Thévenot *et al.*, 2006 [34]; Meloni *et al.*, 2014 [60]; Mureddu *et al.*, 2014 [57]), while serotype 4b is more rarely seen (Greenwood *et al.*, 1991 [91]; Hayes *et al.*, 1991 [92]). In Italy, serotype 1/2a is increasing its importance in the epidemiology of listeriosis. An increase of cases due to serotype 1/2a and a decline in cases due to serotype 4b have been reported by several authors (Gianfranceschi *et al.*,

2009 [93]; Pontello *et al.*, 2012 [94]). Despite the low prevalence rates of serotype 4b, previous surveys have shown that Lineage I strains of serotype 4b belonging to a clonal group (DUP-ID 1038) linked to several listeriosis outbreaks (De Cesare *et al.*, 2007 [24]) were recently recovered in Mediterranean-style dry fermented sausages produced in Italy (Meloni *et al.*, 2009 [87]).

Table 2. Prevalence of *Listeria monocytogenes* in naturally-contaminated Mediterranean-style dry fermented sausages *.

Type of Mediterranean-style dry fermented sausage	Prevalence%	Concentration data	pH of the final product	a_w of the final product
Fermented sausage **	10	<3 CFU/g	4.7–5.4	0.78–0.90
Italian salami **	13.3	Presence in 25 g	4.8–5.2	0.85–0.90
Soudjouk **	7	Presence in 25 g	4.9–6.7	*nd*
Fermented sausage **	3.25	Presence in 25 g	*nd*	*nd*
Fermented sausage **	20	Presence in 25 g	*nd*	*nd*
Fermented sausage **	19.05	Presence in 25 g	*nd*	*nd*
Fermented sausage **	44	Presence in 25 g	*nd*	*nd*
Fermented sausage **	20	Presence in 25 g	*nd*	*nd*
Salami **	16.67	Presence in 25 g	*nd*	*nd*
Salsiccia **	11.54	Presence in 25 g	*nd*	*nd*
Fermented sausage **	20	Presence in 25 g	*nd*	*nd*
Salami **	10	Presence in 25 g	*nd*	*nd*
Salami **	16	Presence in 25 g	*nd*	*nd*
Salami **	5	20 CFU/g	*nd*	*nd*
Fermented sausage **	10	Presence in 25 g	*nd*	*nd*
Salami **	40	Presence in 25 g	*nd*	*nd*
Spanish-style sausage **	3.70	Presence in 25 g	*nd*	*nd*
Salsiccia Sarda ***	20	Presence in 25 g	5.32	0.90
Salsiccia Sarda ****	8	Presence in 25 g	5.37	0.91

* Adapted and modified from Skandaminis and Nychas, 2007 [95]. *nd:* no data. Based on: ** Skandaminis and Nychas, 2007 [95]; *** Meloni *et al.*, 2009 [87]; **** Meloni *et al.*, 2014 [60].

6. Conclusions

The outcome of the previous paragraphs can be summarized from a safety standpoint as follows: Mediterranean-style fermented sausages may be contaminated with *L. monocytogenes* from various sources, including raw meat, slaughterhouse environments, production processes and post-processing conditions. In order to prevent these contamination sources, good manufacturing practices, correct sampling schemes, adequate cleaning and disinfection procedures and HACCP principles have to be applied. The use of starter cultures and the correct drying to lower the water activity can minimize the potential for growth of *L. monocytogenes* in Mediterranean-style fermented sausages. However, the frequent detection of *L. monocytogenes* at the end of ripening of these products highlights the need for food business operators to apply hurdle technology properly and to control the contamination routes of *L. monocytogenes* in meat processing plants.

Conflicts of Interest

The author declares no conflict of interest.

References

1. Meloni, D. Focusing on the main morphological and physiological characteristics of the food-borne pathogen *Listeria monocytogenes*. *J. Vet. Sci. Res.* **2014**, *1*, 1–2.
2. Sauders, B.D.; Wiedmann, M. Ecology of *Listeria* species and *L. monocytogenes* in the natural environment. In *Listeria, listeriosis, and Food Safety*; Ryser, E.T., Marth, E.H., Eds.; Marcel Dekker: New York, NY, USA, 2007; pp. 21–53.
3. Todd, E.C.D.; Notermans, S. Surveillance of listeriosis and its causative pathogen, *Listeria monocytogenes*. *Food Control* **2011**, *22*, 1484–1490.
4. European Food Safety Authority (EFSA); European Centre for Disease Prevention and Control (ECDC). The European Union Summary Report on Trends and Sources of Zoonoses, Zoonotic Agents and Food-borne Outbreaks in 2012. *EFSA J.* **2014**, *12*, 3547.
5. International Life Sciences Insistute (ILSI) Research Foundation; Risk Science Institute. Achieving continuous improvement in reductions in foodborne listeriosis—A risk-based approach. *J. Food Prot.* **2005**, *68*, 1932–1994.
6. Arevalos-Sánchez, M.; Regalado, C.; Martin, S.E.; Domínguez-Domínguez, J.; GarcíaAlmendárez, B.E. Effect of neutral electrolyzed water and nisin on *Listeria monocytogenes* biofilms and on listeriolys in O activity. *Food Control* **2012**, *24*, 116–122.
7. Farber, J.M.; Peterkin, P. *Listeria monocytogenes*, a food-borne pathogen. *Microbiol. Mol. Biol. Rev.* **1991**, *55*, 476–571.
8. Gram, L.; Bagge-Ravne, D.; Yin, N.Y.; Gymoese, P.; Fonnesbech Vogel, B. Influence of food soiling matrix on cleaning and disinfection efficiency on surface attached *Listeria monocytogenes*. *Food Control* **2007**, *18*, 1165–11671.
9. Gandhi, M.; Chikindas, M.L. *Listeria*: A foodborne pathogen that knows how to survive. *Int. J. Food Microbiol.* **2007**, *113*, 1–15.
10. Cruz, C.D.; Fletcher, G.C. Assessing manufacturers' recommended concentrations of commercial sanitizers on inactivation of *Listeria monocytogenes*. *Food Control* **2012**, *26*, 194–199.
11. Fonnesbech Vogel, B.; Huss, H.H.; Ojeniyi, B.; Ahrens, P.; Gram, L. Elucidation of *Listeria monocytogenes* contamination routes in cold-smoked salmon processing plants detected by DNA-based typing methods. *Appl. Environ. Microbiol.* **2001**, *67*, 2586–2595.
12. Nesbakken, T.; Kapperud, G.; Caugant, D.A. Pathways of *Listeria monocytogenes* contamination in the meat processing industry. *Int. J. Food Microbiol.* **1996**, *31*, 161–171.
13. Chasseignaux, E.; Gerault, P.; Toquin, M.; Salvat, G.; Colin, P.; Ermel, G. Ecology of *Listeria monocytogenes* in the environment of raw poultry meat and raw pork meat processing plants. *FEMS Microbiol. Lett.* **2002**, *210*, 271–275
14. Soni, D.K.; Singh, M.; Vijai Singh, D.; Dubey, S.K. Virulence and genotypic characterization of *Listeria monocytogenes* isolated from vegetable and soil samples. *BMC Microbiol.* **2014**, *14*, 241.

15. Liu, D.; Ainsworth, A.J.; Austin, F.W.; Lawrence, M.L. Characterization of virulent and avirulent *Listeria monocytogenes* strains by PCR amplification of putative transcriptional regulator and internalin genes. *J. Med. Microbiol.* **2003**, *52*, 1066–1070.
16. Velge, P.; Roche, S.M. Variability of *Listeria monocytogenes* virulence: A result of the evolution between saprophytism and virulence? *Future Microbiol.* **2010**, *5*, 1799–1821.
17. Jensen, A.; Thomsen, L.E.; Jørgensen, R.L.; Larsen, M.H.; Roldgaard, B.B.; Christensen, B.B.; Vogel, B.F.; Gram, L.; Ingmer, H. Processing plant persistent strains of *Listeria monocytogenes* appear to have a lower virulence potential than clinical strains in selected virulence models. *Int. J. Food Microbiol.* **2008**, *123*, 254–261.
18. Roberts, A.J.; Williams, S.K.; Wiedmann, M.; Nightingale, K.K. Some *Listeria monocytogenes* outbreak strains demonstrate significantly reduced invasion, inlA transcript levels, and swarming motility *in vitro*. *Appl. Environ. Microbiol.* **2009**, *75*, 5647–5658.
19. Rasmussen, O.F.; Skouboe, P.; Dons, L.; Rosen, L.; Olsen, J.E. *Listeria monocytogenes* exists in at least three evolutionary lines: Evidence from flagellin, invasive associated protein and listeriolysin O gene. *Microbiology* **1995**, *141*, 2053–2061.
20. Wiedmann, M.; Bruce, J.K.; Keating, C.; Johnson, A.; McDonough, P.L.; Batt, C.A. Ribotypes and virulence gene polymorphism suggest three distinct *Listeria monocytogenes* lineages with differences in pathogenic potential. *Infect. Immun.* **2997**, *65*, 2707–2716.
21. Wiedmann, M. Molecular subtyping methods for *Listeria monocytogenes*. *J. AOAC Int.* **2002**, *85*, 524–531.
22. OIE Terrestrial Manual 2014, CHAPTER 2.9.7. *Listeria monocytogenes*. Available online: http://www.oie.int/fileadmin/Home/eng/Health_standards/tahm/2.09.07_LISTERIA_MONO.pdf (accessed on 4 February 2015).
23. Piana, A.; Are, R.; Orrù, M.; Saba, F.; Dettori, M.; Maida, I.; Sotgiu, G.; Rais, C.; Mura, M.S. *Listeria monocytogenes* meningoencephalitis: Molecular methods for diagnosis and for monitoring the response to chemotherapy. *Ital. J. Public Health* **2005**, *2*, 29–34.
24. De Cesare, A.; Mioni, R.; Manfreda, G. Prevalence of *Listeria monocytogenes* in fresh and fermented sausages and ribotyping of contaminating strains. *Int. J. Food Microbiol.* **2007**, *120*, 124–130.
25. Denny, J.; McLauchlin, J. Human *Listeria monocytogenes* infections in Europe-an opportunity for improved European surveillance. *Euro Surveill.* **2008**, *13*, 8082.
26. Romanova, N.; Favrin, S.; Griffiths, M.W. Sensitivity of *Listeria monocytogenes* to sanitizers used in the meat processing industry. *Appl. Environ. Microbiol.* **2002**, *68*, 6405–6409.
27. Van Coillie, E.; Werbrouck, H.; Heyndrickx, M.; Herman, L.; Rjipens, N. Prevalence and typing of *Listeria monocytogenes* in Ready-to Eat food products on the Belgian market. *J. Food Prot.* **2004**, *67*, 2480–2487.
28. Shen, Y.; Liu, Y.; Zhang, Y.; Cripe, J.; Conway, W.; Meng, J.; Hall, G.; Bhagwat, A.A. Isolation and characterization of *Listeria monocytogenes* isolates from ready-to-eat foods in Florida. *Appl. Environ. Microbiol.* **2006**, *72*, 5073–5076.
29. Aureli, P.; Fiorucci, G.C.; Caroli, D.; Marchiaro, G.; Novara, O.; Leone, L.; Salmaso, S. An outbreak of febrile gastroenteritis associated with corn contaminated with *Listeria monocytogenes*. *New Engl. J. Med.* **2000**, *342*, 1236–1241.

30. Gillespie, I.A.; McLauchlin, J.; Grant, K.; Little, C.L.; Mithani, V.; Penman, C.; Lane, X.C.; Regan, M. Changing pattern of human listerios, England and Wales, 2001–2004. *Emerg. Infect. Dis.* **2006**, *12*, 1361–1366.
31. Public Health Agency Canada. Lessons Learned: Public Health Agency of Canada's Recall Response to the 2008 Listeriosis Outbreak. Available online: http://www.phac-aspc.gc.ca/fs-sa/listeria/2008-lessons-lecons-eng.php (accessed on 4 February 2015).
32. Interpretive Summary: Quantitative Assessment of Relative Risk to Public Health from Foodborne *Listeria monocytogenes* among Selected Categories of Ready-to-Eat Foods. Available online: http://www.fda.gov/downloads/food/foodscienceresearch/ucm197329.pdf (accessed on 4 February 2015).
33. European Commission (EC). Regulation (EC) No 2073/2005 of 15 November 2005 on microbiological criteria for foodstuffs. *Off. J. Eur. Union* **2005**, *L 338*, 1.
34. Thévenot, D.; Delignette-Muller, M.L.; Christieans, S.; Leroy, S.; Kodjo, A.; Vernozy-Roland, C. Serological and molecular ecology of *Listeria monocytogenes* isolates collected from 13 French pork meat salting-curing plants and their products. *Int. J. Food Microbiol.* **2006**, *112*, 153–161.
35. López, V.; Villator, D.; Oertiz, S.; López, P.; Navas, J.; Dávila, J.C.; Martínez-Suarez, J.V. Molecular tracking of *Listeria monocytogenes* in an Iberian pig abattoir and processing plant. *Meat Sci.* **2008**, *78*, 130–134.
36. Sammarco, M.L.; Ripabelli, C.; Ruberto, A.; Iannitto, G.; Grasso, G.M. Prevalence of *Salmonellae*, *Listeriae*, and *Yersinae* in the slaughterhouse environment and on work surfaces, equipment, and workers. *J. Food Prot.* **1997**, *60*, 367–371.
37. Korsak, N.; Daube, G.; Ghafir, Y.; Chahed, A.; Jolly, S.; Vindevogel, H. An efficient sampling technique used to detect four foodborne pathogens on pork and beef carcasses in nine Belgian abattoirs. *J. Food Prot.* **1998**, *61*, 535–541.
38. Bonardi, S.; Brindani, F.; Maggi, E. Isolation of *Listeria monocytogenes* and *Listeria* spp. from pigs at slaughter in Italy. *Ann. Fac. Medic. Vet. Parma* **2002**, *22*, 205–210.
39. Autio, T.; Sateri, T.; Fredriksson-Ahomaa, M.; Rahkio, M.; Lunden, J.; Korkeala, H. *Listeria monocytogenes* contamination pattern in pig slaughterhouses. *J. Food Prot.* **2000**, *63*, 1438–1442.
40. Belœil, P.A.; Chauvin, C.; Toquin, M.T.; Fablet, C.; Le Nŏtre, Y.; Salvat, G. *Listeria monocytogenes* contamination of finishing pigs: An exploratory epidemiological survey in France. *Vet. Res.* **2003**, *34*, 737–748.
41. Kanuganti, S.R.; Wesley, I.V.; Reddy, P.G.; McKean, J.; Hurd, H.S. Detection of *Listeria monocytogenes* in pigs and pork. *J. Food Prot.* **2002**, *65*, 1470–1474.
42. Gobat, P.F.; Jemmi, T. Epidemiological studies on *Listeria* spp. in slaughterhouses. *Fleischwirtsch* **1991**, *1*, 44–49.
43. Nesbakken, T.; Nerbrink, E.; Rotterud, O.J.; Borch, E. Reduction of *Yersinia enterocolitica* and *Listeria* spp. on pig carcasses by enclosure of the rectum during slaughter. *Int. J. Food Microbiol.* **1994**, *23*, 197–208.
44. Saide-Albornoz, J.J.; Knipe, C.L.; Murano, E.A.; Beran, G.W. Contamination of pork carcasses during slaughter, fabrication, and chilled storage. *J. Food Prot.* **1995**, *58*, 993–997.
45. Borch, E.; Nesbakken, T.; Christensen, H. Hazard identification in swine slaughter with respect to foodborne bacteria. *Int. J. Food Microbiol.* **1996**, *30*, 9–25.

46. Adesyiun, A.A.; Krishnan, C. Occurrence of *Yersinia enterocolitica* O:3. *Listeria monocytogenes* O:4. and thermophilic *Campylobacter* spp. in slaughter pigs and carcasses in Trinidad. *Food Microbiol.* **1995**, *12*, 99–107.
47. Thévenot, D.; Delignette-Muller, M.L.; Christieans, S.; Vernozy-Roland, C. Prevalence of *Listeria monocytogenes* in 13 dried sausage processing plants and their products. *Int. J. Food Microbiol.* **2005**, *102*, 85–94.
48. Meloni, D.; Piras, F.; Mureddu, A.; Fois, F.; Consolati, S.G.; Lamon, S.; Mazzette, R. *Listeria monocytogenes* in five Sardinian swine slaughterhouses: Prevalence, Serotype and Genotype Characterization. *J. Food Prot.* **2013**, *76*, 1863–1867.
49. Buncic, S. The incidence of *Listeria monocytogenes* in slaughtered animals, in meat, and in meat products in Yugoslavia. *Int. J. Food Microbiol.* **1991**, *12*, 173–180.
50. Ripamonti, B.; Zecchini, A.; Colombo, F.; Stella, S.; Cantoni, C. Localizzazione di *Listeria monocytogenes* e *Salmonella* spp. nei macelli. *Ing. Aliment.* **2002**, *18*, 7–13.
51. Autio, T.; Keto-Timonen, R.; Lunden, J.; Bjorkroth, J.; Korkeala, H. Characterisation of persistent and sporadic *Listeria monocytogenes* strains by pulsed-field gel electrophoresis (PFGE) and amplified fragment length polymorphism (AFLP). *Syst. Appl. Microbiol.* **2003**, *26*, 539–545.
52. Fabbi, M.; Andreoli, G.; De Giuli, L.; Carretto, E. *Listeria monocytogenes*. In *Trattato Sulle Infezioni e Tossinfezioni Alimentari*; Rondanelli, E.G., Fabbi, M., Marone, P., Eds.; Selecta Medica: Pavia, Italy, 2005; pp. 357–389.
53. Hof, H.; Rocourt, J. Is any strain isolated detected in food a health risk? *Int. J. Food Microbiol.* **1992**, *16*, 683–692.
54. Giovannacci, I.; Ragimbeau, C.; Queguiner, S.; Salvat, G.; Vendeuvre, J.L.; Carlier, V.; Ermel, G. *Listeria monocytogenes* in pork slaughtering and cutting plants. Use of RAPD, PFGE and PCR-REA for tracing and molecular epidemiology. *Int. J. Food Microbiol.* **1999**, *53*, 127–140.
55. Kathariou, S. *Listeria monocytogenes* virulence and pathogenicity, a food safety perspective. *J. Food Prot.* **2002**, *65*, 1811–1829.
56. Jay, J. Prevalence of *Listeria* spp. in meat and poultry products. *Food Control* **1996**, *7*, 209–214.
57. Mureddu, A.; Mazza, R.; Fois, F.; Meloni, D.; Bacciu, R.; Piras, F.; Mazzette, R. *Listeria monocytogenes* persistence in ready-to-eat sausages and in processing plants. *Ital. J. Food Saf.* **2014**, *3*, 12–15.
58. Mafu, A.A.; Roy, D.; Goulet, J.; Magny, P. Attachment of *Listeria monocytogenes* to stainless steel, glass, polypropylene and rubber surfaces after short contact times. *J. Food Prot.* **1990**, *53*, 742–746.
59. Meloni, D.; Piras, F.; Mureddu, A.; Mazza, R.; Nucera, D.; Mazzette, R. Sources of *Listeria monocytogenes* contamination in traditional fermented sausages processing plants in Italy. *Ital. J. Food Sci.* **2012**, *24*, 214–222.
60. Meloni, D.; Consolati, S.G.; Mazza, R.; Mureddu, A.; Fois, F.; Piras, F.; Mazzette, R. Presence and molecular characterization of the major serovars of *Listeria monocytogenes* in ten Sardinian fermented sausage processing plants. *Meat Sci.* **2014**, *97*, 443–450.
61. Tompkin, R.B. Control of *Listeria monocytogenes* in the food-processing environment. *J. Food Prot.* **2002**, *65*, 709–725.

62. Verghese, B.; Lok, M.; Wen, J.; Alessandria, V.; Chen, Y.; Kathariou, S.; Knabel, S. comK prophage junction fragments as markers for *Listeria monocytogenes* genotypes unique to individual meat and poultry processing plants and a model for rapid niche-specific adaptation, biofilm formation and persistence. *Appl. Environ. Microbiol.* **2011**, *77*, 3279–3292.
63. Zhao, T.; Doyle, M.P.; Zhao, P. Control of *Listeria monocytogenes* in a biofilm by competitive-exclusion microorganisms. *Appl. Environ. Microbiol.* **2004**, *7*, 3996–4003.
64. Samelis, J.; Metaxopoulos, J. Incidence and principal sources of *Listeria* spp. and *Listeria monocytogenes* contamination in processed meats and a meat processing plant. *Food Microbiol.* **1999**, *16*, 465–477.
65. Ordóñez, J.A.; de la Hoz, L. Mediterranean products. In *Handbook of Fermented Meat And Poultry*; Toldrà, F., Ed.; Blackwell Publishing: Ames, IA, USA, 2007; pp. 333–347.
66. Frey, W. *Fabricacion Fiable de Embutidos*; Acribia, Ed.; Zaragoza, España, 1985; pp. 1–39.
67. Greco, M.; Mazzette, R.; De Santis, E.P.L.; Corona, A.; Cosseddu, A.M. Evolution and identification of lactic acid bacteria isolated during the ripening of Sardinian sausages. *Meat Sci.* **2005**, *69*, 733–739.
68. Jofré, A.; Aymerich, T.; Garriga, M. Improvement of the food safety of low acid fermented sausages by enterocins A and B and high pressure. *Food Control* **2009**, *20*, 179–184.
69. Barbuti, S.; Parolari, G. Validation of manufacturing process to control pathogenic bacteria in typical dry fermented products. *Meat Sci.* **2002**, *62*, 323–329.
70. Leistner, L. Principles and applications of hurdle technology. In *New Methods of Food Preservation*; Gould, G.W., Ed.; Springer: Berlin, Germany, 1995; pp. 1–21.
71. Leistner, L.; Gould, G.W. *Hurdle Technologies: Combination Treatments for Food Stability, Safety and Quality*; Kluwer: New York, NY, USA, 2002.
72. Selgas, D.; Garcia de Fernando, G.; Ordonez, J.A. Lipolytic and proteolytic activity of micrococci isolated from dry fermented sausages. *Fleischwirsch. Int.* **1994**, *2*, 51–54.
73. Greco, M.; Mazzette, R.; de Santis, E.P.L.; Bean, V.; Nieddu, M.P.; Pisanu, S. Micrococcaceae of "Sardinian sausage": Note II. Characterization of isolated species and evolution during production. In Proceedings of the 53rd National Congress of Italian Society of Veterinary Science (SISVet), Montecatini terme (PT), Italy, 17–19 September 1999; pp. 331–332.
74. Mazzette, R.; De Santis, E.P.L.; Greco, M.; Virdis, S.; Cosseddu, A.M. Micrococcaceae of ''Sardinian sausage'': Note I. Biochemical and metabolic characteristics. In Proceedings of the 53rd National Congress of Italian Society of Veterinary Science (SISVet), Montecatini terme (PT), Italy, 17–19 September 1999; pp. 329–330.
75. Papa, F.; Giudici, P.; Grazia, L.; Torriani, S. Lactic-acid bacteria selection for better quality salami. *Ind. Aliment.* **1993**, *32*, 258–261.
76. Torriani, S.; Di Bucchianico, R.; Patarini, F.; Zabeo, G.; Dell'aglio, F. Presenza e caratterizzazione biotecnologia di batteri lattici e *Micrococcaceae* negli insaccati abruzzesi. *Ind. Conserve* **1994**, *69*, 3–9.
77. Grazia, L.; Ranieri, S.; Zambonelli, C.; Chiavari, C. Action of homofermentative and heterofermentative lactobacilli on themolding of salami. *Ind. Aliment.* **1998**, *37*, 852–855.

78. Leroy, F.; de Vuyst, L. Temperature and pH conditions that prevail during fermentation of sausages are optimal for production of the antilisterial bacteriocin sakacin K. *Appl. Environ. Microbiol.* **1999**, *115*, 974–981.
79. Hebert, E.M.; Raya, R.R.; Taillez, P.; de Giori, G.S. Characterization of natural isolates of *Lactobacillus* strains to be used as starter cultures in dairy fermentation. *Int. J. Food Microbiol.* **2000**, *59*, 19–27.
80. Lucke, F.K. Utilisation of microbes to process and preserve meat. *Meat Sci.* **2000**, *56*, 105–115.
81. Leistner, L. Shelf-stable products and intermediate moisture foods based on meat. In *Water Activity: Theory and Applications to Food*; Rockland, L.B.; Beuchat, L.R., Institute of Food Technologists, Eds.; Marcel Dekker Inc.: New York, NY, USA, 1987.
82. Balev, D.; Vulkova, T.; Dragoev, S.; Zlatanov, M.; Bahtchevanska, S. A comparative study on the effect of some antioxidants on the lipid and pigment oxidation in dry-fermented sausages. *Int. J. Food Sci. Technol.* **2005**, *40*, 977–983.
83. Colak, H.; Hampikyan, H.; Ulusoy, B.; Bingol, E.B. Presence of *Listeria monocytogenes* in Turkish style fermented sausage (sucuk). *Food Control* **2007**, *18*, 30–32.
84. Cantoni, C.; Aubert, S.D.; Valenti, M.; Comi, G.; Aubert, S. *Listeria* spp. in cheese and dry sausages. *Ind. Aliment.* **1989**, *28*, 1068–1070.
85. Cordano, A.M.; Rocourt, J. Occurrence of *Listeria monocytogenes* in food in Chile. *Int. J. Food Microbiol.* **2001**, *70*, 175–178.
86. Levine, P.; Rose, B.; Green, S.; Ransom, G.; Hill, W. Pathogen testing of ready-to-eat meat and poultry products collected at federally inspected establishment in the United States, 1990 to 1999. *J. Food Prot.* **2001**, *64*, 1188–1193.
87. Meloni, D.; Galluzzo, P.; Mureddu, A.; Piras, F.; Griffiths, M.; Mazzette, R. *Listeria monocytogenes* in RTE foods marketed in Italy: Prevalence and automated EcoRI ribotyping of the isolates. *Int. J. Food Microbiol.* **2009**, *129*, 166–173.
88. Doménech, E.; Jimenez-Belenguer, A.; Amoros, J.A.; Ferrus, M.A.; Escriche, I. Prevalence and antimicrobial resistance of *Listeria monocytogenes* and *Salmonella* strains isolated in ready-to-eat foods in Eastern Spain. *Food Control* **2015**, *47*, 120–125.
89. Hospital, X.F.; Hierro, E.; Fernández, M. Survival of *Listeria innocua* in dry fermented sausages and changes in the typical microbiota and volatile profile as affected by the concentration of nitrate and nitrite. *Int. J. Food Microbiol.* **2012**, *153*, 395–401.
90. McLauchlin, J.; Mitchell, R.T.; Smerdon, W.J.; Jewell, K. *Listeria monocytogenes* and listeriosis: A review of hazard characterization for use in microbiological risk assessment of foods. *Int. J. Food Microbiol.* **2004**, *92*, 15–33.
91. Greenwood, M.; Roberts, H.D.; Burden, P. The occurrence of *Listeria* species in milk and dairy products: A national survey in England and Wales. *Int. J. Food Microbiol.* **1991**, *12*, 197–206.
92. Hayes, P.S.; Graves, L.M.; Ajello, G.W.; Swaminathan, B.; Weaver, R.E.; Wenger, J.D.; Schuchat, A.; Broome. C.V. Comparison of cold-enrichment and U.S. Department of Agriculture methods for isolating *Listeria* spp. from naturally contaminated foods. *Appl. Environ. Microbiol.* **1991**, *57*, 2109–2113.

93. Gianfranceschi, M.V.; D'Ottavio, M.C.; Gattuso, A.; Bella, A.; Aureli, P. Distribution of serotypes and pulsotypes of *Listeria monocytogenes* from human, food and environmental isolates (Italy, 2002–2005). *Food Microbiol.* **2009**, *26*, 520–526.
94. Pontello, M.; Guaita, A.; Sala, G.; Cipolla, M.; Gattuso, A.; Sonnessa, M.; Gianfranceschi, M.V. *Listeria monocytogenes* serotypes in human infections (Italy, 2000–2010). *Ann. Ist. Super. Sanita* **2012**, *48*, 146–150.
95. Skandaminis, P.; Nychas, J.E. Pathogens: Risks and control. In *Handbook of Fermented Meat and Poultry*; Toldrà, F., Ed.; Blackwell Publishing: Ames, IA, USA, 2007; pp. 427–454.

6

Effect of Food and Vitamin D Supplements on the Serum $25(OH)D_3$ Concentration in Children during Winter Months

Ellen van der Gaag [1,*] and Lisanne Brekhoff [2]

[1] Hospital Group Twente, Geerdinksweg 141, 7555 DL Hengelo, The Netherlands

[2] Medical School University of Groningen, Hanzeplein 1, Groningen, The Netherlands; E-Mail: l.brekhoff@student.rug.nl

* Author to whom correspondence should be addressed; E-Mail: e.gaagvander@zgt.nl

External Editor: Christopher J. Smith

Abstract: Aim: To determine the contribution of food and vitamin D supplements on the serum 25-hydroxyvitamin D_3 ($25(OH)D_3$) concentration between October and April in a northern country (almost absent vitamin D synthesis by sunlight). Methods: Children aged 1–18 years were selected who visited the general pediatrician with a complaint whereby serum $25(OH)D_3$ concentration was determined. The intake of vitamin D was calculated based on a dietary questionnaire. Results: 51.1% of the 174 children had a serum $25(OH)D_3$ concentration below 50 nmol/L, 9.2% had a serum $25(OH)D_3$ concentration below 30 nmol/L. Adolescents showed lower concentrations compared to younger children. There was a positive correlation between the total amount of vitamin D obtained from food and the serum $25(OH)D_3$ concentration ($r = 0.218$, $p = 0.004$). The intake of milk contributed more to the serum $25(OH)D_3$ concentration compared to the intake of artificial supplementation, butter or fish. Conclusions: In the absence of vitamin D synthesis by sunlight, vitamin D obtained from food has a significant influence on the serum $25(OH)D_3$ concentration in children. Vitamin D supplements can be described as trivial. This means we should pay more attention to food as a natural source of vitamin D.

Keywords: children; dietary vitamin D; nutrients; vitamin D deficiency; vitamin D supplements

1. Introduction

The main source of vitamin D_3 for people is exposure to sunlight [1,2]. In northern countries however, such as The Netherlands, the skin only synthesizes vitamin D_3 between April and October [1,3]. During the colder months, there is much less vitamin D_3 synthesis in the skin because of the latitude and due to sun rays entering the atmosphere at a more oblique angle. Therefore, ultraviolet-B (UVB) photons have to pass through the ozone at a greater distance and more UVB photons are efficiently absorbed by ozone [4].

Less than 10% of vitamin D is obtained from dietary sources and vitamin D supplements [4]. The major foods that naturally contain vitamin D_3 are (wild caught) salmon, oily fish, cod liver oil and dairy products. Some foods are fortified with vitamin D, but in Europe there are many countries that do not have standard fortification of food groups [1,4–6]. In The Netherlands the foods fortified with vitamin D include margarine, low fat margarine, baby food and baby/grow up milk, but no natural milk [7,8].

In The Netherlands, the recommended serum 25-hydroxyvitamin D_3 (25(OH)D_3) concentration for children is greater than 30 nmol/L [6]. In other continents or countries, such as The United States, Germany, Austria and Switzerland, the recommended serum 25(OH)D_3 concentration is above 50 nmol/L [4,6].

There is limited data available on vitamin D deficiency (VDD) among the European pediatric population [9]. At the end of the winter, the prevalence of VDD is at its highest because the summer supplies are utilized [1,3]. Weggemans *et al.* [3] reported that in The Netherlands, 5% of the light-skinned children and 15%–30% of the dark-skinned children have VDD at the end of winter.

Because of the importance of vitamin D for many functions of human body and due to the limited vitamin D synthesis in the skin between October and April, it is important to have another source of vitamin D during these months.

Since vitamin D is also obtained from dietary sources and vitamin D supplements, this study evaluated the effects of the intake of food and vitamin D supplements on the serum 25(OH)D_3 concentration in children in The Netherlands between October and April.

2. Experimental Section

2.1. Study Design

The study was an epidemiologic pilot study conducted at the ZGT Hengelo, The Netherlands, covering the period from February 2011 to April 2013 (with exclusion of the months from May to September).

2.2. Study Population

Children aged 1–18 years were selected who visited the general pediatrician with general pediatric complaints (like tiredness, abdominal pain, respiratory infections) whereby serum 25(OH)D_3 concentration was determined. They were referred by a general practitioner.

2.3. Measurements

A custom-made isotope dilution LC-MS/MS method is used for the analysis of vitamin D (25-OH-vitamin D_3). The method starts with protein precipitation of the serum proteins with acetonitril containing deuterated internal standard (D_6-25-OH vitamin D_3). This step also releases vitamin D from its binding proteins. After centrifugation and removing and evaporating the supernatant, the sample is reconstituted in 70% methanol followed by a centrifugation step. Reversed phase (C8) chromatographic separation was performed on a Shimadzu HPLC system ('s-Hertogenbosch, The Netherlands), consisting of two LC 20AD pumps, a SIL-20AC autosampler, and a CTO 20A column oven. Mass spectrometric detection was carried out on a Q-Trap 3200 mass spectrometer from Applied Biosystems (Foster City, CA, USA), in the MRM (multiple reaction monitoring) mode using APCI (atmospheric pressure chemical ionization) as an ionization method. At two concentrations (42 and 206 nmol/L) the intra-assay CVs (coefficients of variation) were calculated to be 3.6% and 2.7%, respectively and the inter-assay CVs 4.0% and 3.3%, respectively. The method is linear up to 1500 nmol/L and has a limit of quantification of 3 nmol/L. This is an excellent alignment with NIST (SRM 972) standards. Comparison to other laboratories was continuously observed by participating in the DEQAS (The Vitamin D External Quality Assessment Scheme) proficiency testing.

Information was collected on the vitamin D intake with use of a food questionnaire. The investigator noted the kind and amount of food the children had eaten during the past week. The mean was calculated to estimate the daily intake. To circumvent investigator bias the same investigator was used to take the food questionnaires for all the children. The intake was noted of the type and quantity of milk in milliliters, the type of butter the children put on their bread, the number of slices of bread that was buttered, the type and quantity of fish eaten and the use of vitamin D supplements. The medical complaints that brought them to the doctor was obtained, the amount of time the child spent playing outdoors every day, and the month of the year in which the serum $25(OH)D_3$ concentration was determined. The amount of vitamin D per day obtained from food and vitamin D supplements was calculated. For example, to calculate the vitamin D content from fish, the parents were asked to specify the amount and type of fish the child ate that week. Since there are big differences in the amount of vitamin D between oily fish and other types of fish, the Nederlands Voedingsstoffenbestand (NEVO) tables [10] were consulted to calculate the actual individual intake.

When a food was missing in the NEVO table, the super market was visited and food labels were read to determine the amount of vitamin D in the food. Each portion of butter the children put on one slice of bread was assumed to be 5 g, and one portion of fish was assumed to be 55 g.

2.4. Statistical Analysis

Statistical analysis was carried out using IBM SPSS Statistics 20. For normally-distributed variables, means and standard deviations were calculated and for non-normal distributions, medians and interquartile ranges (IQR) were calculated. A p value of less than 0.05 was considered significant.

Linear regression analysis was used to determine the specific effect of the age of the child on the total amount of vitamin D from food. In other words, could the age of the child be a factor of any influence on the $25(OH)D_3$ concentration. It was also used to determine the separate effect of the variables of gender, and the total amount of vitamin D obtained from food, milk, butter, fish and vitamin D supplements on the serum $25(OH)D_3$ concentration. The correlation coefficients were calculated using the Pearson test. Thereafter, multiple regression analysis was used to determine the effect of all the significant variables on the serum $25(OH)D_3$ concentration.

To determine the effect of different types of milk on the serum 25(OH)D_3 concentration, the children were classified into five groups: children who did not drink milk, children who drank yogurt drinks and chocolate milk, children who drank semi-skimmed milk, children who drank whole milk, and the residual group. To investigate the influence of different types of butter on the serum 25(OH)D_3 concentration, the children were classified into four groups: children who did not use butter, children who used low fat margarine, children who used margarine and children who used dairy butter. To determine the effect of the different types of milk and butter on the serum 25(OH)D_3 concentration, dummy variables were made and linear regression analysis was carried out. Subsequently, correlation coefficients were calculated using the Pearson test and thereafter multiple regression analysis was used to determine the effect of all the significant variables on the serum 25(OH)D_3 concentration.

2.5. Ethical Committee

No intervention was given. Existing laboratory results from standard patient care were used and therefore ethical consideration by an ethics committee was not necessary. For the use of the questionnaires, all patients agreed to participate in the collection of the data. The data was recorded and subsequently coded anonymously by the investigator.

3. Results and Discussion

3.1. Descriptives

There were 174 participating children and the median age was 8.5 years (IQR, 5.0–13.1). Of these children, 51.1% were female (Table 1). The medical complaints differed per child. 16.1% of the children were referred for abdominal pain, 18.4% suffered from tiredness, 6.9% had bronchial hyper reactivity and 5.7% had recurrent respiratory tract infections. Other problems were constipation, obesity, diarrhea, dizziness or asthma.

Table 1. Baseline characteristics.

Characteristic	Patients (n = 174)
Age, median (IQR) (years)	8.5 (5.0–13.1)
Gender, number (%)	
Female	89 (51.1%)
Male	85 (48.9%)
Serum 25(OH)D concentration, mean ± SD (nmol/L)	52.0 ± 18.4
Vitamine D obtained with the total amount of food per day, mean ± SD (μg)	1.3 ± 1.9
Vitamin D obtained with milk per day, mean ± SD (μg)	0.5 ± 1.8
Vitamin D obtained with butter per day, mean ± SD (μg)	0.7 ± 0.5
Vitamin D obtained with fish, mean ± SD (μg)	0.1 ± 0.2
Vitamin D obtained with vitamin D supplements, mean ± SD (μg)	3.4 ± 4.8

IQR = Interquartile range; SD = standard deviation.

The mean serum 25(OH)D_3 concentration was 52.0 ± 18.4 nmol/L (Table 1). Sixteen (9.2%) of 174 children had a serum 25(OH)D_3 concentration below 30 nmol/L, and 89 (51.1%) had a serum

$25(OH)D_3$ concentration below 50 nmol/L. At adolescence age, the serum $25(OH)D_3$ concentration decreased (Figure 1). Table 2 shows the mean daily intake of vitamin D per nutrient of three age groups.

Figure 1. Mean serum $25(OH)D_3$ concentration per age.

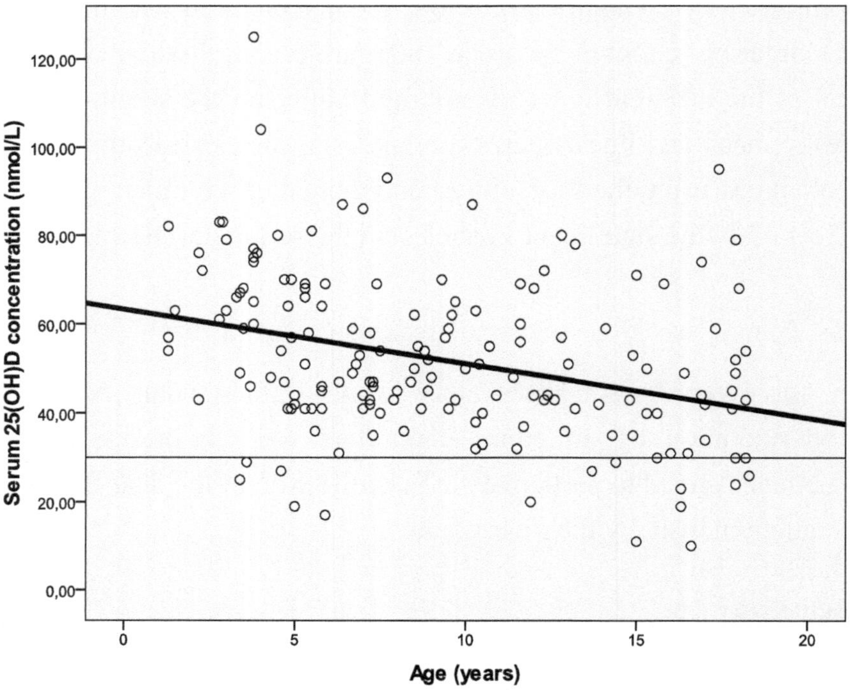

Table 2. Mean and standard deviation of daily intake of vitamin D per nutrient for three age groups.

Groups	Butter (µg)	Milk (µg)	Fish (µg)	Supplementation (µg)	Total (µg)
Toddlers (1–4 years old)	0.7 ± 0.3	3.2 ± 3.7	0.0 ± 0.1	5.8 ± 4.8	9.7
School-aged (4–12 years old)	0.8 ± 0.5	0.3 ± 1.5	0.1 ± 0.2	3.8 ± 4.6	5.0
Adolescents (>12 years old)	0.6 ± 0.6	0.0 ± 0.2	0.2 ± 0.3	1.8 ± 4.8	2.6

3.2. Correlations

Linear regression analysis showed a negative correlation between the age and the total amount of vitamin D obtained from food ($r = 0.260$, $p = 0.001$). There was a positive correlation between the total amount of vitamin D obtained from food and the serum $25(OH)D_3$ concentration ($r = 0.218$, $p = 0.004$). A further positive correlation was detected between milk intake and the serum $25(OH)D_3$ concentration ($r = 0.186$, $p = 0.014$). The other variables (butter, fish or supplements) had no significant effect on the serum $25(OH)D_3$ concentration (Table 3). Pearson correlation analysis showed a significant positive correlation between the total amount of vitamin D obtained from food and milk intake ($r = 0.951$, $p = 0.000$). Therefore, the variable milk intake was excluded in the multiple regression analysis since milk intake had a smaller impact on the serum $25(OH)D_3$ concentration compared to the total amount of vitamin D obtained from food (95% CI 2.1 (0.7–3.5), $p = 0.004$).

If the total amount of vitamin D obtained from food increased with 1 µg, the serum $25(OH)D_3$ concentration increased with 2.1 nmol/L. The R-square for the model with the total amount of vitamin D obtained from food was 0.048.

Table 3. Results linear regression.

Variables	Linear regression B * (95% CI **); ß, coefficient
Gender	−1.7 (−7.2–3.8), $p = 0.535$; ß = −0.047
Total amount of vitamin D obtained from food	2.1 (0.7–3.5), $p = 0.004$ ***; ß = 0.218
Milk	1.9 (0.4–3.4), $p = 0.014$ ***; ß = 0.186
Butter	4.2 (−0.9–9.3), $p = 0.106$; ß = 0.124
Fish	4.8 (−9.2–18.8), $p = 0.497$; ß = 0.052
Vitamin D supplements	0.3 (−0.3–0.9), $p = 0.315$; ß = 0.077

* Regression coefficient; * 95% confidence interval; *** significant.

3.3. Linear Regression

The linear regression analysis showed no statistically significant correlation for the different types of milk and the serum 25(OH)D_3 concentrations (Table 4). Positive relations can be made between the intake of whole milk and the serum 25(OH)D_3 concentration ($r = 0.116$, $p = 0.13$). For the residual milk group (toddler milk, soy milk, buttermilk, follow-up milk and yogurt) there was also a positive but not significant relation seen with the serum 25(OH)D_3 concentrations ($r = 0.111$, $p = 0.14$) (Table 4). Yogurt drinks and semi skimmed milk containing no vitamin D showed small negative relations. None of these findings were significant due to the relatively small sample size. The type of butter was of no influence on the serum 25(OH)D_3 concentration. Fortified margarine ($r = -2.2$, $p = 0.66$), fortified diet margarine ($r = 4.5$, $p = 0.13$) and non-fortified full fat butter ($r = 0.4$, $p = 0.94$) all showed non-significant relations between their vitamin D content and the serum 25(OH)D_3. This is surprising considering that the fortified products contain six times more vitamin D compared to the non-fortified full fat butter.

Table 4. Results linear regression of different types of milk and serum 25(OH)D_3 concentration.

Variables	Linear regression B * (95% CI **); ß, coefficient
Yogurt drinks and chocolate milk	−0.6 (−6.7–5.6), $p = 0.856$; ß = −0.014
Semi-skimmed milk	−3.1 (−9.4–3.1), $p = 0.325$; ß = −0.075
Whole milk	7.0 (−2.0–16.0), $p = 0.127$; ß = 0.116
Residual group	5.0 (−1.7–11.7), $p = 0.143$; ß = 0.111

* Regression coefficient; ** 95% confidence interval.

3.4. Food or Supplements?

This study investigated the effects of food and vitamin D supplements on the serum 25(OH)D_3 concentrations between October and April in children aged between 1 and 18 years old. The influence of the total amount of vitamin D obtained from food was examined. In particular, the intake of milk, butter, fish and vitamin D supplements and the different types of milk and butter on the serum 25(OH)D_3 concentration was closely examined. In our study, vitamin D intake by food showed a stronger relation with serum 25(OH)D_3 compared to that of intake by supplements.

The results of this epidemiologic pilot study indicate that between October and April, the total amount of vitamin D obtained from food is positive related to serum 25(OH)D_3 concentration. The type of food with the greatest impact was milk. A positive trend was seen between the intake of whole milk and the residual group (toddler milk, soy milk, buttermilk, follow-on milk and yogurt), and the serum 25(OH)D_3 concentration. This trend was however not statistically significant.

The effect of food is remarkable; when the total amount of vitamin D obtained from food increased with 1 μg, the serum 25(OH)D_3 concentration increased with 2.1 nmol/L. The contribution of food can be considered very valuable during winter months.

Due to the limited food fortification with vitamin D and the limited intake of vitamin D supplements in The Netherlands, and considering the different climate zones compared to other countries, it is difficult to compare this research with research from foreign countries. There is generally also very little research on the influence of food and vitamin D supplements on the serum 25(OH)D_3 concentration in The Netherlands between October and April in pediatric patients.

3.5. Intake and Age

The Dutch National Food Consumption Survey investigated the actual intake of vitamin D in a random sample of the Dutch population (n = 3819). The advised dietary intake of 2.5 μg vitamin D per day for school aged children and adolescents was found in their cohort with mean intakes of 2.3–3.1 μg/day depending on the age groups. No data was available for children aged below seven years [6]. Our study shows that toddlers have an adequate vitamin D intake but that school-aged children and adolescents do not have an adequate vitamin D intake from dietary sources alone (Table 2). Other studies also show decreasing 25(OH)D_3 levels with age [11–14]. When supplementation was added, adequate intake was reached. This could also be seen in the decrease of vitamin D levels related to a particular age group. Furthermore, The Dutch National Food Consumption Survey reported that most of the vitamin D is obtained from basic foods and little vitamin D is obtained from fortified foods or vitamin D supplements [6]. However, our research showed that milk was associated with an increased serum 25(OH)D_3 concentration. A possible explanation for this difference is that The Dutch National Food Consumption Survey does not contain the same age group tables as our research.

3.6. Food Sources

During the winter months 25(OH)D_3 levels are above zero. It is not likely, however, that a serum 25(OH)D_3 concentration of 30–50 nmol/L can be reached only with the intake of the foods mentioned above. Other sources for vitamin D should therefore be identified. One explanation could be that we did not include the amount of meat that was eaten as meat which also contains vitamin D_3. According to the Dutch Food composition database (NEVO tables) [10], fish contains 10–100 times more vitamin D_3 compared to meat, this made us initially decide not to include the meat products. In a recent study however, a re-analysis of eight other studies also found unidentified sources for serum 25(OH)D_3 [15]. They measured serum 25(OH)D_3 and accurately back-calculated basal vitamin D_3 input. They estimated the basal vitamin D_3 intake should be 50 μg, irrespective of the source. When the actual vitamin D input from food sources (3–4 μg) and solar sources (8–12 μg) were calculated, a deficit exists of 40 μg between intake and blood concentrations. They, and also other resources, suggested meat sources could be a possible candidate to fill the calculated input gap [15,16]. Since we performed an inventory pilot study, the meat intake should be taken into account in future studies.

The actual amount of vitamin D in milk was not very high but it was an important factor for the final vitamin D levels. This could possibly be due to the use of vitamin D fortification of toddler milk although we did also see this trend in children who did not take toddler milk (Table 4). Ovesen *et al.* [17] also recognized this mechanism, and elucidated it by the content of the metabolite 25-hydroxyvitamin D_3 in food products which could be more biologically active compared to vitamin D_3 [17]. Most animal sources contain vitamin D_3 (which needs to be hydroxylated in the liver), as well as the hydroxylated metabolite, 25(OH)D_3. The bioactivity of 25(OH)D_3 in food could be up to five times higher compared to vitamin D_3. Consensus on this factor is, however, lacking [18]. This activity factor is not included in dietary intake surveys, so true vitamin D intake will typically be underestimated from dietary sources. Additionally, most food composition databases used in different countries (Denmark, France, Germany, Switzerland, Canada or The United States) do not differentiate between vitamin D_2 and vitamin D_3. Nor do they give any information about the inclusion or exclusion of the vitamin D metabolites or how the data were obtained as reviewed by Schmid *et al.* [15]. Perhaps this mechanism also plays a role in the butter mechanism. The actual vitamin D content in natural full fat butter was lower compared to margarine. However, the effect on the increase of serum levels by the intake of natural butter was greater than could be expected beforehand. This points out that more factors contribute to the final 25(OH)D_3 levels in serum. This cannot only be calculated alone based on values mentioned in different databases.

3.7. Limitations

A limitation of this study is the use of a food questionnaire. The parents were asked about the food intake over the past week. The mean was calculated to estimate the daily vitamin D intake. A food recall is less reliable compared to that of an actual food diary. To minimize the differences, the same investigator conducted all the food questionnaires. This enabled an even comparison of all the given answers. The parent dependent variable of being able to securely remember the intake could not be taken into account. Existing food composition databases were used to calculate the actual intake of vitamin D_3 from food. In these databases there was no information on the different types of vitamin D (vitamin D_2, vitamin D_3 or 25(OH)D_3).

Finally, the children in this study all came with a complaint to the pediatrician meaning that they were initially not completely healthy. However, the complaints did not have an impact on vitamin D metabolism so the effect on the serum 25(OH)D_3 concentration was minimal.

The results of this study are important because between October and April there is almost no vitamin D synthesis in the skin and many children have a low serum 25(OH)D_3 concentration during these months. Vitamin D has several important effects in the body, making it important to maintain an adequate serum 25(OH)D_3 concentration.

4. Conclusions

Between October and April food has an unexpected significant contribution to the serum 25(OH)D_3 concentration. Supplements with vitamin D were trivial compared to vitamin D rich food sources. Furthermore, the mean serum 25(OH)D_3 concentration of many children is below the normal range. Due to the important function of vitamin D, it is advisable to recommend nutrition that is rich in vitamin D for children between October and April. Milk is especially important. As the risk for lower 25(OH)D_3

concentrations increases with age into adolescence, more research is needed on the influence of several factors on the serum 25(OH)D_3 concentration in adolescents and special attention is needed for this group.

Acknowledgments

The authors thank Job van der Palen, epidemiologist, for his statistical help. They also thank Boony Thio and Hans Krabbe for their help in finishing the paper.

Author Contributions

Ellen van der Gaag included the patients, Lisanne Brekhoff evaluated the data. Both authors contributed to the writing of the manuscript.

Conflicts of Interest

The authors declare no conflict of interest.

References

1. Holick, M.F. Vitamin D: A D-lightful solution for health. *J. Investig. Med.* **2011**, *59*, 872–880.
2. Webb, A.R. Who, what, where and when-influences on cutaneous vitamin D synthesis. *Prog. Biophys. Mol. Biol.* **2006**, *92*, 17–25.
3. Weggemans, R.M.; Schaafsma, G.; Kromhout, D. Towards an adequate intake of vitamin D. An advisory report of the Health Council of the Netherlands. *Eur. J. Clin. Nutr.* **2009**, *63*, 1455–1457.
4. Webb, A.R.; Kline, L.; Holick, M.F. Influence of season and latitude on the cutaneous synthesis of vitamin D3: Exposure to winter sunlight in Boston and Edmonton will not promote vitamin D3 synthesis in human skin. *J. Clin. Endocrinol. Metab.* **1988**, *67*, 373–378
5. Hennessy, A.; Walton, J.; Flynn, A. The impact of voluntary food fortification on micronutrient intakes and status in European countries. *Proc. Nutr. Soc.* **2013**, *72*, 433–440.
6. Van Rossum, C.T.M.; Fransen, H.P.; Verkaik-Kloosterman, J.; Buurma-Rethans, E.J.M.; Ocke M.C. Dutch National Food Consumption Survey 2007–2010: Diet of Children and Adults Aged 7 to 69 Years. Available online: http://www.rivm.nl/bibliotheek/rapporten/350050006.html (accessed on 12 July 2014).
7. Van der Meer, I.M.; Boeke, A.J.; Lips, P.; Grootjans-Geerts, I.; Wuister, J.D.; Devillé, W.L.; Wielders, J.P.; Bouter, L.M.; Middelkoop, B.J. Fatty fish and supplements are the greatest modifiable contributors to the serum 25-hydroxyvitamin D concentration in a multiethnic population. *Clin. Endocrinol. (Oxf.)* **2008**, *68*, 466–472.
8. Gezondheidsraad. Evaluatie van de Voedingsnormen voor Vitamine D. Available online: http://www.gezondheidsraad.nl/sites/default/files/201215evaluatievoedingsnormenvitamineD.pdf (accessed on 6 April 2013).
9. Braegger, C.; Campoy, C.; Colomb, V.; Decsi, T.; Domellof, M.; Fewtrell, M.; Hojsak, I.; Mihastsch, W.; Molgaard, C.; Shamir, R.; *et al.* Vitamin D in the healthy paediatric population. *J. Pediatr. Gastroenterol. Nutr.* **2013**, *56*, 692–701.

10. Nederlands Voedingsstoffenbestand (NEVO). Available online: http://nevo-online.rivm.nl/ (accessed on 5 April 2013).
11. Roth, D.E.; Martz, P.; Yeo, R.; Prosser, C.; Bell, M.; Jones, A.B. Are national vitamin D guidelines sufficient to maintain adequate blood levels in children? *Can. J. Public Health* **2005**, *96*, 443–449.
12. Rajakumar, K.; Holick, M.F.; Jeong, K.; Moore, C.G.; Chen, T.C.; Olabopo, F.; Haralam, M.A.; Nucci, A.; Thomas, S.B.; Greenspan, S.L. Impact of season and diet on vitamin D status of African American and Caucasian children. *Clin. Pediatr. (Phila)* **2011**, *50*, 493–502.
13. Absoud, M.; Cummins, C.; Lim, M.J.; Wassmer, E.; Shaw, N. Prevalence and predictors of vitamin D insufficiency in children: A Great Britain population based study. *PLoS One* **2011**, *6*, e22179.
14. Carpenter, T.O.; Herreros, F.; Zhang, J.H.; Ellis, B.K.; Simpson, C.; Torrealba-Fox, E.; Kim, G.J.; Savoye, M.; Held, N.A.; Cole, D.E. Demographic, dietary, and biochemical determinants of vitamin D status in inner-city children. *Am. J. Clin. Nutr.* **2012**, *95*, 137–146.
15. Heaney, R.P.; Armas, L.A.; French, C. All-source basal vitamin D inputs are greater than previously thought and cutaneous inputs are smaller. *J. Nutr.* **2013**, *143*, 571–575.
16. Schmid, A.; Walther, B. Natural vitamin D content in animal products. *Adv. Nutr.* **2013**, *4*, 453–462.
17. Ovesen, L.; Brot, C.; Jakobsen, J. Food contents and biological activity of 25-hydroxyvitamin D: A vitamin D metabolite to be reckoned with? *Ann. Nutr. Metab.* **2003**, *47*, 107–113.
18. Jakobsen, J. Bioavailability and bioactivity of vitamin D_3 active compounds—Which potency should be used for 25-hydroxyvitamin D_3? *Int. Congr. Ser.* **2007**, *1297*, 133–142.

7

Antioxidant Generation during Coffee Roasting: A Comparison and Interpretation from Three Complementary Assays

Sebastian E. W. Opitz [1], Samo Smrke [1], Bernard A. Goodman [2,*], Marco Keller [3], Stefan Schenker [3] and Chahan Yeretzian [1,*]

1 Institute of Chemistry and Biological Chemistry, Zurich University of Applied Sciences, Einsiedlerstrasse 31, CH-8820 Wädenswil, Switzerland; E-Mails: opit@zhaw.ch (S.E.W.O.); smrk@zhaw.ch (S.S.)

2 State Key Laboratory for Conservation and Utilization of Subtropical Agro-Bioresources, Guangxi University, Nanning, 530004 Guangxi, China

3 Bühler AG, Gupfenstrasse 5, 9240 Uzwil, Switzerland; E-Mails: marco.keller@buhlergroup.com (M.K.); stefan.schenker@buhlergroup.com (S.S.)

* Authors to whom correspondence should be addressed; E-Mails: bernard_a_goodman@yahoo.com (B.A.G.); yere@zhaw.ch (C.Y.)

External Editor: Christopher J. Smith

Abstract: Coffee is a major source of dietary antioxidants; some are present in the green bean, whereas others are generated during roasting. However, there is no single accepted analytical method for their routine determination. This paper describes the adaption of three complementary assays (Folin-Ciocalteu (FC), ABTS and ORAC) for the routine assessment of antioxidant capacity of beverages, their validation, and use for determining the antioxidant capacities of extracts from coffee beans at different stages in the roasting process. All assays showed a progressive increase in antioxidant capacity during roasting to a light roast state, consistent with the production of melanoidins having a higher antioxidant effect than the degradation of CGAs. However, the three assays gave different numbers for the total antioxidant capacity of green beans relative to gallic acid (GA), although the range of values was much smaller when chlorogenic acid (CGA) was used as reference. Therefore, although all three assays indicated that there was an increase in antioxidant activity during coffee roasting, and the large differences in responses to GA and CGA illustrate their different sensitivities to different types of antioxidant molecule.

Keywords: coffee; antioxidant assays; Folin-Ciocalteu; ABTS; ORAC, flow injection analysis

1. Introduction

Epidemiological studies have revealed that coffee consumption is linked to several beneficial health effects [1,2]. Furthermore, it has been suggested that these are associated with the high antioxidant capacity of the beverage, which is greater than in most other food products [3–6], and represents a major source of antioxidants in the diets of many people [7–9].

In many plant organs, phenolic compounds are the most abundant antioxidant molecules and the phenolic content is often considered to represent a good approximation of the total antioxidant capacities of foods and beverages [10,11]. Thus, beneficial health effects have been strongly linked to (poly)phenolic components, and in tea, for example, flavonoids account for much of the antioxidant capacity [12], and have been specifically linked to its health-related properties [13].

In unroasted coffee beans, the chlorogenic acids (CGAs) are a major source of the total antioxidant capacity [14,15] with a high level of bioavailability [16], although as esters of primarily caffeic and quinic acids, they belong to a different sub-family of phenols from those in teas. Also, CGAs are at least partially degraded during the roasting process [17,18], which also results in the production of different types of antioxidant molecules, most notably the ill-defined products from the Maillard and caramelization reactions [19,20]. The water-soluble, macromolecular products from these reactions, known as melanoidins, have been linked to many chemical and biological properties, and are responsible for the taste and color of coffee [20]. Thus with this beverage, measurements of polyphenol contents are unlikely to give a complete picture of its antioxidant properties.

A wide range of assays have been reported for the measurement of antioxidant capacities of food products [1–25], but many are based on different types of chemical reaction and they do not all provide equivalent information. There are also many reports of the antioxidant capacity of coffee, in particular its dependence on the roast degree, but different assays have been used, and there are inconsistencies between results reported by different groups [17–19,26–31]; for example, conditions reported for the highest antioxidant potential have varied over a wide range of roasting conditions from green coffee through light to medium to dark roasted coffee, and the results are seemingly dependent on the method used. Thus, in order to get an accurate description of the antioxidant properties of coffee brews, and to produce information that may be relevant to the health effects of coffee consumption, it is necessary to have an understanding of the chemical basis of the various assays, which should be able to detect both phenolic and non-phenolic antioxidant molecules. In addition, it is important that assays and their experimental protocols are widely accepted in order to compare results obtained in different studies. Finally, although cellular assays [32] can provide biological support for any chemical findings, ultimately health properties can only be firmly established by true *in vivo* studies, which are subject to many confounding factors. Thus at the present time we are forced to rely to a large extent on chemical assays for understanding the antioxidant properties of food products.

A major aim of our current research program is to develop an analytical platform for assessing the antioxidant properties of coffee brews that is both easy to operate and can generate highly reproducible results, whilst at the same time providing values that are relevant to quality perception and biological properties of the beverage [33]. In the present paper, we describe the determination of antioxidant properties of brews prepared from green coffee beans and beans roasted at three different speeds and to different temperatures in an attempt to understand the critical stages in the roasting process for the formation and destruction of antioxidant molecules. In order to do this we have adapted the experimental protocols for three commonly-used complementary assays for determining the antioxidant capacity of food products to the routine and easy assessment of the antioxidant properties of coffee using automated flow injection techniques. Two of these assays are based on electron transfer reactions, namely the Folin-Ciocalteu (FC) assay for total phenols [34], and the ABTS assay, which is based on oxidation of the reagent to the cationic chromophore, 2,2'-azinobis-(3-ethylbenzothiazoline-6-sulphonate) ($ABTS^{\bullet+}$) [35]. The third assay, which is based on hydrogen atom transfer, is the oxygen radical absorbance capacity (ORAC), in which the antioxidant and the substrate fluorescein compete for thermally generated radicals after their formation by decomposition of 2,2'-azobis-2-amidinopropane, dihydrochloride (AAPH) [36]. Therefore, in addition to presenting their applications to the measurement of antioxidant properties of extracts of coffee at various stages of the roasting process, we also describe the optimization of these analytical methods for reproducibility, accuracy and speed of analysis and critically assess the information that is produced by them.

2. Experimental Section

2.1. Materials

Gallic acid monohydrate (purity > 99%) (GA), 2,2'-azobis-2-amidinopropane-dihydrochloride (AAPH), 2,2'-azinobis-(3-ethylbenzothiazoline-6-sulphonate) (ABTS), Folin-Ciocalteu reagent, salts and solvents were purchased from Sigma-Aldrich (Buchs SG, Switzerland). 5-Caffeoyl quinic acid (5-CQA) was obtained from Acros Organics (Geel, Belgium).

2.2. Coffee Roasting and Preparation

All coffee measurements were performed on a single batch of *Coffea arabica* L. beans from Costa Rica (washed Costa Rica, strictly hard bean). Coffee samples were roasted in a 20 kg batch roaster (Roastmaster™ 20, Bühler AG, Uzwil, Switzerland) according to three profiles, described as fast, medium and slow (Table 1). Within these profiles, each sample was prepared separately by roasting 8 kg of green coffee beans to a certain temperature along the profile. Beans were roasted and removed at the following temperatures: (1) 150, 170, 190, 203, 209 and 215 °C for fast, (2) 130, 150, 170, 190, 203, 209 and 215 °C for medium, and (3) 110, 130, 150, 170, 190, 203, 209 and 215 and 220 °C for slow roasting conditions. The lower roasting temperatures were performed twice, whereas the higher roasting temperatures (209, 215, and 220 °C) were only performed once. After a storage period of a minimum of 10 days in a refrigerator at 4 °C, the coffee samples were ground with a ball mill (MM 400, Retsch GmbH, Haan, Germany) for 1 min at 30 Hz. For green coffee beans, as well as those that were only partially roasted (to 110 or 130 °C), the samples were cooled in liquid nitrogen before grinding in order to obtain finely ground powders. The mass loss and the humidity content were determined by gravimetric analysis, the latter as loss of water in % after roast and ground (R & G) coffee was dried at 120 °C for 10 min with a Halogen Moisture Analyzer (HG 63, Mettler Toledo, Greifensee, Switzerland) (Table 1). Finally, the color of the roasted and ground coffee was determined with a Colortest II (NeuhausNeotec GmbH, Reinbek, Germany), which showed that the samples removed at 203 °C corresponded to a light roast. For analysis, the coffee samples were brewed in triplicate using a 200 mL French press (Bodum, Triengen, Switzerland) in which 12 g of ground coffee powder was infused with approximately 200 mL (mean value 190.5 ± 0.5 g) of hot water at 92 °C [18].

Table 1: Summary of roasting data for Arabica coffee samples for different profiles and roasting degrees ($n = 1$).

Roast profile	Final roast temperature °C	Roast time min:sec	Color of R & G coffee Colortest II	Mass loss %	Humidity %
Fast	150	03:00 ± 4	166.8 ± 0.7	5.8	ND
	170	03:52 ± 5	154.0 ± 1.0	8.7	4.17
	190	04:43 ± 2	130.2 ± 7.78	12.0	3.32
	203	05:15 ± 0	97.2 ± 13.44	14.7	2.26
	209	05:27	71.3	16.4	2.19
	215	05:37	56.3	18.6	1.84
Medium	130	03:50 ± 6	ND	4.8	ND
	150	05:07 ± 10	161.5 ± 2.1	7.0	ND
	170	06:45 ± 2	160.7 ± 1.9	9.4	4.06
	190	08:23 ± 19	139.5 ± 3.1	12.3	3.16
	203	09:36 ± 21	109.4 ± 6.6	15.3	2.03
	209	10:29	93.7	15.9	1.43
	215	11:05	78.7	17.6	1.26
Slow	110	03:07 ± 17	ND	2.8	ND
	130	04:54 ± 9	ND	5.0	ND
	150	06:45 ± 2	162.3 ± 1.4	7.4	ND
	170	09:06 ± 1	162.5 ± 0.2	9.5	4.16
	190	12:04 ± 4	140.7 ± 0.0	12.6	3.06
	203	14:51 ± 24	106.8 ± 2.1	15.3	1.92
	209	15:48	91.7	15.9	1.74
	215	16:36	78.0	16.8	1.32
	220	17:51	67.3	19.2	1.19

The coffee was roasted either at a fast, a medium or a slow roast profile and each coffee for a given roast temperature was roasted separately. Beans were removed at the presented temperatures, which correspond to the roasting time. Mass loss, humidity and color of the ground coffee were measured, where applicable. ND = not determined.

2.3. Antioxidant Assays

Each sample was analyzed by each of the three following assays that are commonly used for assessment of antioxidant values. In each case, the coffee extracts were diluted to adjust the concentration to the specific dynamic range of the assays. Samples for the FC assay were diluted 1:50, for the ABTS assay 1:200, and 1:400 for the ORAC assay, and analyzed with a FIA lab 3200 instrument (FiALab Instruments Inc., Bellevue, WA, USA). A Cetac ASX-260 autosampler (Omaha, NE, USA) together with a 1 mL syringe pump were used for automatic introduction of the samples for all assays. In the FC assay the antioxidant capacity is measured with a flow injection analysis (FIA) setup including a six-port injection valve. Both FC reagent and NaOH solutions are pumped simultaneously and the sample/standard is injected with the six-port valve, which is filled by the autosampler. Constantly pumped FC reagent carries the sample from the injection loop towards the T-piece, where the solution mixes with the NaOH solution. After further mixing and dispersion in the reaction coil, the absorbance of the blue colored complex product is measured in the detector cell. In the ABTS assay the antioxidant capacity is measured via a sequential injection analysis (SIA) setup. After the transfer tube from the autosampler is purged with sample/standard, the carrier syringe is pre-filled with carrier solution, followed by sequential stacking of reagents ($ABTS^{\bullet+}$, sample and carrier) in the holding tube. By delaying subsequent actions for 30 s, the reaction time of the stacked reagents is prolonged before the $ABTS^{\bullet+}$ absorbance is measured in the detector cell. In the ORAC assay the antioxidant capacity is measured via a sequential injection analysis (SIA) setup using a stop flow technique in the detector cell. After the transfer tube from the autosampler is purged with sample/standard, the carrier syringe is pre-filled with carrier solution, followed by sequential stacking of reagents (AAPH, fluorescein, sample and carrier) in the holding tube. Then the content of the holding coil is dispensed through the detector cell, stopped at the peak of the fluorescein signal and the degradation of fluorescein is monitored for 1 h.

All samples were filtered before analysis using 0.45 μm polyethylene terephthalate filters (Machery-Nagel, Düren, Germany). Results are expressed as gallic acid equivalents (GAE), which were used as a standard reference material; this was preferred to the more commonly-used Trolox, because of solubility problems of the latter in our system.

2.3.1. Folin-Ciocalteu (FC) Assay

This assay makes use of the fact that phenolic compounds undergo autoxidation in an alkaline medium, and that these products then react with the FC reagent, which consists of a mixture of phosphomolybdate and phosphotungstate, to form "molybdenum blue" [34], which is then measured. Our automated adaption of the FC assay consists of a flow injection analysis (FIA) setup, in which the sample or antioxidant standard is injected (injection loop 100 μL) into the flow stream (flow rate 30 μL/s) of the FC reagent (0.2 M with respect to acid) (Figure 1a). After mixing with NaOH solution (0.25 M, flow rate 30 μL/s) to raise the solution pH (to ~pH 10), dispersion in the reaction coil (1 m tubing length) mixes the components, and the molybdenum blue reaction product is measured photometrically at 765 nm.

Figure 1. Schemes for antioxidant measurements by (**a**) FC, (**b**) ABTS and (**c**) ORAC assays.

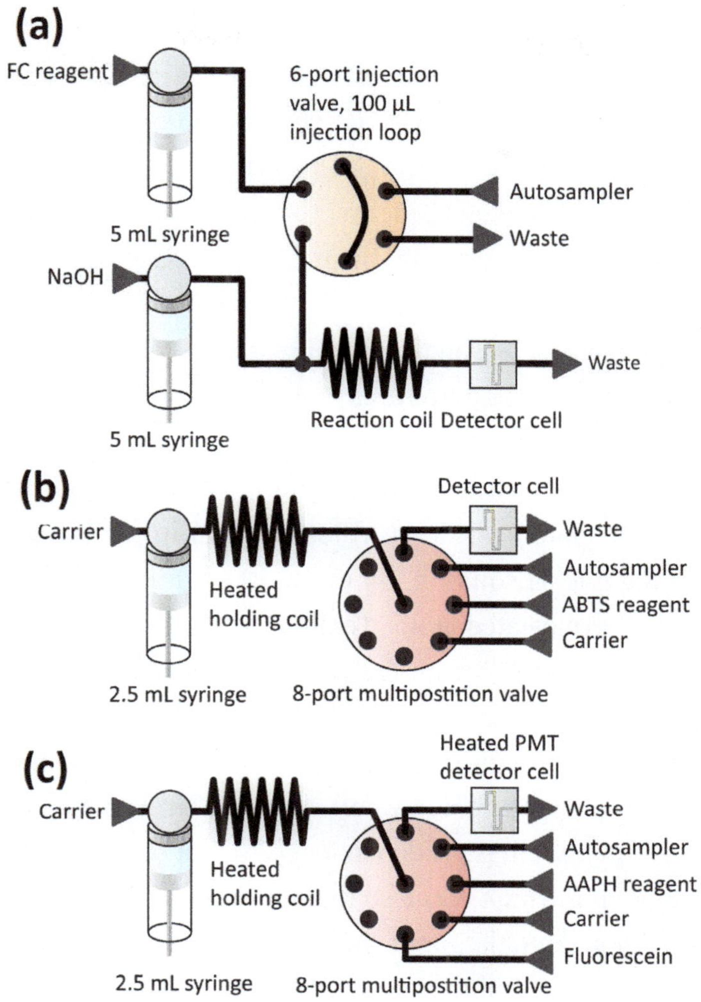

2.3.2. ABTS (2,2'-Azinobis-(3-Ethylbenzothiazoline-6-Sulphonate)) Assay

In the ABTS assay, which is also known as the Trolox equivalent antioxidant capacity (TEAC) assay, the green-blue stable radical cationic chromophore, $ABTS^{•+}$ is reduced by antioxidants and a subsequent drop in color intensity is observed. The radical cation has to be generated prior to analysis, because it is not sufficiently stable to be available in bulk. Several different approaches have been published for $ABTS^{•+}$ generation and in the present work we used oxidation by potassium persulfate as originally described by Re *et al.* [35]. The assay was performed using a sequential injection analysis (SIA) setup to minimize consumption of the reagent (Figure 1b). The sample or antioxidant standard (injection volume 40 µL), $ABTS^{•+}$solution (injection volume 20 µL) and carrier solution (0.1 M Na_2HPO_4, pH = 7.4) were aspirated sequentially into a heated holding coil, which was thermostated at 35 °C. By aspirating and dispensing the components through the holding coil, an even dispersion was achieved; the instrument was stopped after the sampling process was completed to achieve a total reaction time of 60 s [35,37]. After dispensing the mixture to the detector cell (flow rate 30 µL/s), the absorbance of the $ABTS^{•+}$ solution was measured photometrically at 734 nm. Before starting the experiment, the concentration of $ABTS^{•+}$ solution was set to an initial absorbance of 0.75 AU as the preparation of the $ABTS^{•+}$ solution was not accurately reproducible.

2.3.3. Oxygen Radical Absorbance Capacity (ORAC) Assay

The ORAC assay measures the effect of a test sample on the decrease in fluorescence of fluorescein under a constant flux of free radicals that originate from AAPH and result in the formation of a non-fluorescent product [36]. When antioxidants are present, a proportion of the radicals reacts directly with an antioxidant molecule abstracting a hydrogen atom to form a diamagnetic product and a "stable" antioxidant-derived radical, which does not react with fluorescein. However, although a proportion of the reactive radicals still reacts with fluorescein, there is a secondary reaction between antioxidants and fluorescein radicals that results in the regeneration of fluorescein. This latter reaction results in a lag phase in the fluorescein degradation [38], and the results from the assay represent a combination of both types of reaction. By quantifying the integrated area under the curve of the fluorescence decay, this assay combines several ways of quantification of antioxidant capacity, e.g., lag time, initial rate, and total inhibition in a single value [22]. Our adaption of the original assay [36] uses a SIA setup in which the sample or antioxidant standard (injection volume of 25 μL), fluorescein solution (25 μL of 0.37 μg/mL solution), and the AAPH solution (50 μL of 50 mg/mL solution) are aspirated sequentially into a holding coil, which is thermostated at 35 °C (Figure 1c). However, the incubation step at 37 °C described by Ou *et al.* [36] was omitted from the AAPH preparation procedure. A crucial step in our assay is the control of the initial formation of radicals from AAPH. Therefore, the AAPH was stored in ice and the tubing purged each time with cold AAPH before use. Introducing the components and the carrier (0.1 M Na_2HPO_4, pH 7.4) into the holding coil produced an even dispersion in a buffered solution, as the fluorescence signal of fluorescein is pH dependent [36]. Subsequently the reagents-sample mixture was dispensed to the detector cell, which was also thermostated at 35 °C. At the peak of the fluorescein signal, the flow was stopped and degradation of fluorescein was monitored for 1 h in a photomultiplier tube. A 5 mm light emission diode combined with a blue color filter was used as an excitation source and an interference green filter was placed before the photomultiplier tube to filter the emitted light. Area under the curve (AUC)-quantification was then applied to the fluorescein decay curve to produce a measure of the antioxidant value.

2.4. Validation of Assays

Each assay was validated before being used in routine measurements of coffee antioxidant capacity. The validation was performed with GA as the reference compound for each assay, and consisted of determinations of repeatability within a day (intraday), reproducibility between days (interday), the limit of detection (LoD), limit of quantification (LoQ), linearity range, coefficient of determination and recovery percentage.

After preparation of a GA stock solution in degassed water, working standard solutions were prepared at appropriate concentrations for the linearity range of the assays. Intraday repeatability was assessed by repetitive injections of a single standard GA sample. Interday reproducibility was performed over a period of several days and involved the repeated preparation of a calibration curve of freshly prepared standards and reagents. The LoD and LoQ were investigated by using GA standard solutions at low concentrations, and the dynamic range for the analyses was determined from the linearity of the calibration curve and the LoD. Assessment of recovery was made by spiking samples of a freeze-dried soluble coffee (dissolved in 100 mL of degassed water) with a GA standard solution.

In addition to these standard validation criteria, robustness tests were carried out on certain parameters of some assays. Because the $ABTS^{\bullet+}$ reagent degrades slowly, the stability of the $ABTS^{\bullet+}$ radical is a critical factor for accurate measurements, and this was tested by injecting a blank sample and a 3 mg/L GA standard every hour during measurements. Further, the initial $ABTS^{\bullet+}$ concentrations were varied to investigate the robustness of the initial absorbance. Also, although results obtained by the ABTS assay can be expressed either as an absorbance decrease, which is an absolute difference in absorbance between a blank and a sample signal, or as an inhibition percentage (Inh. %), we only present data for the latter, which represents the ratio between the observed absorbance after the reaction with an antioxidant and that of the maximum value of the blank (water).

For the ORAC assay, AAPH and fluorescein concentrations were varied to test the response of selected concentrations of the standards. For AAPH, 90% and 110% of the chosen AAPH concentration were tested and three different GA standard concentrations (10, 15 and 20 mg/L) were measured. From the mean value of the corresponding AUC values, the response was also calculated as %. The same standard concentrations were also used to test the effects of changing the fluorescein concentrations between 90% and 110% of the chosen value.

3. Results and Discussion

3.1. Methodology Validation

Prior to measurements, the whole FIA lab system was thoroughly purged with freshly degassed ethanol and only freshly degassed water was used for preparation of samples and solutions. This turned out to be necessary in order to avoid the formation of air bubbles in the FIA system and to optimize the performance. For the FC assay, the dynamic range was determined to lie between 2 and 60 mg/L of GA, and a correlation coefficient of 0.9996 was obtained for a calibration curve in this concentration range. Measurements of the antioxidative response of GA standard solutions in the range 2 to 8 mg/L gave a LoD of 1.6 mg/L GA and a LoQ of 3.2 mg/L GA. In addition, when solutions of freeze-dried coffee samples (160 mg/L) were spiked with 30 mg/L GA, a recovery of 93% was calculated. The intraday repeatability was 0.9% RSD and the interday reproducibility (over three days) was 2% RSD (Table 2). Compared to the other two assays, this assay showed the highest reproducibility, the lowest sensitivity, and the highest dynamic range (Table 2).

Table 2. Validation of analytical results for the FC, ABTS and ORAC assay.

Validation criterion	FC	ABTS	ORAC
Intraday RSD (%)	0.9 at 30 mg/L GA	0.4 at 3 mg/L GA	4.7 at 10 mg/L GA
Interday RSD (%)	2.1 at 30 mg/L GA	4.4 at 3 mg/L GA	8.2 at 10 mg/L GA
LoD (mg/L GA)	1.6	0.12	0.43
LoQ (mg/L GA)	3.2	0.4	2.1
Linearity range (mg/L GA)	2–60	0.5–5	5–25
R^2	0.9996	0.9987	0.9993
Recovery (%)	93 (of 30 mg/L GA)	89 (of 2 mg/L GA)	78 (of 20 mg/L GA)

All validation measurements were performed with gallic acid (GA). LoD = limit of detection, LoQ = limit of quantification, RSD = relative standard deviation.

The dynamic range for the ABTS assay was 0.5 to 5 mg/L, considerably lower than with the FC assay, and the calibration curve for these concentrations showed a correlation coefficient of 0.9987 (Table 2). Here the maximum value of 5 mg/L could not be exceeded since that resulted in zero absorbance of $ABTS^{\bullet+}$. The LoD was determined as 0.12 mg/L GA and LoQ as 0.4 mg/L GA. Interday reproducibility was measured over three days with a deviation of 4.4% RSD and the intraday repeatability was 0.4% RSD. A recovery of 89% was determined for freeze-dried coffee samples (40 mg/L) spiked with 2 mg/L of GA (Table 2). Inhibition percentage was used for quantification, as the recovery values for absorbance drop were considerably lower. For the ABTS assay, repeatability was excellent and reproducibility was acceptable, although better values (1.3% to 1.95% RSD) have been reported for a FIA instead of a SIA setup [37,39]. On testing the stability of $ABTS^{\bullet+}$ for an experimental time of 8 h, the slope of degradation (AU in 10 h) was found to be −0.026 with an R^2 of 0.82. The variation of the results was small with a %RSD of 0.6 (n = 8). A similar decrease in slope (−0.023 AU in 10 h) was observed over 13 h with a R^2 of 0.71 and a variation of 0.8% RSD (n = 13). Variations in the initial $ABTS^{\bullet+}$ concentration were studied and the effect on the deviation in measured inhibition percentage was estimated. Here with a higher absorbance of 0.88 AU, the deviation was −15.8% and for the lower value of 0.60 AU it was +18.2% compared to the values obtained for 0.7 AU. From these values the maximum allowable deviation of the initial $ABTS^{\bullet+}$ absorbance was calculated as ±0.025 absorption units (AU), which would be within a 3% error margin.

For the ORAC assay, the linearity range was between 5 and 25 mg/L GA and a calibration curve within that range produced a correlation coefficient of 0.9993 (Table 2); the LoD was 0.43 mg/L GA and LoQ 2.1 mg/L GA. The intraday repeatability was calculated as 4.7% RSD and the interday reproducibility over five days was 8.2% RSD. Recovery of 78% was determined for freeze-dried coffee solutions (2000 mg/L) spiked with 15 mg/L of GA (Table 2). The reproducibility of the ORAC measurements, which were made in series with SIA analyses, was in the range of other studies. The literature reports values for the intermediate precision relative standard deviations (RSD_{int}) of 0.7 to 13% and reproducibility relative standard deviations (RSD_R) of 4.4 to 13.8% [36,40,41], which were achieved by parallel measurements using a fluorescence plate reader. Hence, the reproducibility of the assay was still acceptable despite the rather long reaction times. The stability of AAPH was tested hourly by injecting a 20 mg/L GA standard. For 30 h and 30 injections the RSD was 3%, for 40 h and 40 injections it was 4.3%. Also initial concentrations of AAPH as well as fluorescein were varied to assess the importance of precise initial concentrations. Relative concentrations of 90% and 110% were tested for the two substances, which resulted in corresponding responses of 89% and 111% when varying fluorescein, and 107% and 86% when varying AAPH. The main drawback of this assay is the amount of time required to measure samples, but the validation procedure gave a time window of ≥40 h during which the assay conditions remained stable. A major factor affecting the performance of the assay is the stability of radical generator AAPH, which is temperature sensitive [42]. As we needed a valid time period of 40 h, it was important to ensure the same AAPH quality for this time range. Therefore, it was stored in ice water and was purged before each run to have fresh AAPH solution for radical generation. Despite the long reaction time of 1 h, results with acceptable reproducibility were obtained.

3.2. Effects of Coffee Roasting Conditions on Antioxidant Contents

The results from the assays of the extracts from coffee samples roasted to various stages are presented in Figure 2. Large differences in the antioxidant values were obtained from the different methods, which increases in the order ABTS < FC < ORAC, when compared to GA as reference standard; the mean values for the green coffee, for example, were 373, 1885, and 5083 mg/L GA equivalents, respectively. Clearly GA and the antioxidant components of green coffee behave differently with respect to the three assays, even though phenolic compounds (the CGAs) are the major antioxidants in green coffees.

The water content of the green beans was 10.9%, but this decreased with increasing roast temperature/roasting time within each roasting profile (Table 1). However, when comparing the data between roasting profiles, the color, mass loss and water (humidity) contents were comparable at similar sampling temperatures (130 to 209 °C). The trends with respect to roasting progression were similar for each of the analytical methods, and antioxidant values for all roasted samples were greater than those of the green beans, except the first slightly roasted coffee sample of the slow roast (110 °C). Thus during roasting to a light degree (up to 200 °C), molecules with antioxidant properties are produced in greater amounts than those that are destroyed. Although part of this increase in measured antioxidant capacity can be explained by mass loss during roasting, this loss is insufficient to fully explain the increase (see Table 1). Furthermore, while we attempted to produce samples with consistent extraction efficiency by grinding them with a ball mill, we cannot rule out the possibility that the more voluminous and porous roasted and ground coffee beans resulted in higher extraction efficiency than the harder green coffee beans. Antioxidant values tended to increase with increasing temperature and reached an optimum at around 200 °C [43]. Especially with the FC method, large decreases towards darker roasts were observed. Since the FC method is often considered to be especially sensitive to phenolic compounds, these latter results can be explained by the greater degradation of CGAs at higher temperatures. The effect is strongest with the slow roast, where the energy input is maintained over the longest time span, and the change in magnitude of the antioxidant activity is less pronounced with the ABTS and ORAC assay, which also indicates that these methods have different sensitivities to other types of antioxidants in the roasted coffee [20].

Figure 2. Antioxidant capacity (as gallic acid equivalents, GAE, mean ± SD) of green and roasted coffee measured using (**a**) the Folin-Ciocalteu, (**b**) the ABTS, and (**c**) the ORAC assays. Three roast profiles (fast, f; medium, m and slow, s) were conducted and coffee beans were removed at the given temperatures. The values are depicted with colors from green to brown with increasing roast degree; six roast degrees for the fast profile, seven for the medium profile, and nine for the slow profile were roasted (roast replicate $n = 2$, except for the higher roast temperatures 209, 215, 220 °C, which were only roasted once (brewing replicates $n = 3$).

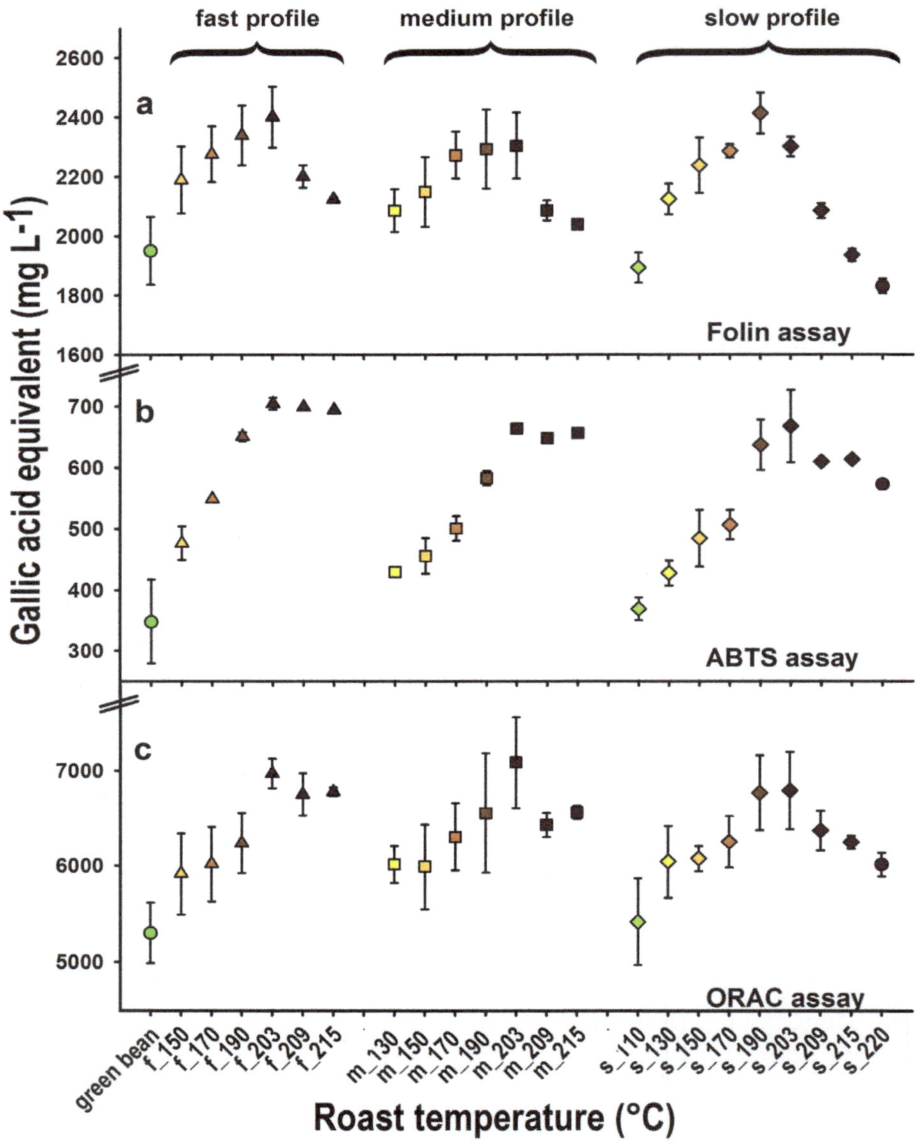

Finally, with each assay there was little influence of the speed of roasting on the antioxidant values observed for each temperature, with the exception of the samples heated to the highest temperature for medium and slow roasting times in the FC assay, where the decrease is increasing with roast length. If, as suggested in the previous paragraph, these lower results with the FC assay correspond to increased destruction of the CGAs, it would appear that the speed of roasting could be an important parameter for the preservation of the CGA component in roasted coffee samples. This effect of speed on the preservation of antioxidant activity is similar with the ABTS assay, as also here the highest increase is measured with the fast roast profile.

3.3. Significance of the Numbers Produced by the Different Methods

The FC assay [34] is often considered to be a measure of "total phenolic content", and since the CGAs represent the main polyphenols in coffee [14], it is tempting to describe the FC values for green coffee as approximating to the CGA concentration. However, this cannot be the situation for the roasted coffee samples, since roasting results in appreciable increases in antioxidant values even though it is known that progressive degradation of CGAs occurs during the roasting process [17,18]. Furthermore, as reported by Prior *et al.* [22], the FC reagent is not specific for phenolic compounds and it can also be reduced by many non-phenolic compounds, e.g., melanoidins [19]; thus it may be more accurately described as measuring the total reducing capacity of a sample [23]. The high pH (pH 10) used in the FC assay results in deprotonation of one phenolic –OH group, a process important for the autoxidation reaction of polyphenols. Thus phenoxy or semiquinone radicals are readily formed at high pH, and it seems likely that such radicals are the reactive compounds in this assay. However, although semiquinone radicals tend to be reasonably stable, at high pH, secondary reactions, such as dimerization, degradation and further oxidation reactions involving generation of some –OH groups can lead to higher activity than would be expected from oxidation of simple catechol groups [44]. Furthermore, cyclic voltammetry studies have shown that oxidation of 5-CQA is not reversible at high pH [45]. Also, phenoxy radicals are often quite reactive and may thus react with other components in the complex coffee matrix. Thus there are potentially reactions occurring in the assay that could lead to either over- or underestimation of the actual phenolic contents of the coffee brews.

In the ABTS assay, the test sample is added after the $ABTS^{\bullet+}$ has been generated, and the amount of radical remaining after a certain reaction time is determined. As with the FC assay, the ABTS assay is not specific for any particular type of antioxidant, since any compound with a redox potential lower than that of $ABTS^{\bullet+}$ ($E = 0.68$ V) may react with the radical [46].

However, the competitive mechanism of the ORAC assay provides a fundamentally different approach to antioxidant measurement compared to the FC and ABTS assays. By integrating the response over time, this assay includes the reducing effects of secondary components in the sample for which there may be an appreciable lag time before they show a response. Nevertheless, the long time required for the assay represents a major drawback for the routine use of the ORAC assay, which also has technical limitations related to the formation of the peroxyl radicals. For example, radical forming compounds such as AAPH are sensitive to temperature and can undergo spontaneous decay. In air-saturated solutions, it is thought that the AAPH radical reacts rapidly with O_2 to give a more stable radical, and the loss of fluorescence is then an indication of the extent of damage resulting from reaction with the radicals. However, a big advantage of this assay is the combined quantification of initial reaction rate as well as the total amount of inhibition, which includes slow-reacting or secondary antioxidant products [36]. This combination makes it a useful assay to determine antioxidant capacities of samples with a complex matrix and mixtures of antioxidant active compounds such as coffee. A lag phase could be the consequence of a reaction between fluorescein radicals and antioxidant compounds which leads to regeneration of the original fluorescein [38]. A considerable factor affecting the ORAC assay is the presence of metal ions, which can be responsible for a possible underestimation of the numbers generated [47].

Although there has been increasing use in recent years of measures of antioxidant capacities of food products as markers of their contributions to healthy diets [32], the wide disparity in the numbers obtained by different methods in the present work shows that caution should be exercised in their interpretation. The more than 10-fold range in values for the antioxidant content (expressed as GA equivalents) of green (unroasted) coffee in the present work is a reflection of the different chemical processes involved in the three assays used. If reactions of GA, which was used as a reference standard throughout, were truly representative of phenol chemistry, then the diverse numbers could illustrate variability in the relative sensitivity of the various methods to phenolic and non-phenolic antioxidants. However, the results with green coffee suggest that there are also appreciable differences in the chemical behavior of GA and the CGAs (the major antioxidants in green coffee) under the conditions of the different assays. As a consequence, we compared the responses of GA and 5-CQA in the three assays. Considerable differences were observed, and on a molar basis GA was more reactive than 5-CQA in the ABTS assay, but less reactive in the FC and the ORAC assays. When the antioxidant capacity of green coffee was recalculated as CGA equivalents (CGAE), mean values of 3874, 3080 and 2671 mg/L were obtained for the ABTS, FC and ORAC assays, respectively. These are much closer than the corresponding numbers (373, 1885, and 5083 mg/L) for GA equivalents, and are also consistent with CGAs being the main antioxidants in green coffee.

The differences in the responses of the standard compounds GA and 5-CQA in the ABTS assay are pronounced with GA values being ~five times higher for the same concentration. This is broadly in line with other publications, where GA has also been reported to be around 2.7 to 4.3 times more potent as a reductant than CGA or caffeic acid [18,39,48,49]. However, the analyses of green coffee gave values >10 times higher for CGAE than GAE values. The reaction time was then considered as a possible factor affecting the behavior of GA and 5-CQA in the ABTS assay, since the antioxidant capacity determined for phenolic compounds using this method has been reported to be sensitive to the choice of analytical parameters [50]. However, in time delay experiments with concentrations that give similar responses in the ABTS assay (2 mg/L GA and 15 mg/L 5-CQA), we found that both phenols produced similar absorption intensity curves (data not shown). Furthermore, Tian and Schaich [51] have reported that 5-CQA reacts essentially instantaneously with $ABTS^{\bullet+}$, so the cause of the different behavior of GA and 5-CQA in this assay is probably chemical rather than physical.

The differences in the responses of GA and 5-CQA in the FC and ORAC assays were less pronounced, than in the ABTS assay, but not insignificant. In the FC assay, the value for GA was ~0.8 times that of 5-CQA, which is similar to the result reported by Chun and Kim [52], although on the basis of the number of hydroxyl groups, the antioxidant capacity of GA should be higher [18,53]. The number of electrons involved in oxidation increased for CGA with increasing pH, which could also be an explanation for the observed higher antioxidant potential of 5-CQA in the present study [45]. With the ORAC assay, the response of 5-CQA was 2.5 times that of GA, and an even greater ratio (3.5) has been reported by Yeh and Yen [49]. These latter results could be the consequence of secondary reactions which also contribute to the antioxidant capacity determined by this method [53], and in the coffee extracts there could be further complications arising from the presence of transition metal species [48]. It should be noted that the ORAC assay was run for ≥30 min to avoid any underestimation of antioxidant capacity due to slow reacting antioxidants.

Recent studies of coffee brews using high performance size exclusion chromatography coupled to on-line assays have suggested that CGAs and melanoidins (the components responsible for the largest contributions to the antioxidant capacity of coffee brews) produce different responses in the FC and ABTS methods [18]. In that work it was concluded that the ABTS assay was relatively more sensitive towards melanoidins than the FC assay, which is also supported by the present results. Furthermore, the contributions to the antioxidant capacity derived from CGAs decreased with a higher roasting degree, whilst there was only a minor increase in the contribution from melanoidins. For the coffee samples roasted to a light roast, only a small amount of degradation of CGAs was observed with the FC assay. However, this was accompanied by a larger increase in the melanoidin contribution, a result that suggests that melanoidins produced in the early stage of roasting have higher antioxidant potential than those produced at later stages. This hypothesis will be tested in future experiments, as well as the extent of contributions from other components in the coffee brews.

Although this research shows the progressive formation of antioxidant molecules during the coffee roasting process, it also highlights a fundamental problem in interpreting results from antioxidant assays. Different molecules have different relative responses in different assays, and even phenols can have widely different relative responses in commonly-used assays. Because of this, the choice of reference standard can have a major influence on the results, as illustrated by the numbers from the various assays expressed as GA or CGA equivalents. Problems relating to the selection of reference standards in antioxidant activity assessment have been discussed in detail by Nenadis *et al.* [54]. However, apart from recommending that uric and ascorbic acids should not be used for this purpose, they concluded that there was no single ideal reference standard, whilst lending some support to the use of Trolox. Also, in the six assays considered in that paper, the GA/Trolox response ratio varied between 0.52 and 3.76. Furthermore, if the antioxidant values for green coffee beans in the present work are expressed relative to Trolox using the relative activities of GA and Trolox reported by Nenadis *et al.* [54], values of 1200, 7090, and 5340 mg/L are obtained for the ABTS, FC and ORAC assays, respectively.

In addition to these analytical problems, it is also unlikely that all antioxidants in foods behave in the same way after consumption. Thus, whereas some may be taken up intact during food digestion, others may be altered, and yet others may be unavailable, e.g. liposolubility. Therefore, questions must be raised as to the relevance of analyses such as those presented here to the health of consumers, and until we have more information, caution should be exercised in interpreting antioxidant values beyond simple shelf-life prediction. Furthermore, even the use of such antioxidant assays for shelf-life prediction requires calibration with actual storage-induced changes, possibly on a product-by-product basis.

4. Conclusions

The results from this work show clearly the progressive generation of antioxidant molecules during the roasting of coffee, and that this occurs more rapidly than any degradation of the initial antioxidant capacity during the early stages of the roasting process. After an optimum value at around 200 °C, the antioxidant activity decreased due to degradation processes. Similar trends were observed with three different types of assay that used different types of reagents, and hence putatively had different responses to antioxidant molecules in the coffee; the active reagents were first a metal complex, second a stable radical, and third a radical with moderate stability. However, these three analytical methods produced widely different numbers relative to the GA standard, although much more similar numbers were obtained when the results were expressed as CGA equivalents. Reproducibility of replicate samples was good, and each method showed a consistent set of trends in antioxidant values with roasting conditions. Thus there are differences in the chemical nature of the components measured with the individual methods, a conclusion which indicates that the concept of a single antioxidant number is meaningless, especially in the context of the "healthiness" of a food product. Profiles of antioxidant properties from complementary methods are potentially more meaningful, but the extent to which differences in biological behavior are reflected by the results from the various assays is still unknown. However, although the analytical results described in the present paper give a representation of various aspects of the antioxidative capacity in coffee; it represents just a first step towards developing knowledge about the health effects of the beverage. Nevertheless, it demonstrates the necessity of using a combination of assays for the determination of antioxidant properties of food products. The methods presented here are sufficiently simple that they can be applied routinely in food quality analysis, although at the present time we still do not have a definitive interpretation of the meaning of the numbers generated.

Acknowledgments

We acknowledge Babette Klopprogge for preparing coffee brews and antioxidant measurements, Sergio Petrozzi for technical assistance in establishing the experimental setup for FIA analysis. We also thank the Commission for Technology and Innovation (CTI) of Switzerland for their financial contribution to the CTI-project "Modulating the Coffee Roasting Process to Optimize the Antioxidant Potential of Coffee" (Project no. 13897.1 PFIW-IW).

Author Contributions

In alphabetical order by section: design of the project: B.A. Goodman, S. Schenker, and C. Yeretzian; design of experiments: S.E.W. Opitz, S. Smrke and C. Yeretzian; validation of experiments: S.E.W. Opitz and S. Smrke; coffee roasting: M. Keller and S. Schenker; measurements of coffee brews: S.E.W. Opitz; writing of the paper: B.A. Goodman, S.E.W. Opitz, and C. Yeretzian

Conflicts of Interest

The authors declare no conflict of interest.

References

1. Goodman, B.A. *Coffee Consumption and Health*; Nova Science Publishers: New York, NY, USA, 2012.
2. Chu, Y.F. *Coffee: Emerging Health Effects and Disease Prevention*; Wiley-Blackwell, Hoboken, NJ, USA, 2012.
3. Richelle, M.; Tavazzi, I.; Offord, E. Comparison of the antioxidant activity of commonly consumed polyphenolic beverages (coffee, cocoa, and tea) prepared per cup serving. *J. Agric. Food Chem.* **2001**, *49*, 3438–3442.

4. Nicoli, M.C.; Anese, M.; Manzocco, L.; Lerici, C.R. Antioxidant properties of coffee brews in relation to the roasting degree. *Lebensm. Wiss. Technol.* **1997**, *30*, 292–297
5. Anese, M.; Nicoli, M.C. Antioxidant properties of ready-to-drink coffee brews. *J. Agric. Food Chem.* **2003**, *51*, 942–946
6. Vicente, S.J.V.; Ishimoto, Y.; Torres, E.A.F.S. Coffee modulates transcription factor Nrf2 and highly increases the activity of antioxidant enzymes in rats. *J. Agric. Food Chem.* **2014**, *62*, 116–122.
7. Pulido, R.; Hernández-García, M.; Saura-Calixto, F. Contribution of beverages to the intake of lipophilic and hydrophilic antioxidants in the Spanish diet. *Eur. J. Clin. Nutr.* **2003**, *57*, 1275–1282.
8. Svilaas, A.; Kaur Sakhi, A.; Frost Andersen, L.; Svilaas, T.; Ström, E.C.; Jacobs, D.R., Jr.; Ose, L.; Blomhoff, R. Intakes of antioxidants in coffee, wine, and vegetables are correlated with plasma carotenoids in humans. *J. Nutr.* **2004**, *134*, 562–567
9. Fukushima, Y.; Ohie, T.; Yonekawa, Y.; Yonemoto, K.; Aizawa, H.; Mori, Y.; Watanabe, M.; Takeuchi, M.; Hasegawa, M.; Taguchi, C.; Kondo, K. Coffee and green tea as a large source of antioxidant polyphenols in the Japanese population. *J. Agric. Food Chem.* **2009**, *57*, 1253–1259.
10. Scalbert, A.; Johnson, I.T.; Saltmarsh, M. Polyphenols: Antioxidants and beyond. *Am. J. Clin. Nutr.* **2005**, *81*, 215S–217S.
11. Perez-Jimenez, J.; Neveu, V.; Vos, F.; Scalbert, A. Systematic analysis of the content of 502 polyphenols in 452 foods and beverages: An application of the phenol-explorer database. *J. Agric. Food Chem.* **2010**, *58*, 4959–4969.
12. Chen, C.W.; Ho, C.T. Antioxidant properties of polyphenols extracted from green and black teas. *J. Food Lipids* **1995**, *2*, 35–46.
13. Frei, B.; Higdon, J.V.; Antioxidant activity of tea polyphenols *in vivo*: Evidence from animal studies. *J. Nutr.* **2003**, *133*, 3275–3284.
14. Clifford, M.N. Chlorogenic acids and other cinnamates—Nature, occurrence and dietary burden. *J. Sci. Food Agric.* **1999**, *70*, 362–372.
15. Madhava Naidu, M.; Sulochanamma, G.; Sampathu, S.R.; Srinivas, P. Studies on extraction and antioxidant potential of green coffee. *Food Chem.* **2008**, *107*, 377–384.
16. Farah, A.; Monteiro, M.; Donangelo, C.M.; Lafay, S. Chlorogenic acids from green coffee extract are highly bioavailable in humans. *J. Nutr.* **2008**, *138*, 2309–2315.
17. Moon, J.K.; Yoo, H.S.; Shibamoto, T. Role of roasting conditions in the level of chlorogenic acid content in coffee beans: Correlation with coffee acidity. *J. Agric. Food Chem.* **2009**, *57*, 5365–5369.
18. Smrke, S.; Opitz, S.E.W.; Vovk, I.; Yeretzian, C. How does roasting affect the antioxidants of a coffee brew? Exploring the antioxidant capacity of coffee via on-line antioxidant assays coupled with size exclusion chromatography. *Food Funct.* **2013**, *4*, 1082–1092.
19. Delgado-Andrade, C.; Morales, F.J. Unraveling the contribution of melanoidins to the antioxidant activity of coffee brews. *J. Agric. Food Chem.* **2005**, *53*, 1403–1407.
20. Wang, H.Y.; Qian, H.; Yao, W.R. Melanoidins produced by the Maillard reaction: Structure and biological activity. *Food Chem.* **2011**, *128*, 573–584.
21. Prior, R.L.; Cao, G.H. *In vivo* total antioxidant capacity: Comparison of different analytical methods. *Free Radic. Biol. Med.* **1999**, *27*, 1173–1181.

22. Prior, R.L.; Wu, X.L.; Schaich, K. Standardized methods for the determination of antioxidant capacity and phenolics in foods and dietary supplements. *J. Agric. Food Chem.* **2005**, *53*, 4290–4302.
23. Huang, D.; Ou, B.; Prior, R.L. The chemistry behind antioxidant capacity assays. *J. Agric. Food Chem.* **2005**, *53*, 1841–1856.
24. Magalhaes, L.M.; Segundo, M.A.; Reis, S.; Lima, J.L.F.C. Methodological aspects about *in vitro* evaluation of antioxidant properties. *Anal. Chim. Acta* **2008**, *613*, 1–19.
25. Karadag, A.; Ozcelik, B.; Saner, S. Review of methods to determine antioxidant capacities. *Food Anal. Methods* **2009**, *2*, 41–60.
26. Daglia, M.; Papetti, A.; Gregotti, C.; Berte, F.; Gazzani, G. *In vitro* antioxidant and *ex vivo* protective activities of green and roasted coffee. *J. Agric. Food Chem.* **2000**, *48*, 1449–1454.
27. Del Castillo, M.D.; Ames, J.M.; Gordon, M.H. Effect of roasting on the antioxidant activity of coffee brews. *J. Agric. Food Chem.* **2002**, *50*, 3698–3703.
28. Nicoli, M.C.; Innocente, N.; Pittia, P.; Lerici, C.R. Staling of roasted coffee: Volatile release and oxidation reactions during storage. *ASIC 15th Int. Sci. Colloq. Coffee* **1993**, *15*, 557–566.
29. Borrelli, R.C.; Visconti, A.; Mennella, C.; Anese, M.; Fogliano, V. Chemical characterization and antioxidant properties of coffee melanoidins. *J. Agric. Food Chem.* **2002**, *50*, 6527–6533.
30. Cämmerer, B.; Kroh, L.W. Antioxidant activity of coffee brews. *Eur. Food Res. Technol.* **2006**, *223*, 469–474.
31. Da Silveira Duarte, S.M.; de Abreu, C.M.P.; de Menezes, H.C.; dos Santos, M.H.; Gouvea, C.M.C.P. Effect of processing and roasting on the antioxidant activity of coffee brews. *Food Sci. Technol. (Campinas)* **2005**, *25*, 387–393.
32. López-Alarcón, C.; Denicola, A. Evaluating the antioxidant capacity of natural products: A review on chemical and cellular-based assays. *Anal. Chim. Acta* **2013**, *763*, 1–10.
33. Opitz, S.E.W.; Smrke, S.; Goodman, B.A.; Yeretzian, C. Methodology for the measurement of antioxidant capacity of coffee: A validated platform composed of three complementary antioxidant assays. In *Processing and Impact on Antioxidants in Beverages*; Preedy, V., Ed.; Academic Press: Oxford, UK, 2014; pp. 253–264.
34. Singleton, V.L.; Orthofer, R.; Lamuela-Raventos, R.M. Analysis of total phenols and other oxidation substrates and antioxidants by means of Folin-Ciocalteu reagent. *Methods Enzymol.* **1999**, *299*, 152–178.
35. Re, R.; Pellegrini, N.; Proteggente, A.; Pannala, A.; Yang, M.; Rice-Evans, C. Antioxidant activity applying an improved ABTS radical cation decolorization assay. *Free Radic. Biol. Med.* **1999**, *26*, 1231–1237.
36. Ou, B.; Hampsch-Woodill, M.; Prior, R.L. Development and validation of an improved oxygen radical absorbance capacity assay using fluorescein as the fluorescent probe. *J. Agric. Food Chem.* **2001**, *49*, 4619–4626.
37. Pellegrini, N.; Serafini, M.; Colombi, B.; del Rio, D.; Salvatore, S.; Bianchi, M.; Brighenti, F. Total antioxidant capacity of plant foods, beverages and oils consumed in Italy assessed by three different *in vitro* assays. *J. Nutr.* **2003**, *133*, 2812–2819.
38. Bisby, R.H.; Brooke, R.; Navaratnam, S. Effect of antioxidant oxidation potential in the oxygen radical absorption capacity (ORAC) assay. *Food Chem.* **2008**, *108*, 1002–1007.

39. Iveković, D.; Milardović, S.; Roboz, M.; Grabarić, B.S. Evaluation of the antioxidant activity by flow injection analysis method with electrochemically generated ABTS radical cation. *Analyst* **2005**, *130*, 708–714.
40. Watanabe, J.; Oki, T.; Takebayashi, J.; Yamasaki, K.; Takano-Ishikawa, Y.; Hino, A.; Yasui, A. Method validation by interlaboratory studies of improved hydrophilic oxygen radical absorbance capacity methods for the determination of antioxidant capacities of antioxidant solutions and food extracts. *Anal. Sci.* **2012**, *28*, 159–165.
41. Huang, D.J.; Ou, B.X.; Hampsch-Woodill, M.; Flanagan, J.A.; Prior, R.L. High-throughput assay of oxygen radical absorbance capacity (ORAC) using a multichannel liquid handling system coupled with a microplateflourescence reader in 96-well format. *J. Agric. Food Chem.* **2002**, *50*, 4437–4444.
42. Prior, R.L.; Hoang, H.; Gu, L.W.; Wu, X.L.; Bacchiocca, M.; Howard, L.; Hampsch-Woodill, M.; Huang, D.J.; Ou, B.X.; Jacob, R. Assays for hydrophilic and lipophilic antioxidant capacity (oxygen radical absorbance capacity (ORAC(FL)) of plasma and other biological and food samples. *J. Agric. Food Chem.* **2003**, *51*, 3273–3279.
43. Vignoli, J.A.; Viegas, M.C.; Bassoli, D.G.; de Toledo Benassi, M. Roasting process affects differently the bioactive compounds and the antioxidant activity of arabica and robusta coffees. *Food Res. Int.* **2014**, *61*, 279–285.
44. Hotta, H.; Sakamoto, H.; Nagano, S.; Osakai, T.; Tsujino, Y. Unusually large numbers of electrons for the oxidation of polyphenolic antioxidants. *Biochim. Biophys. Acta* **2001**, *1526*, 159–167.
45. Namazian, M.; Zare, H.R. Electrochemistry of chlorogenic acid: Experimental and theoretical studies. *Electrochim. Acta* **2005**, *50*, 4350–4355.
46. Scott, S.L.; Chen, W.J.; Bakac, A.; Espenson, J.H. Spectroscopic parameters, electrode-potentials, acid ionization-constants, and electron-exchange rates of the 2,2'-azinobis(3-ethylbenzothiazoline-6-sulfonate) radicals and ions. *J. Phys. Chem.* **1993**, *97*, 6710–6714.
47. Nkhili, E.; Brat, P. Reexamination of the ORAC assay: Effect of metal ions. *Anal. Bioanal. Chem.* **2011**, *400*, 1451–1458.
48. Ozgen, M.; Reese, R.N.; Tulio, A.Z.; Scheerens, J.C.; Miller, A.R. Modified 2,2-azino-bis-3-ethylbenzothiazoline-6-sulfonic acid (ABTS) method to measure antioxidant capacity of selected small fruits and comparison to ferric reducing antioxidant power (FRAP) and 2,2'-diphenyl-1-picrylhydrazyl (DPPH) methods. *J. Agric. Food Chem.* **2006**, *54*, 1151–1157.
49. Yeh, C.T.; Yen, G.C. Effects of phenolic acids on human phenolsulfotransferases in relation to their antioxidant activity. *J. Agric. Food Chem.* **2003**, *51*, 1474–1479.
50. Pannala, A.S.; Chan, T.S.; O'Brien, P.J.; Rice-Evans, C.A. Flavonoid B-ring chemistry and antioxidant activity: Fast reaction kinetics. *Biochem. Biophys. Res. Commun.* **2001**, *282*, 1161–1168.
51. Tian, X.; Schaich, K. Effects of molecular structure on kinetics and dynamics of the Trolox equivalent antioxidant capacity assay with $ABTS^{+\cdot}$. *J. Agric. Food Chem.* **2013**, *61*, 5511–5519.
52. Chun, O.K.; Kim, D.O. Consideration on equivalent chemicals in total phenolic assay of chlorogenic acid-rich plums. *Food Res. Int.* **2004**, *37*, 337–342.
53. Hotta, H.; Nagano, S.; Ueda, M.; Tsujino, Y.; Koyama, J.; Osakai, T. Higher radical scavenging activities of polyphenolic antioxidants can be ascribed to chemical reactions following their oxidation. *Biochim. Biophys. Acta* **2002**, *1572*, 123–132.

54. Nenadis, N.; Lazaridou, O.; Tsimidou, M.Z. Use of reference compounds in antioxidant activity assessment. *J. Agric. Food Chem.* **2007**, *55*, 5452–5460.

8

Biosensors for the Detection of Food Pathogens

Palmiro Poltronieri [1,*], Valeria Mezzolla [1], Elisabetta Primiceri [2,3] and Giuseppe Maruccio [2,3]

[1] Institute of Sciences of Food Productions, National Research Council, ISPA-CNR, Via Lecce-Monteroni, 73100 Lecce, Italy; E-Mail: valeria.mezzolla@gmail.com

[2] NNL, Institute of Nanoscience-CNR, Via per Arnesano, I-73100 Lecce, Italy; E-Mails: elisabetta.primiceri@unisalento.it (E.P.); giuseppe.maruccio@unisalento.it (G.M.)

[3] Department of Mathematics and Physics "Ennio De Giorgi", University of Salento, Via per Arnesano, I-73100 Lecce, Italy

* Author to whom correspondence should be addressed; E-Mail: palmiro.poltronieri@ispa.cnr.it

---

Abstract: Food pathogens frequently cause foodborne diseases. There is a need to rapidly identify the source of the bacteria in order to contain their spread and epidemics. A pre-enrichment culture or a direct culture on agar plate are standard microbiological methods. In this review, we present an update on alternative molecular methods to nucleic acid-based detection for species identification. Biosensor-based methods rely on the recognition of antigen targets or receptors by antibodies, aptamers or high-affinity ligands. The captured antigens may be then directly or indirectly detected through an antibody or high-affinity and high-specificity recognition molecule. Various different detection methods are discussed, from label-free sensors and immunosensors to fluorescence-based ones. Each method shows advantages and disadvantages in terms of equipment, sensitivity, simplicity and cost-effectiveness. Finally, lab-on-a-chip (LOC) devices are introduced briefly, with the potential to be fast, sensitive and useful for on-site bacteria detection in food processing laboratories to check potential contamination by sample monitoring combined with a rapid pre-enrichment step.

Keywords: food pathogens; rapid identification; molecular recognition; detection methods; labelling; antibody chips; lab-on-a-chip

---

1. Introduction

Numerous cases of foodborne illness among humans are caused by pathogens delivered with foods. In addition to human health risks, microbial contamination can result in food spoilage. The most common foodborne infections are caused by *Campylobacter* spp., *Salmonella* spp. and *Escherichia coli* O157:H7. Other bacterial species, such as *Listeria monocytogenes*, *Bacillus cereus*, *Bacillus anthracis*, the enterohemorrhagic strains of *Escherichia coli* O103:H25, O26, O111, O115, O128, O145 and the recently isolated O104:H4 harboring an antibiotic resistance plasmid, *Staphylococcus aureus* strains coding for thermostable enterotoxins, *Clostridium botulinum*, *Clostridium perfringens*, *Yersinia enterocolitica*, *Vibrio* spp., *Aeromonas* spp., *Shigella* spp. and *Mycobacterium bovis*, viruses and protozoa are transmitted through food consumption (source: Center for Disease Control, USA) [1].

Raw meat, milk, seeds and vegetables may be the source of bacteria, which are transferred through cross-contamination during food preparation, reaching the food supply chain. Certain types of foods (milk, cream, meat) are required to be free of pathogens, such as *L. monocytogenes* and *S. aureus*. The threshold for the acceptance of a pathogen depends on its ability to grow at refrigerated conditions, at which psychrotrophic bacteria, such as *L. monocytogenes*, are able to proliferate. On the contrary, the presence of *Legionella pneumophila* in water pipelines in hotels and thermal baths is considered a health threat when the bacteria density reaches several thousands of cells in one liter of water [2].

The standard microbiology method in food pathogen detection is colony counting on an agar plate, a procedure that requires several days for revealing the presence of a pathogen. This review will focus on methods for the detection of bacteria in food samples incubated in a pre-enrichment broth for times shorter than the standard 24–48 h, in order to individuate the bacteria, even if the exponential growth phase has not reached its maximum peak. A concentration step is required to collect bacteria from larger volumes into a small area, such as through filtration or using immuno-magnetic nanoparticles or beads loaded with species-specific antibodies or epitope binding proteins. Many publications have exploited the immunomagnetic separation of bacteria from a solution or food resuspension. Nowadays, the Pathatrix Auto System (Life Technologies, Carlsbad, CA, USA) has been validated and approved by AOAC for *Salmonella* spp. and *Listeria* spp. detection. Using these concentrators, bacteria from an enriched culture of 15 mL incubated for 15 h can be processed through DNA purification and amplification or immunodetection [3].

The highest sensitivity in species identification of bacteria has been achieved with molecular methods based both on polymerase chain reaction (real-time PCR, digital PCR) and isothermal amplification methods, like rolling circle amplification (RCA), recombinase polymerase amplification (RPA) and loop-mediated isothermal amplification (LAMP). In addition to the quantitative evaluation of the amplified DNA by quantitative real-time PCR, other methods have been developed, which are based on the hybridization of target DNA with highly selective probes bound to a surface. The detection limit of DNA methods range between 10 to 100 colony forming units/mL of sample [4,5].

While several methods have been applied for species identification after bacteria have reached the exponential phase in culture broth, there is a need to overcome remaining bottlenecks in the microbiological analysis of foods. For example, the food matrix is a constraint for the elution of bacteria and liberates PCR inhibitors that may affect the subsequent analyses. In addition, the bacteria are stressed and require longer times to grow.

Two identification methods can be applied in food safety based on ISO certification: plating and counting on selective agar and real-time PCR with dedicated DNA extraction kits (Bio-Rad) applied to bacteria from enrichment broths at 24 or 48 h of incubation. This molecular method has proven to be sensitive enough to detect 10 to 100 bacteria (from volumes up to 50 mL when combined with immunomagnetic separation) [4–6].

Methods, such as lateral flow-based dip-sticks with species-specific antibodies, and generally all immunosensors are not sensitive enough, and for this reason, they are used for species identification in enrichment broths at 24–48 h of incubation.

Nowadays, the developed immunoassays are based upon bacterial species-specific antibodies, aptamers [7,8] and immuno-recognition of bacterial antigens (such as bacteriophage tailspike protein) [9]. These methods require standard conditions for the optimum binding of proteins or other highly-specific affinity compounds on: beads (Luminex, Austin, TX, USA), glass slides, gold surfaces, microplates and membranes suitable for chromatographic separation of antigen-antibody complexes, combined with dipsticks, microfluidic channels on paper (µPADs) [10] or lateral flow immuno-assays (LFIA), in which the capture antibody is conjugated with a detection molecule exploiting colorimetric methods, chemiluminescence or gold nanoparticles [11]. We found that the sensitivity of this last method applied in *Salmonella* spp. detection was approximately 10^5 CFU/mL, thus making it unsuitable for detection in pre-enrichment broth at early stages of growth (18–24 h).

Presently many biosensor-based methods are still labor-intensive, expensive, and not easily implementable for in-field applications.

2. Detection Methods in Label-Free Sensors and Immunosensors

An ideal biosensor should detect target molecules directly without the use of labelled ligands or multiple washing steps. There are several bottlenecks to be solved in order to perform an efficient analysis. The first is the use of proper surfaces. The second is to efficiently and uniformly print the capture probe (antibody, binding proteins, aptamers) and to have a constant probe density in order to obtain the detection of targets with high reproducibility. The third is to obtain a high sensibility even at a low concentration of the targets. A number of methods are used for the rapid detection of biomolecules in solution. These include immunoassays, chromatographic methods, magnetic and biological biosensor methods (using screen-printed biosensors with immobilized enzymes), DNA biosensors and antibody-based detection methods.

2.1. Surface Plasmon Resonance (SPR)

SPR biosensors have been used in the direct and indirect detection of pathogenic microorganisms [12]. The technique was largely studied in label-free immunosensors for the detection of bacteria [13–15]. SPR can be used for the setup of immunosensors applied to the detection of food pathogens in enrichment broth, in liquids or in food dilutions.

The SPR technique for biosensing allows real-time monitoring of chemical and bio-chemical interactions occurring at the interface between a thin gold film and a dielectric interface or transparent material, such as the liquid analyte. SPR principles have been deeply described and discussed in other reviews [16]. SPR is an optical technique that uses the evanescent wave phenomenon to measure changes in refractive index very close to the sensor surface. The evanescent wave produced by an incident, monochromatic light beam is able to interact with free electrons (plasmons) in the metal film at a special angle (α) of incident light (SPR angle). The angular position is highly sensitive to any change at the metal-dielectric interface. Many real-time monitoring SPR methods exploit the prism-based Kretschmann configuration that uses the excitation of a surface-bound electromagnetic wave from the metal side. The occurrence of a binding event between the investigated antigen and antibody can be monitored by following the shift of the angular minimum toward higher angles or the variation in reflectance at a fixed angle (corresponding to the maximum slope of the curve). Thus, it is possible to record the SPR reflectance curve at different angles of incidence of light. In a report, a *L. pneumophila* cell suspension at a concentration of 10,000 CFU/mL was detected using an SPR immunosensor provided with a miniaturized microfluidic system [2].

2.1.1. SPR Imaging

SPR microscopy is suitable for the evaluation of end-point quantification. In grating-coupled SPR, macroscopically-thick metal films are deposited on inexpensive and disposable grating. When the wave is excited from the transparent material side, in the Otto-configuration, the sample is placed on a topographically-modified surface, such as a grating or a grid printed by spot deposition in a microarray format [17]. Such a method does not require a prism. The surface, even if not in gold, does not need to be uniformly thick. Grating-coupled surface plasmon resonance imaging (GCSPRI) has been applied to multiplexed detection of microbes, toxins and viruses. GCSPRI instrument showed the ability to simultaneously measure binding at over 1000 unique, discrete regions of interest (ROIs) by utilizing a compact microarray of antibodies or other specific capture molecules immobilized on the sensor chip.

2.1.2. Enhancement of Sensitivity by Combining SPR with A Labelling or Capturing Method

Standard SPR detection requires usually high concentrations of the screened target antigens. This may lead to non-specific binding and a limit of detection of bacteria in the range of 10^5–10^6 CFU/mL. For this reason, improvement of the SPR detection limit has been achieved with an additional labelling step.

To enhance the signal from bacteria bound to the antibodies on the gold surface [18], antibody-functionalized gold nanoparticles (immuno-AuNP) were allowed to flow into the SPR cell, producing an increase in the reflectance units already at an *L. monocytogenes* concentration of 1000 CFU/mL [4]. The sensitivity was significantly enhanced, based on a mass effect.

A new, indirect method capable of enhancing SPR sensitivity is label-enhanced surface plasmon fluorescence. An alternative to a fluorophore, to avoid quenching problems due to the gold surface, a more efficient method based on the absorbance of light is quantitative label-enhanced SPR [19]. Label-enhanced SPR can be run simultaneously and in parallel to label-free SPR. The specificity is very high, and the influence of environment factors, such as temperature and buffer composition variations, is reduced.

In the subtractive inhibition assay, *E. coli* O157:H7 cells and goat polyclonal antibodies for *E. coli* O157:H7 were incubated for a short amount of time, and then, the *E. coli* O157:H7 cells, to which antibodies were bound, were removed by a stepwise centrifugation process. The remaining free unbound antibodies were detected through an interaction with rabbit anti-goat IgG polyclonal antibodies immobilized on the sensor chip using a BIAcore 3000 biosensor. The results showed that the signal was inversely correlated with the concentration of *E. coli* O157:H7 cells in a range from 3.0×10^4 to 3.0×10^8 CFU/mL. Compared to direct SPR, the detection limit of the subtractive inhibition assay method was reduced by one order of magnitude [20].

Lectins have been used to capture the bacteria on the gold surface, in the development of surface plasmon resonance biosensors. In this prototype, the adhesion of bacteria depends on the affinity between specific lectins and surface carbohydrates [21].

2.2. Electrochemical Impedance Spectroscopy (EIS)

Electrochemical impedance spectroscopy (EIS) has been extensively applied as a method of choice for biosensors [22], combined with disposable, screen-printed chips [23]. Each electrode can be disposed after use in order to prevent carry-over contamination. Due to its high sensitivity and easy setup, EIS has been extensively applied for biosensor fabrication, particularly in the field of microbiology. Since the 1990s, impedance methods have been used for bacterial identification. These methods record the changes in the sensor electrical impedance induced by bacterial metabolism and cell growth as a result of the release of ionic metabolites from the living cells (carbon dioxide and organic acids produced by catabolism and ion exchange through the cell membrane). One application of this technique has been applied to differentiate between viable and dead cells [24,25]. The method has been used also by Yang and coworkers [26,27], who applied a three-electrode configuration for the impedimetric detection of viable *S. typhimurium*. Since no changes happen at the electrodes surface, the system can be described by a simplified equivalent circuit composed of a resistor (R_s) associated with the solution between the two electrodes and a double-layer capacitor for each electrode (C_{dl}). These two components determine the total (frequency-dependent) impedance. At high frequencies, the medium resistance is essentially the main contributor to the total impedance; whereas at low frequencies, the double-layer capacitance dominates. It was shown that both the capacitance and resistance measurements can be used for the enumeration of bacteria [26].

In the past few years, advances in microfabrication technologies allowed a miniaturization of the impedimetric biosensors into a chip format offering new opportunities in the field of bacterial detection. Several advantages are provided, since dedicated chips can be mass produced at a low cost, so each electrode can be disposed of after use, avoiding cross-contamination [26]. In this respect, interdigitated microelectrodes (IME) were used for the quantitative detection of *S. typhimurium* in pure culture and milk samples by means of impedance measurements [27]. Specifically, low frequencies (<100 Hz) were suitable to monitor the impedance changes associated with bacterial growth, and a significant change in impedance at 10 Hz was recorded because of the increase in the double-layer capacitance, while a simple measurement of the resistance is not sufficient, since bacteria cells attached to the electrode surface cause an increase in resistance, while a decrease in medium resistance limits the evaluation of total impedance. The detection time of the IME impedance sensor for

S. typhimurium at a given initial cell number was shorter than that by the classic impedimetry and was close to that of electrochemical methods. A detection limit as low as 0.5 CFU/mL with a detection time of more than 10 h was reported [27].

In addition, microfabricated systems allow for the functionalization of the electrode surface with molecules able to capture specific bacterial cells. In this case, the equivalent circuit consists of the ohmic resistance of the electrolytic solution (R_s), a double-layer capacitance (C_{dl}), an interfacial electron-transfer resistance (R_{et}) and a Warburg impedance (Z_w), which is associated with the diffusion of the redox probe. Among them, C_{dl} and R_{et} depend on the electrical properties of the solution/electrode interface and can be used to monitor the attachment of molecules [28] and cells [29]. As a result of specific biorecognition reactions, this method allows the detection of a specific type of bacteria or different bacteria present at the same time, even in complex matrices, thanks to a strong reduction of unspecific contributions to the signal. Moreover, the time needed for the analysis can be strongly reduced (less than 2 h) compared with growth-based impedance methods (which requires at least 24 h). For example, in 2002, Ruan *et al.* [30] developed an impedance biosensor for the detection of *Escherichia coli* O157:H7 based on the immobilization of bacteria to electrodes functionalized with specific antibodies. A secondary antibody conjugated with alkaline phosphatase as the labeled enzyme was used to amplify the signals. These biosensors were able to detect the target bacteria with a detection limit of 6×10^3 cells/mL. Yang *et al.* [31] were able to avoid the use of secondary labelled antibodies and develop label-free immunosensors for the detection of the same bacterial strain (*E. coli* O157:H7) by exploiting the better sensitivity of indium tin oxide (ITO) interdigitated electrodes, due to the compatibility in terms of size between the electrodes and bacterial cells. In fact, when bacterial cells are recognized by the specific antibodies immobilized on the electrodes, they are attached to the surface and induce an increase in the electron transfer resistance, measurable even without any additional amplification step. More recently, Radhakrishnan and colleagues [32] were able to detect *L. monocytogenes* by electrochemical impedance spectroscopy with very high sensitivity using a mouse monoclonal antibody immobilized onto a gold electrode. They reported detection limits of 5 CFU/mL and 4 CFU/mL for ideal solutions and filtered tomato extract, respectively.

To increase the sensitivity of the assays, the use of innovative materials for electrode fabrication was also investigated, and in this respect, nanomaterials, such as nanoparticles, nanowires, nanotubes and graphene, seem to be very promising. For example, Yang and colleagues [33] developed an immunosensor for the detection of *Salmonella* spp. exploiting the immobilization of gold nanoparticles onto a glassy carbon and have demonstrated improved performances in terms of sensitivity and stability. On the other hand, Wang [34] proposed the use of TiO_2 nanowires connected by gold microelectrodes and functionalized with anti-*Listeria* antibodies for the rapid and sensitive detection of *Listeria monocytogenes*. In this case, both a high sensitivity and a high specificity in combination with a fast response time were reported.

Recently, advances in microfabrication and microfluidics and the development of lab-on-a-chip technology (LOC) has offered new possibilities in many fields and also in microbiology and food security monitoring. In fact, LOC is the most suitable approach to address the challenge of bacterial isolation, thanks to the higher surface-to-volume ratio in a microchannel (which increases the active area of a microdevice for cell capturing) and the possibility to integrate other components, such as micropillars, micropores, microfilters and mixers, enhancing capture efficiency. Therefore, the future will be based on integrated platforms in which biosensors will be integrated with additional modules in order to perform all of the analytical procedures on the same chip.

For example, Jiang [35] has moved towards such an integrated platform by presenting a low-cost, miniaturized and sensitive bacteria sensor based on electrical impedance spectroscopy coupled with a smartphone platform. The proposed platform is composed of interdigitated electrodes on a micro-hole silicon substrate and a microfluidic chamber based on a nano-porous filter paper, which is used to pre-concentrate bacteria in sample solutions, which lowered the detection limit to 10 bacterial cells/mL.

Recent advancements in electrochemical impedance spectroscopy (EIS) and in differential pulse voltammetry (DPV) have been applied to detect DNA and protein-protein interactions on arrays [36].

2.3. Other Direct Methods

Semiconductor technologies have been also applied successfully in biosensor arrays. In particular, field-effect transistor technologies can play an important role: thin film transistors (TFTs) can be applied either as potentiometric sensors for the detection of different biomolecular interactions or as circuit elements for the conditioning and read-out of signals of promising label-free electrical detection techniques [37]. The use of nanostructured devices in chemical/biological sensors in place of conventional sensing technologies has the advantages of high sensitivity, low decreased energy consumption and potentially highly miniaturized integration. In this respect, owing to their particular structure, excellent electrical properties and high chemical stability, carbon nanotube and graphene-based electrical devices have been widely developed for high performance label-free chemical/biological sensors [38].

Surface acoustic wave biosensors and quartz crystal microbalances (QCM) have been applied to detect *E. coli* O157 strains in diluted samples [18]. Functionalized god nanoparticles were added to enhance the change in detection signals. Pathogenic *E. coli* was detected in food specimens using an amperometric immunosensor based on a self-assembled gold surface with immobilized gold nanoparticles and chitosan-multiwalled carbon nanotubes-SiO_2-thionine, in a layer-by-layer film deposition process. The electrode connected with an amperemeter allowed for the visualization of bacterial cells, ranging from 400 up to 10^5 CFU/mL [39]. The capacitative immunosensors based on QCM were shown to detect *E. coli* cells with a detection limit of 100 CFU/mL in inoculated food samples, using an electrochemical detector [39]. The specific capture of viable pathogens was followed by capacitative detection on a quartz crystal microbalance sensor followed by detection with antibody-functionalized gold nanoparticles [2]. In a different quartz crystal-based setting, the change in capacitance caused by the bacteria was directly measured with good sensitivity [40].

On-chip surface-enhanced Raman spectroscopy (SERS) [41,42] has been applied to pathogen detection. Surface-enhanced Raman spectroscopy (SERS) is a Raman spectroscopic (RS) technique that provides a greatly enhanced Raman signal from Raman-active analyte molecules that have been adsorbed onto certain specially prepared metal surfaces. Multispectral technologies, such as Raman, have been applied to individuate, through spectral profiles, single compounds, as well as multi-analyte in complex matrixes [41,42]. Electromagnetic spectrum wavelengths useful for multispectral technologies fall below the range of 10^{-8} meters. SERS technology is both surface selective and highly sensitive. Using a hybrid electrokinetic mechanism, it is possible to concentrate bacteria at the stagnation area on the SERS-active roughened electrode, easily identifying different pathogens by their characteristic spectra. Since this method does not need capture antibodies to selectively concentrate the investigated pathogen on the surface, it is not an immunosensor. On-chip SERS is based on the electromagnetic effect due to the excitation of localized surface plasmons. The observed signal enhancement is a consequence of the specimen physically adsorbed onto the surface.

Fourier transformation-infrared spectrometry (FT-IR) was used in label-free IR fingerprinting [43]. A benchtop FT-IR spectrometer and a portable mid-IR spectrometer were used to characterize *Salmonella* spp. and *E. coli* O157 strains in complex food matrices, such as diluted milk and spinach extract, after immuno-purification using magnetic beads.

Another biosensor system has been based on thermally-blocked magnetic nanoparticles. Magnetic nanoparticle detection systems are based on an AC susceptometer measuring a wide frequency range, from 1 Hz up to 500 kHz. The magnetic flux change is measured by a detection-coil system connected to the AC source, delivering the current to the excitation coil that gives the excitation field. Signals are fed into a low-noise lock-in amplifier. This biosensor can detect the concentration of different molecules in a liquid [44]. The software elaboration of the magnetic field measurements in Tesla units results in a graphical resolution of plots with the frequency on the *x*-axis and magnetic susceptibility on the *y*-axis. Recently, this method was applied to measuring the presence of *Mycobacterium tuberculosis* in sputum samples, with a sensitivity in the range of 100–1000 CFU/mL, to individuate the acute phase of tuberculosis and to avoid the spreading of infection to healthy individuals [45].

The recent development of molecular electronics also offers new opportunities for implementing miniaturized and ultrasensitive transducers. So far, these methods have not been applied to bacteria detection but no constrains exist in this respect. At the molecular level, sensitivity is huge and, in one study, a nanobiosensor able to detect even single hybridization events was demonstrated by using arrays of nanojunctions and oligonucleotides conjugated to gold nanoparticles [46]. In these devices, target-probe binding events on the electrodes result in a quantized change in conductivity as nanoparticles close the gap. More recently, a protein transistor was made of an antibody molecule and two nanoparticles connected with two electrodes [47]. By means of CdSe quantum dots attached to the antibody, it was shown that the protein transistor can be gated by an applied optical field in addition to the electric signal detection. The authors demonstrated a reproducible approach to create a stable molecule/electrode interface [48] and successively applied this technology for DNA sequencing [49].

Additional methods exploited in label-free biosensors have been described and discussed by other authors, such as optical ellipsometry (OE), oblique-incidence reflectivity difference (OI-RD), resonant cavity biosensors and resonant waveguide grating (RWG) [16]. Optical biosensors have also been studied with an integrated Mach-Zehnder interferometer for the detection of *L. monocytogenes* [50]. Reflectometric interference spectroscopy (RIfS) has been applied by ForteBio (Menlo Park, CA, USA) in their Octet system for the monitoring of several different target molecules at the same time [16].

However, most of these methods, as in other biosensor systems, require costly detectors, so that applications to food pathogen detection are limited.

3. Label-Based Detection Methods

Biosensor devices for pathogen detection generally consist of several elements, including a biological capture molecule (affinity probes or antibodies), a labelled antibody interacting with the bacteria captured from the solution (in the case of beads, a double label is already present on each different bead type) and a signal detection system.

Fluorescent methods make use of a detection system that reads the fluorescence signal of excited molecules, such as cyanines and Alexa Fluor dyes, or CdSe/ZnS quantum dots (QDs); the QDs possess high stability over time, but are excited on average by 50% of the molecules present on the sensor. Other sources of fluorescence are in the red and infrared wavelengths. Examples of such dyes are the HiLyte dyes, with high brightness, photo-stability and intense fluorescence, which are insensitive to pH changes. Quantum dots have been used to visualize *Salmonella* spp. enriched from a sample using immunomagnetic separation [6]. QDs exhibited a fluorescence emission peak at 620 nm, with a red shift of the conjugates of QD-dBSA-biotin (QD-dBSA, 625 nm) and QD-Ab (630 nm).

Some biosensors can be automated; many methods also have a multiplex capability, and some have high-throughput performance. Examples of such methods include electrochemiluminescent (ECL) assays performed in 96-well plates and cytometry-based assays that use antibody-coated microspheres. ECL detection is based on electrochemical stimulation of reporter molecules, which undergo electron transfer reactions, resulting in the emission of light. These reporter molecules, such as ruthenium(II) trisbipyridal ($Ru(bpy)_3^{2+}$), are attached to biological conjugates (e.g., antibodies) and, when stimulated by an electrode, allow detection at very low concentrations [51]. Instrument-specific 96-well assay plates are available for use with fully-automated, commercially available ECL detection systems (MesoScale Diagnostics, Gaithersburg, MD, USA).

Cytometric bead arrays (CBAs) make use of a fluidic method in which microspheres, individually coded using different ratios of red and infrared fluorophores, serve as the solid support for a sandwich-type immunoassay (xMAP technology, Luminex Corp., Austin, TX, USA) [52]. Capture molecules are attached to the microspheres and bind the target, while an orange-fluorescent reporter molecule (R-phycoerythrin (R-PE)) is used to generate the signal. The reporter for immunoassays is an R-PE-tagged antibody or a biotinylated antibody used in combination with streptavidin-conjugated R-PE. The bead/target/reporter complexes are read by a specialized flow cytometer, which determines the identity of each individual bead (based on the combined red/infrared fluorescent molecules) and the target-associated signal on each bead (based on the orange fluorescence). Hundreds of different sets of coded beads can be distinguished, allowing detection of up to 100 targets in one sample. Superparamagnetic fluorescent microspheres can also be used to incorporate immunomagnetic separation (IMS) procedures [52].

Lateral flow immuno-separation is a well-established method exploited in many commercial kits for bacterial identification after enrichment broth culture. Several types of substrates, membranes and detection systems are available, such as colorimetric or fluorescent quantification. In lateral flow immuno-assays for the identification of *Salmonella* spp., the detection antibody was labelled with gold nanoparticles (AuNP) and detected by densitometry using UV wavelength [53]. In another protocol, mastitis bacteria were detected using biotinylated antibodies and streptavidin coupled to carbon nanotube labels [54] visualized by flatbed scanning of the pixel gray intensity.

3.1. Protein Chips, Antibody Chips and Aptamer Chips

Planar protein arrays in microplates [55], antibodies bound to microarray glass slides [4] and middle-throughput wells in microplates have been combined with magnetic separation and fluorescent silica nanoparticles in solution [56] and applied to the detection of pathogens and their toxins in the immunoassay format.

The protein chip methodology is based on the binding to a glass surface of an antibody or an aptamer recognizing the bacterial species, for the capture of the bacteria to the surface. In our experiments (unpublished data), we tested either capture aptamers bound to a streptavidin-coated surface or capture antibodies directly bound to the epoxy-modified glass slides. Since the bacterial wall harbors several protein antigens in different locations in the membrane, it could be possible to use the same antibody either as a capture antibody or also as a detection probe. In the protein chip experiment, an activated surface, such as clean glass treated with a suitable chemistry, such as epoxy-groups, for the binding of amino residues (by a Schiff reaction and a reduction step to stabilize the Schiff base), is spotted with capture antibodies, streptavidin or proteins with high affinity for a bacterial species. The capture step can be either species-specific or genus-specific, if the detecting antibody shows high specificity for *L. monocytogenes* and does not bind to *Listeria innocua*.

Detection antibodies can be directly labelled or detected using specific proteins (protein A, anti-IgG antibodies) that have been efficiently labelled. The fluorophores used more for such applications are Cyanine3/Alexa Fluor 555 with excitation/emission values at 550/568 nm, and Cyanine5/Alexa Fluor 647, with excitation at 650 nm and emission at 668 nm. Furthermore, species-specific biotinylated aptamers can be used for the detection step, combined with fluorescence-labelled streptavidin.

The detection method used can differ depending on whether the assay is in solution or on a surface, making use of laser scanners, cytofluorometer, fluoro-imagers or fluorescence microplate readers [4,57]. However, a portable and simple detection system is not yet available. Yet, improvements have been made to increase the sensitivity of fluorescence detection using a low-cost CMOS webcam and dedicated software readers [56]. The authors reported that by using a multi-wavelength LED illuminator and suitable filters, the photo camera acquired a video that was processed for signal-to-noise reduction by the software, achieving a limit-of-detection of 30 μM fluorescein in independent spots. In another report, a portable microarray slide reader measuring 19 cm in length has been developed that performs as well as commercial laser scanners, but that is much cheaper [58].

Approaches for Enhancement of Signal

A further approach to obtain an increase in sensitivity is to exploit multiple fluorescent labelling of the molecular recognition elements that bind to target bacteria. Using an antibody chip with spotted *Salmonella* spp. cells in serial dilutions, the detection method of choice is to incubate the slide with fluorescently (Alexa or cyanine)-labelled anti-*Salmonella* antibodies. We found that the sensitivity of this protocol was approximately 10^4 CFU/mL. It was shown that lactoferrin (Lf) and lactoferricin peptides exert antimicrobial activity through binding to bacterial membranes. Recently, the binding of fluorescent Lf to bacterial membranes was shown in *Salmonella* cells bound onto glass slides. The fluorescence of Lf binding was at levels similar to the results obtained with boro-dipyrromethene boron fluoride (BODIPY), a lipid probe with excitation/emission at 530–550 nm, and to the anti-*Salmonella* antibody [53]. These findings indicate the feasibility to increase the detection of *Salmonella* spp. and related species of bacteria on protein chips by additive fluorescence detection using, at the same time, the anti-*Salmonella* antibody, BODIPY and labelled lactoferrin.

It is possible to exploit two different monoclonal antibodies recognizing two proximal epitopes on the same surface antigen. The antibodies should be provided with an oligonucleotide tag, to exploit the amplification of the signal using the proximity probe ligation assay (PLA) [59]. In this method, it is possible to detect even single molecules in different formats, from *in situ* hybridization to confocal microscopy to capture antibodies fixed on glass slides or microplates. These methods resemble the use of fluorescently-labelled nucleic acid dendrimers to amplify the hybridization signals of DNA sequences bound to oligonucleotide arrays.

3.2. Fluorescence Resonance Energy Transfer (FRET)

Real-time monitoring of ligand-receptor interactions was performed on bacteria bound to carbohydrates exposed onto liposome surfaces using fluorescence resonance energy transfer (FRET). Ligand-receptor interactions were investigated at the bilayer surface using electronic absorption spectroscopy and fluorescence resonance energy transfer [60]. Since fluorescence is intrinsically more sensitive than colorimetry, the detection limit of the assay is in the sub-nanomolar range or lower. Polydiacetylene (PDA) liposomes were made with Lissamine Rhodamine (LR)-tagged phospholipids inserted in the bilayer, where LR acted as an acceptor (maximum absorption ~560 nm and maximum emission ~583 nm). Further, polydiacetylene (PDA) acts as a universal acceptor in FRET, which means that multiple sensors can be developed with PDA (acceptor) functionalized with donors and different receptors attached on the surface of PDA liposomes. With the addition of *E. coli* to PDA solution, the blue absorption peak at 540 nm was decreased and the intensities of the red absorption peaks (centered at 490 and 540 nm) were increased. The FRET emission response, on the other hand, depends on the FRET efficiency and on the distance between the donor and the acceptor (mean optimum radius at 50 nm).

3.3. Total Internal Reflection Fluorescence (TIRF)

Total internal reflection fluorescence (TIRF) microscopy has been used to observe the fluorescence of single molecules. The region visualized, a few hundred nanometers wide, allows the visualization of the specimen zone surrounding the evanescent wave under a coverslip layered with a drop of immersion oil, achieving a selective description of surfaces at a high axial resolution [61]. Fluorophore conjugates are excited by means of an evanescent field generated upon total reflection of laser light at the surface. The short penetration depth avoids the excitation of fluorophores distant from the surface.

4. Conclusions

Several different types of biosensors, based on direct detection and indirect methods, such as lateral flow immunoassays and protein chips, have been reviewed in food pathogens detection. The assay performance, including detection limit and assay time, were also compared. The problem of the portability of the instrumentation has been approached with the development of cheaper, small-sized scanners and with software for the analysis of video images using mobile devices. Each biosensor technique has its own advantages and disadvantages in terms of the equipment required, sensitivity, simplicity and cost-effectiveness. Lab-on-a-chip (LOC) devices have a strong potential to be used in the field, since they can be miniaturized and automated; also being potentially fast and very sensitive. There are still several issues to be solved before in-field applications, including the pre-treatment of a sample, such as the enrichment of bacteria in culture broth, the proper storage of reagents, the full integration into a battery-powered system, the detection limit and the enhancement of the sensitivity of each method.

Acknowledgments

Palmiro Poltronieri and Valeria Mezzolla are supported by the Italian projects "Strumenti Innovativi per il Miglioramento della Sicurezza Alimentare" (S.I.Mi.S.A. PON_02.00186) and "Conoscenze Integrate per la Sostenibilità ed Innovazione Agroalimentare" (C.I.S.I.A.).

Author Contributions

Palmiro Poltronieri and Valeria Mezzolla have contributed with the protein chip experimental setup and label-independent SPR methods, respectively. Giuseppe Maruccio and Elisabetta Primiceri have contributed with the sections on electronic detection and impedance methods.

Conflicts of Interest

The authors declare no conflict of interest.

References

1. Scallan, E.; Hoekstra, R.M.; Angulo, F.J.; Tauxe, R.V.; Widdowson, M.-A.; Roy, S.L.; Jones, J.L.; Griffin, P.M. Foodborne illness acquired in the United States—Major pathogens. *Emerg. Infect. Dis.* **2011**, *7*, 7–15.
2. De Lorenzis, E.; Manera, M.G.; Cimaglia, F.; Montagna, G.; Chiesa, M.; Poltronieri, P.; Santino, A.; Rella, R. SPR based immunosensor for detection of *Legionella pneumophila* in water samples. *Optics Comm.* **2013**, *294*, 420–426.
3. Pathatrix® Auto System. Available online: http://www.lifetechnologies.com/cn/zh/home/industrial/food-safety/pathatrix-auto-system.html (accessed on 24 February 2014).
4. Poltronieri, P.; de Blasi, M.D.; D'Urso, O.F. Detection of *Listeria monocytogenes* through RealTime PCR and biosensor methods. *Plant Soil Environ.* **2009**, *9*, 363–369.
5. D'Urso, O.F.; Poltronieri, P.; Marsigliante, S.; Storelli, C.; Hernandez, M.; Rodriguez Lazaro, D. A filtration-based real-time PCR method for the quantitative detection of viable *Salmonella enterica* and *Listeria monocytogenes* in food samples. *Food Microbiol.* **2009**, *26*, 311–316.
6. Kuang, H.; Cui, G.; Chen, X.; Yin, H.; Yong, Q.; Xu, L.; Peng, C.; Wang, L.; Xu, C. A one-step homogeneous sandwich immunosensor for *Salmonella* detection based on magnetic nanoparticles (MNPs) and quantum dots (QDs). *Int. J. Mol. Sci.* **2013**, *14*, 8603–8610.

7. Amaya-González, S.; de-los-Santos-Alvarez, N.; Miranda-Ordieres, A.J.; Lobo-Castañón, M.J. Aptamer-based analysis: A promising alternative for food safety control. *Sensors* **2013**, *13*, 16292–16311.
8. Paniel, N.; Baudart, J.; Hayat, A.; Barthelmebs, L. Aptasensor and genosensor methods for detection of microbes in real world samples. *Methods* **2013**, *64*, 229–240.
9. Dutt, S.; Tanha, J.; Evoy, S.; Singh, A. Immobilization of P22 bacteriophage Tailspike protein on Si surface for optimized *Salmonella* capture. *J. Anal. Bioanal. Tech.* **2013**, *S7*, 7.
10. Liana, D.D.; Raguse, B.; Gooding, J.J.; Chow, E. Recent advances in paper-based sensors. *Sensors* **2012**, *12*, 11505–11526.
11. Cimaglia, F.; Rosu, V.; Chiesa, M.; Poltronieri, P.; Aliverti, A.; Santino, A.; Sechi, L.A. Quantum dot nanoparticle-based lateral flow assay for rapid detection of *Mycobacterium* species using anti-FprA antibodies. *Nanotechnol. Dev.* **2012**, *2*, e5.
12. Bergwerff, A.A.; van Knapen, F. Surface plasmon resonance biosensors for detection of pathogenic microorganisms: Strategies to secure food and environmental safety. *J. AOAC Int.* **2006**, *89*, 826–831.
13. Deisingh, A.K.; Thompson, M. Biosensors for the detection of bacteria. *Can. J. Microbiol.* **2004**, *50*, 69–77.
14. Mazumdar, S.D.; Barlen, B.; Kramer, T.; Keusgen, M. A rapid serological assay for prediction of *Salmonella* infection status in slaughter pigs using surface plasmon resonance. *J. Microbiol. Meth.* **2008**, *75*, 545–550.
15. Dudak, F.C.; Boyaci, I.H. Rapid and label-free bacteria detection by surface plasmon resonance (SPR) biosensors. *Biotechnol. J.* **2009**, *4*, 1003–1011.
16. Sun, Y.-S. Optical biosensors for label-free detection of biomolecular interactions. *Instrum. Sci. Technol.* **2014**, *42*, 109–127.
17. Marusov, G.; Sweatt, A.; Pietrosimone, K.; Benson, D.; Geary, S.J.; Silbart, L.K.; Challa, S.; Lagoy, J.; Lawrence, D.A.; Lynes, M.A. A microarray biosensor for multiplexed detection of microbes using grating-coupled surface plasmon resonance imaging. *Environ. Sci. Technol.* **2012**, *46*, 348–359.
18. Guo, X.; Lin, C.S.; Chen, S.H.; Ye, R.; Wu, V.C. A piezoelectric immunosensor for specific capture and enrichment of viable pathogens by quartz crystal microbalance sensor, followed by detection with antibody-functionalized gold nanoparticles. *Biosens. Bioelectron.* **2012**, *38*, 177–183.
19. Granqvist, N.; Hanning, A.; Eng, L.; Tuppurainen, J.; Viitala, T. Label-enhanced Surface Plasmon Resonance: A new concept for improved performance in optical biosensor analysis. *Sensors* **2013**, *13*, 15348–15363.
20. Wang, Y.; Ye, Z.; Si, C.; Ying, Y. Subtractive inhibition assay for the detection of *E. coli* O157:H7 using surface plasmon resonance. *Sensors* **2011**, *11*, 2728–2739.
21. Wang, Y.; Ye, Z.; Si, C.; Ying, Y. Monitoring of *Escherichia coli* O157:H7 in food samples using lectin based surface plasmon resonance biosensor. *Food Chem.* **2013**, *136*, 1303–1308.
22. Li, N.; Brahmendra, A.; Veloso, A.J.; Prashar, A.; Cheng, X.R.; Hung, V.W.S.; Guyard, C.; Terebiznik, M.; Kerman, K. Disposable immunochips for the detection of *Legionella pneumophila* using electrochemical impedance spectroscopy. *Anal. Chem.* **2012**, *84*, 3485–3488.

23. Silley, P.; Forsythe, S. Impedance microbiology—A rapid change for microbiologists. *J. Appl. Bacteriol.* **1996**, *80*, 233–243.
24. Wawerla, M.; Stollem, A.; Schalchm, B.; Eisgruber, H. Impedance microbiology: Applications in food hygiene. *J. Food Prot.* **1999**, *62*, 1488–1496.
25. Wan, Y.; Lin, Z.; Zhang, D.; Wang, Y.; Hou, B. Impedimetric immunosensor doped with reduced graphene sheets fabricated by controllable electrodeposition for the non-labelled detection of bacteria. *Biosens. Bioelectron.* **2011**, *26*, 1959–1964.
26. Yang, L.; Ruan, C.; Li, Y. Detection of viable *Salmonella typhimurium* by impedance measurement of electrode capacitance and medium resistance. *Biosens. Bioelectron.* **2003**, *19*, 495–502.
27. Yang, L.; Li, Y.; Griffis, C.L.; Johnson, M.G. Interdigited microelectrode (IME) impedance sensor for the detection of viable *Salmonella typhimurium*. *Biosens. Bioelectron.* **2004**, *19*, 1139–1147.
28. Chiriacò, M.S.; Primiceri, E.; Monteduro, A.G.; Bove, A.; Leporatti, S.; Capello, M.; Ferri-Borgogno, S.; Rinaldi, R.; Novelli, F.; Maruccio, G. Towards pancreatic cancer diagnosis by EIS biochips. *Lab Chip* **2012**, *13*, 730–734.
29. Primiceri, E.; Chiriaco, M.S.; Dioguardi, F.; Monteduro, A.G.; D'Amone, E.; Ionescu, R.E.; Rinaldi, R.; Maruccio, G. Automatic transwell assay by an EIS cell chip to monitor cell migration. *Lab Chip* **2011**, *11*, 4081–4086.
30. Ruan, C.; Yang, L.; Li, Y. Immunobiosensor chips doe detection of *Escherichia coli* O157:H7 using electrochemical impedance spectroscopy. *Anal. Chem.* **2002**, *74*, 4814–4820.
31. Yang, L.; Li, Y.; Erf, G.F. Interdigitated array microelectrode-based electrodechemical impedance immunosensor for detection of *Escherichia coli* O157:H7. *Anal. Chem.* **2004**, *76*, 1107–1113.
32. Radhakrishnan, R.; Jahne, M.; Rogers, S.; Suni, I. Detection of *Listeria monocytogenes* by electrochemical impedance spectroscopy. *Electroanalysis* **2013**, *25*, 2231–2237.
33. Yang, G.J.; Huang, J.L.; Meng, W.J.; Shen, M.; Jiao, X.A. A reusable capacitive immunosensor for detection of *Salmonella* spp. based on grafted ethylene diamine and self-assembled gold nanoparticle monolayers. *Anal. Chim. Acta* **2009**, *647*, 159–166.
34. Wang, R.; Dong, W.; Ruan, C.; Kanayeva, D.; Tian, R.; Lassiter, K.; Li, Y. TiO2 nanowire bundle microelectrode based impedance immunosensor for rapid and sensitive detection of *Listeria monocytogenes*. *Nano Lett.* **2008**, *9*, 2625–2631.
35. Jiang, J.; Wang, X.; Chao, R.; Ren, Y.; Hu, C.; Xu, Z.; Liu, G.L. Smartphone based portable bacteria pre-concentrating microfluidicsensor and impedance sensing. *Sens. Actuators B Chem.* **2014**, *193*, 653–659.
36. Regan, E.M.; Hallett, A.J.; Wong, L.C.C.; Saeed, I.Q.; Langdon-Jones, E.E.; Buurma, N.J.; Pope, S.J.A.; Estrela, P. A novel cobalt complex for enhancing amperometric and impedimetric DNA detection. *Electrochim. Acta* **2014**, *128*, 10–15.
37. Li, P.; Migliorato, P.; Estrela, P. Application of field-effect transistors to label-free electrical DNA biosensor arrays. In *Electrochemical DNA Biosensors*; Ozsoz, M., Ed.; Pan Stanford Publishing: San Diego, CA, USA, 2012; pp. 163–204.
38. Hu, P.; Zhang, J.; Li, L.; Wang, Z.; O'Neill, W.; Estrela, P. Carbon nanostructure-based field-effect transistors for label-free chemical/biological sensors. *Sensors* **2010**, *10*, 5133–5159.

39. Li, Y.; Cheng, P.; Gong, J.; Fang, L.; Deng, J.; Liang, W.; Zheng, J. Amperometric immunosensor for the detection of *Escherichia coli* O157:H7 in food specimens. *Anal. Biochem.* **2012**, *421*, 227–233.
40. Li, D.; Feng, Y.; Zhou, L.; Ye, Z.; Wang, J.; Ying, Y.; Ruan, C.; Wang, R.; Li, Y. Label-free capacitive immunosensor based on quartz crystal Au electrode for rapid and sensitive detection of *Escherichia coli* O157:H7. *Anal. Chim. Acta* **2011**, *687*, 89–96.
41. Liu, T.Y.; Tsai, K.T.; Wang, H.H.; Chen, Y.; Chen, Y.H.; Chao, Y.C.; Chang, H.H.; Lin, C.H.; Wang, J.K.; Wang, Y.L. Functionalized arrays of Raman-enhancing nanoparticles for capture and culture-free analysis of bacteria in human blood. *Nat. Commun.* **2011**, *2*, 538.
42. Cheng, I.F.; Chang, H.C.; Chen, T.Y.; Hu, C.; Yang, F.L. Rapid (<5 min) identification of pathogen in human blood by electrokinetic concentration and surface-enhanced Raman spectroscopy. *Sci. Rep.* **2013**, *3*, 2365.
43. Ravindranath, S.P.; Mauer, L.J.; Deb-Roy, C.; Irudayaraj, J. Biofunctionalized magnetic nanoparticle integrated mid-infrared pathogen sensor for food matrixes. *Anal. Chem.* **2009**, *81*, 2840–2846.
44. Fornara, A.; Johansson, P.; Petersson, K.; Gustafsson, S.; Qin, J.; Olsson, E.; Ilver, D.; Krozer, A.; Muhammed, M.; Johansson, C. Tailored magnetic nanoparticles for direct and sensitive detection of biomolecules in biological samples. *Nano Lett.* **2008**, *8*, 3423–3428.
45. Kaittanis, C.; Boukhriss, H.; Santra, S.; Naser, S.A.; Perez, J.M. Rapid and sensitive detection of an intracellular pathogen in human peripheral leukocytes with hybridizing magnetic relaxation nanosensors. *PLoS One* **2012**, *7*, e35326.
46. Maruccio, G.; Primiceri, E.; Marzo, P.; Arima, V.; Della Torre, A.; Rinaldi, R.; Pellegrino, T.; Krahne, R.; Cingolani, R. A nanobiosensor to detect single hybridization events. *Analyst* **2009**, *134*, 2458–2461.
47. Chen, Y.-S.; Hong, M.-Y.; Huang, G.S. A protein transistor made of an antibody molecule and two gold nanoparticles. *Nat. Nanotechnol.* **2012**, *7*, 197–203.
48. Maruccio, G. Molecular electronics: Protein transistors strike gold. *Nat. Nanotechnol.* **2012**, *7*, 147–148.
49. Chen, Y.-S.; Lee, C.-H.; Hung, M.-Y.; Pan, H.-A.; Chiou, J.-C.; Huang, G.S. DNA sequencing using electrical conductance measurements of a DNA polymerase. *Nat. Nanotechnol.* **2013**, *8*, 452–458.
50. Sarkar, D.; Gunda, N.S.; Jamal, I.; Mitra, K. Optical biosensors with an integrated Mach-Zender interferometer for detection of *Listeria monocytogenes*. *Biomed. Microdevices* **2014**, *16*, 509–520.
51. Drummond, T.G.; Hill, M.G.; Barton, J.K. Electrochemical DNA sensors. *Nat. Biotechnol.* **2003**, *21*, 1192–1199.
52. Leach, K.M.; Stroot, J.M.; Lim, D.V. Same-day detection of *Escherichia coli* O157:H7 from spinach by using electrochemiluminescent and cytometric bead array biosensors. *Appl. Environ. Microbiol.* **2010**, *76*, 8044–8052.
53. Cimaglia, F.; Poltronieri, P. Institute of Sciences of Food Productions, National Research Council, ISPA-CNR, Lecce, Italy. Unpublished work, 2014.
54. Mujawar, L.H.; Moers, A.; Norde, W.; van Amerongen, A. Rapid mastitis detection assay on porous nitrocellulose membrane slides. *Anal. Bioanal. Chem.* **2013**, *405*, 7469–7476.

55. Eickhoff, H.; Malik, A. Planar protein arrays in microtiter plates: Development of a new format towards accurate, automation-friendly and affordable (a(3)) diagnostics. *Adv. Biochem. Eng. Biotechnol.* **2013**, *133*, 149–165.
56. Rastogi, S.K.; Hendricks, V.J.; Branen, J.R.; Branen, A.L. Magnetic bead and fluorescent silica nanoparticles based optical immunodetection of staphylococcal enterotoxin B (SEB) in bottled water. *Sens. Transducers J.* **2009**, *7*, 191–202.
57. Balsam, J.; Bruck, H.A.; Kostov, Y.; Rasooly, A. Image stacking approach to increase sensitivity of fluorescence detection using a low cost complementary metal-oxide-semiconductor (CMOS) webcam. *Sens. Actuators B Chem.* **2012**, *171–172*, 141–147.
58. Thompson, D.L.; Pearson, F.; Thomas, C.; Rao, R.; Matthews, D.; Albala, J.S.; Wachsmann-Hogiu, S.; Coleman, M.A. An adaptable, portable microarray reader for biodetection. *Sensors* **2009**, *9*, 2524–2537.
59. Melin, J.; Jartvius, J.; Larsson, C.; Soderberg, O.; Landegren, U.; Nilsson, M. Ligation-based molecular tools for lab-on-chip devices. *New Biotechnol.* **2008**, *25*, 42–47.
60. Dogra, N.; Li, X.; Kohli, P. Investigating ligand-receptor interactions at bilayer surface using electronic absorption spectroscopy and fluorescence resonance energy transfer. *Langmuir* **2012**, *28*, 12989–12998.
61. Schult, K.; Katerkamp, A.; Trau, D.; Grawe, F.; Cammann, K.; Meusel, M. Disposable optical sensor chip for medical diagnostics: new ways in bioanalysis. *Anal. Chem.* **1999**, *71*, 5430–5435.

9

Stability of Capsaicinoids and Antioxidants in Dry Hot Peppers under Different Packaging and Storage Temperatures

Qumer Iqbal [1,†,*], Muhammad Amjad [1], Muhammad Rafique Asi [2], Agustin Ariño [3], Khurram Ziaf [1], Aamir Nawaz [4] and Tanveer Ahmad [1]

[1] Institute of Horticultural Sciences, University of Agriculture, Faisalabad 38000, Pakistan; E-Mails: amjaduaf@gmail.com (M.A.); khurramziaf@uaf.edu.pk (K.Z.); horticulture.tanveer@gmail.com (T.A.)

[2] Nuclear Institute for Agriculture and Biology (NIAB), Faisalabad 38000, Pakistan; E-Mail: asimuhammad@yahoo.co.uk

[3] Veterinary Faculty, University of Zaragoza, Zaragoza 50013, Spain; E-Mail: aarino@unizar.es

[4] Faculty of Agricultural Sciences and Technology, Bahauddin Zakariya University, Multan 60000, Pakistan; E-Mail: aamirkhan_74@hotmail.com

[†] Corresponding author's present address: Nuclear Institute for Agriculture and Biology (NIAB), Faisalabad 38000, Pakistan.

* Author to whom correspondence should be addressed; E-Mail: qumerhort@gmail.com

Academic Editor: Christopher J. Smith

Abstract: The maintenance of the quality and storage life of perishable fruits and vegetables is a major challenge for the food industry. In this study, the effects of different temperatures, packaging materials and storage time on the stability of capsaicinoids and antioxidants, such as total carotenoids, ascorbic acid and total phenolic compounds, were studied in three commercially cultivated hot pepper hybrids, namely Sky Red, Maha and Wonder King. For this purpose, dry whole pods were packed in jute bags and low-density polyethylene bags (LDPE), stored for five months under controlled conditions at 20, 25 or 30 °C and analyzed on Day 0 and at 50-day intervals until Day 150. The three hot pepper hybrids differed significantly with respect to their capsaicinoids and antioxidant concentrations, but the results indicated that with the increase in storage temperature and time, a gradual and steady decrease in these levels was equally observed for all hybrids. Overall, mean concentrations after five months were significantly reduced by 22.6% for ascorbic acid, 19.0% for phenolic compounds, 17% for carotenoids and 12.7% for capsaicinoids. The trends of capsaicinoids and antioxidants evolution were decreasing gradually during storage until Day 150, this effect being more pronounced at higher temperature. Furthermore, the disappearance rates of capsaicinoids and antioxidants were higher in peppers packed in jute bags than in those wrapped with LDPE. In conclusion, despite the sensitivity of capsaicinoids and antioxidants to oxygen, light and moisture, the packaging in natural jute or synthetic LDPE plastic bags, as well as the storage at ambient temperature preserved between 77.4% and 87.3% of the initial amounts of these health- and nutrition-promoting compounds during five months' storage.

Keywords: hot peppers; capsaicinoids; total carotenoids; ascorbic acid; total phenolics; storage

---

1. Introduction

Pepper fruits (*Capsicum annuum*) are important vegetables consumed worldwide as food and as spice condiments, such as curry, chili powder and paprika. The dehydrated fruits in the red ripe stage are globally commercialized and valued for their nutritional, health and sensory attributes. Peppers are regarded to be a good source of nutrients and phytochemicals, such as ascorbic acid, carotenoids and phenolic compounds, with well-known antioxidant properties and potential health benefits [1,2]. Hot cultivars are rich in capsaicinoids, responsible for the specific taste of pepper fruits, which may be also used in pain relievers, due to their pharmacological properties [3]. Additionally, the presence of capsaicinoids in hot peppers has a favorable effect on the stability of carotenoids during thermal drying [4]. The amounts of bioactive compounds depend on the pepper variety, ripening stage and growing conditions. A recent study conducted in the present authors' laboratory showed a considerable increase in ascorbic acid and carotenoid contents during the course of ripening in hot peppers [5].

Although there seems to be a general agreement about the bioactive compounds present in hot peppers, there is little information available about the stability of capsaicinoids and antioxidants during postharvest storage and the consequences for their nutritional value and contribution to health. A number of works have reported on the effect of storage and packing materials on the long-term stability of polyphenols in foods and drinks and noted that they are very sensitive to light, oxygen and heat [6,7]. Talcott *et al.* [8] revealed that the antioxidant property of carrot polyphenols greatly reduced during storage at high temperatures, while Mustapha and Selselet-Attou [9] observed significant reduction in total phenolic contents in dates even at 10 °C after five months' storage. Several earlier studies focused on the stability of capsaicinoids and antioxidants in paprika stored at various temperatures and reported substantial fluctuations during storage. Thus, the concentration of capsaicinoids in paprika was significantly decreased as storage prolonged under ambient storage conditions, and the maximum decrease was recorded for dihydrocapsaicin [10]. Similarly, ascorbic acid concentration in ground paprika decreased significantly by 35% of the original level after four months of storage [11], while carotenoids and flavonoids showed more stability [12]. The evolution of carotenoid content in paprika powder indicated that 83% of the original concentration was retained in product stored in plastic sacs at 4 °C during 12 months [13].

However, analyses of the stability of capsaicinoids and antioxidants in paprika powder cannot be used to draw parallels with whole fruits, since processing conditions may alter enzyme-catalyzed processes, whereas postharvest ripening continues to take place in dry fruits during storage. In Pakistan, dry hot peppers are usually stored in jute bags at retail stores, but in mega stores and trading centers, they are available in polyethylene bags. Storage conditions and time after harvest have a great impact on the bioactive compounds of any produce. Therefore, the objective of the present work was to study the stability of capsaicinoids and antioxidants of dry whole hot peppers in protective packaging of polyethylene or jute during storage for five months at selected conditions of temperature, applying those normally used by hot pepper dealers.

2. Experimental Section

2.1. Pepper Samples

Three commercially-available hot pepper hybrids (*Capsicum annuum* L.), namely Sky Red, Maha and Wonder King, were grown in plastic tunnels under a drip irrigation system at the Vegetable Research Area of the Institute of Horticultural Sciences (University of Agriculture, Faisalabad, Pakistan) and harvested at the red ripe stage with pedicels. The fruits were dried in the sun for 6 to 8 days with an average daily temperature of 39 °C and relative humidity of 36% before the storage experiments. The moisture content of dried samples was determined with the air-forced oven drying method (indirect distillation at 105 °C), according to Method 44-15A of the American Association for Cereal Chemistry [14]. Moisture content in all hot pepper samples was 12%–13% after the sun-drying period.

2.2. Packaging and Storage Conditions

Dried pods of hot pepper hybrids (250 g) were packed in either natural jute bags or synthetic low-density polyethylene bags (9 µm thick, 20 by 32 cm). All samples were stored for 5 months in humidifiers (National MFG. Co., Lincoln, NE, USA) under controlled conditions at 20, 25 or 30 °C with 75% relative humidity, and analyzed on Day 0 and at 50-day intervals until Day 150. Experiments were conducted in triplicate.

2.3. Capsaicin and Dihydrocapsaicin

Hot pepper pods were oven-dried at 60 °C for 2–5 days, cooled and then ground to dried pepper powder. Samples were analyzed by using a chromatographic method previously described [5]. A mixture of sample:acetonitrile in a ratio of 1:10 was placed in 120-mL glass bottles with Teflon-lined lids, capped and placed in a water bath at 80 °C for 4 h and swirled manually every hour. The bottles were removed from the water bath and cooled at room temperature. The supernatant content of samples (2–3 mL) was filtered through a 0.45 µm filter (Millex®-HV filter) using a 5-mL disposable syringe (Millipore, Bedford, MA, USA) into an HPLC sample vial. For the liquid chromatographic analysis of capsaicinoids, an HPLC system LC-10 (Shimadzu, Japan) equipped with an SPD-10A

UV-Vis detector (set at 280-nm wavelength) was used. The analysis was carried out with the isocratic mobile phase (acetonitrile:water, 60:40) at a flow rate of 1 mL/min using a column Discovery C18 (250 × 4.6 mm, 5 μm) supplied by Supelco (Bellefonte, PA, USA). The limit of detection (LOD) was 0.1 μg/g for both capsaicinoids.

2.4. Total Carotenoids

Based on the Association of Official Analytical Chemists (AOAC) official Method 970.64 [15], two grams of hot pepper sample were ground using a mortar and pestle and transferred to a 100-mL flask covered with a stopper. The sample was blended for one minute with a mixture of 30 mL hexane:acetone:ethanol:toluene (10:7:6:7). For hot saponification, 2 mL of 40% methanolic KOH were pipetted into the flask, swirled for one minute and placed in a 56 °C water bath for 20 min. The sample was cooled for one hour in the dark, and then, 30 mL of hexane were pipetted into the flask, dried over anhydrous sodium sulfate made up to volume and shaken vigorously for one minute. Upper phase was 50 mL. Absorbance was measured at 436 nm using the IRMECO UV-Vis spectrophotometer Model U2020 with β-carotene (Sigma-Aldrich, St. Louis, MO, USA) as the standard.

2.5. Ascorbic Acid

Ascorbic acid was quantitatively determined according to the 2,6-dichlorophenolindophenol AOAC official Method 967.21 [16]. A sample of hot peppers (10 g) was blended with 2.5 mL of 20% metaphosphoric acid, and distilled water was then added up to the 100-mL mark. Ten milliliters of the suspension were titrated with freshly-prepared standard of 2,6-dichlorophenolindophenol dye until a light, but distinct rose pink color persisted for 15 s.

2.6. Total Phenolic Compounds

Total phenolic contents of hot peppers were analyzed using the modified Folin-Ciocalteu reagent method as described elsewhere [5]. About 0.5 g of the sample was macerated in 3 mL 80% aqueous acetone with a mortar and pestle. The extracts were placed into tightly-sealed micro-tubes and maintained in darkness at 4 °C overnight. Samples were centrifuged at 1000 rpm for 2 min. A mixture of 135 μL H_2O, 750 μL 1/10 dilution Folin-Ciocalteu reagent (Sigma-Aldrich, St. Louis, MO, USA) and 600 μL 7.5% (w/v) Na_2CO_3 was added to 50 μL of extract in 1.5-mL micro-tubes. After vortexing for 10 s, the mixture was incubated at 45 °C in a water bath for 15 min. Samples were allowed to cool to room temperature before reading the absorbance at 765 nm using the IRMECO UV-Vis spectrophotometer Model U2020. A blank was prepared from 50 μL 80% aqueous acetone. The gallic acid standard curve was prepared from freshly-made 1 mg/mL gallic acid (Acros Organics, Belgium) in 80% aqueous acetone.

2.7. Statistical Analysis

Analysis of variance was computed with the data from each attribute using the STATISTICA Computer Program (Version 2003, StatSoft Inc., Tulsa, OK, USA). The least significant difference test at the 5% level of probability was used to check the differences among mean values according to Hill and Lewicky [16]. There were 72 treatments, and each treatment was replicated thrice.

3. Results and Discussion

3.1. Evolution of Capsaicinoids Contents

Pungency is an important factor for determining the commercial quality in hot peppers, the level of which depends mainly on the concentration of capsaicinoids, such as capsaicin and dihydrocapsaicin [17]. The former is the primary molecule responsible for the heat and burning sensations associated with hot peppers. The initial levels of capsaicin and dihydrocapsaicin were significantly different among hybrids (Table 1), Sky Red (33.7 and 20.6 mg/100 g), Maha (22.5 and 13.6 mg/100 g) and Wonder King (24.4 and 15.6 mg/100 g), respectively (Figure 1). Overall, capsaicin concentrations in polyethylene-packed hot peppers gradually decreased during 150-day storage by 8.8% at 20 °C, 10.4% at 25 °C and 12.9% at 30 °C, while decreasing amounts observed in samples packed in jute bags were 9.2%, 12.2% and 14.0%, respectively (Table 2). Dihydrocapsaicin evolved in a similar trend showing a reduction of 11.1%, 12.8% and 15.4% (20 °C, 25 °C and 30 °C, respectively) in hot peppers packed in polyethylene bags after 150 days of storage, whereas for those stored in jute bags, the percent reductions reached 10%, 14.9% and 16% (Table 2). Then, both capsaicinoids decreased significantly during storage for 150 days ($p < 0.05$), and the disappearance rate increased with storage temperature and tended to be higher in jute bags than in polyethylene wrapping. Jute is a natural fiber that is permeable to air and moisture, and it is biodegradable and easily decomposable in the environment. In contrast, polyethylene is a low cost and low weight, durable, synthetic polymer that provides a barrier to oxygen and water vapor, but it is not biodegradable and causes pollution of the environment.

The trends of capsaicinoids evolution in the three hot peppers hybrids can be observed in Figure 1. The highest reductions of capsaicin were observed in the Maha hybrid, followed by Wonder King and Sky Red, whereas the rate of dihydrocapsaicin disappearance was highest in Wonder King, followed by Maha and Sky Red. These results indicated that the three hot peppers hybrids behave somewhat differently with respect to the fading out of capsaicinoids during storage. It has been reported that differences in biochemical factors (*i.e.*, peroxidase activity) within hot pepper hybrids can influence the stability of capsaicinoids with the advancement of maturity and drying [18].

Overall, our studies indicated that with the increase in storage temperature and time, a gradual and steady decrease in capsaicinoids concentrations was observed in hot peppers. Available research information regarding the effect of temperature and packaging materials on capsaicinoids concentrations is scarce and mainly focused on powder paprika. In correlation with our results, Topuz and Ozdemir [10] observed 16% and 15% decrease in capsaicin and dihydrocapsaicin concentrations, respectively, in powder paprika after six months' storage at ambient temperature.

Table 1. Mean square values from the analysis of variance for capsaicin, dihydrocapsaicin, total carotenoids, ascorbic acid and total phenolic contents (mg/100 g) of hot pepper hybrids under different packaging and storage conditions.

Source of variation	Degree freedom	Capsaicin	Dihydrocapsaicin	Total carotenoids	Ascorbic acid	Total phenolic compounds
Hybrid (H)	2	2,537.866 **	937.711 **	16,090.310 **	1,068.297 **	13,753.680 **
Storage Period (S)	3	87.174 **	51.529 **	2,306.240 **	845.790 **	1,064.536 **
Packaging (P)	1	4.544 **	1.123 *	78.313 **	39.681 **	22.633 **
Temperature (T)	2	9.435 **	6.982 **	276.108 **	223.127 **	143.317 **
H × S	6	0.461 *	0.749 **	21.130 **	15.002 **	62.977 **
H × P	2	0.007 NS	0.020 NS	5.870 NS	0.292 NS	0.176 NS
H × T	4	0.472 *	0.127 NS	10.700 *	8.648 **	4.276 **
S × P	3	0.696 **	0.226 NS	13.888 *	4.884 **	3.217 **
S × T	6	1.215 **	0.874 **	76.090 **	44.643 **	18.147 **
P × T	2	0.475 NS	0.257 NS	15.352 *	0.592 NS	1.139 NS
H × S × P	6	0.066 NS	0.031 NS	1.517 NS	0.619 NS	0.214 NS
H × S × T	12	0.333 *	0.308 *	3.909 NS	3.195 **	1.988 **
H × P × T	4	0.086 NS	0.063 NS	2.204 NS	1.091 NS	2.319 *
S × P × T	6	0.074 NS	0.076 NS	4.375 NS	0.724 NS	0.403 NS
H × S × P × T	12	0.039 NS	0.026 NS	5.063 NS	0.615 NS	0.474 NS
Error	144	0.171	0.166	3.693	0.622	0.781

* Significant at the 0.05 level; ** Significant at the 0.01 level; NS, Not significant.

Table 2. Evolution of capsaicinoids and antioxidants in hot peppers under different storage conditions. Data shown are the mean values ± standard deviation for three hot pepper hybrids (Sky Red, Maha and Wonder King), and ▼ is the percent reduction from 0 to 150 days on a dry weight basis.

Parameter (mg 100 g^{-1})	Polyethylene bag				Jute bag			
	Days	20 °C	25 °C	30 °C	Days	20 °C	25 °C	30 °C
Capsaicin	0		26.9 ± 6.0		0		26.9 ± 6.0	
	50	26.0 ± 5.7	25.8 ± 6.2	25.4 ± 6.1	50	25.8 ± 5.7	25.2 ± 5.9	24.9 ± 5.9
	100	25.2 ± 6.1	24.8 ± 6.0	24.5 ± 5.5	100	25.2 ± 6.0	24.1 ± 6.1	24.1 ± 5.8
	150	24.5 ± 6.1	24.1 ± 5.1	23.4 ± 6.0	150	24.4 ± 6.1	23.6 ± 6.0	23.1 ± 6.0
	▼	8.8%	10.4%	12.9%	▼	9.2%	12.2%	14.0%
Dihydrocapsaicin	0		16.6 ± 3.6		0		16.6 ± 3.6	
	50	16.0 ± 3.8	15.5 ± 3.7	15.0 ± 3.4	50	15.8 ± 3.8	15.2 ± 3.5	15.3 ± 3.9
	100	15.4 ± 3.7	15.2 ± 3.7	14.4 ± 3.5	100	15.3 ± 3.5	14.6 ± 3.6	14.3 ± 3.6
	150	14.8 ± 3.4	14.5 ± 3.9	14.0 ± 3.6	150	14.9 ± 3.1	14.1 ± 3.7	13.9 ± 4.0
	▼	11.1%	12.8%	15.4%	▼	10.0%	14.9%	16.0%
Total carotenoids	0		95.5 ± 14.6		0		95.5 ± 14.6	
	50	91.5 ± 14.2	89.6 ± 15.0	89.4 ± 13.6	50	90.9 ± 13.8	87.9 ± 15.3	88.1 ± 12.8
	100	87.4 ± 15.6	86.3 ± 15.4	86.7 ± 12.0	100	87.3 ± 15.1	83.9 ± 14.7	83.3 ± 15.2
	150	84.6 ± 15.0	80.2 ± 16.6	76.2 ± 16.2	150	84.3 ± 15.0	78.5 ± 16.6	73.7 ± 15.5
	▼	11.5%	16.1%	20.2%	▼	11.7%	17.8%	22.9%

Table 2. *Cont.*

Parameter (mg 100 g^{-1})	Polyethylene bag				Jute bag			
	Days	20 °C	25 °C	30 °C	Days	20 °C	25 °C	30 °C
Ascorbic acid	0		40.3 ± 4.5		0		40.3 ± 4.5	
	50	38.2 ± 4.4	36.1 ± 3.7	35.7 ± 4.0	50	37.1 ± 4.7	35.6 ± 4.8	34.5 ± 4.1
	100	35.9 ± 4.4	34.6 ± 3.1	32.3 ± 2.8	100	35.3 ± 4.2	33.1 ± 3.9	31.4 ± 3.3
	150	35.4 ± 4.7	31.4 ± 3.0	29.2 ± 1.6	150	34.3 ± 4.1	29.7 ± 2.6	26.9 ± 3.1
	▼	12.2%	22.1%	27.6%	▼	14.7%	26.2%	33.2%
Total phenolic compounds	0		56.2 ± 13.6		0		56.2 ± 13.6	
	50	54.3 ± 13.2	52.7 ± 14.2	51.7 ± 15.2	50	53.9 ± 13.6	51.9 ± 14.5	51.2 ± 14.6
	100	51.1 ± 13.3	49.4 ± 13.5	46.9 ± 15.1	100	50.7 ± 14.4	48.1 ± 13.0	45.7 ± 14.0
	150	48.2 ± 12.8	46.7 ± 14.2	44.6 ± 14.2	150	47.6 ± 13.9	45.5 ± 15.1	43.4 ± 13.7
	▼	14.1%	16.9%	20.5%	▼	15.3%	19.0%	22.8%

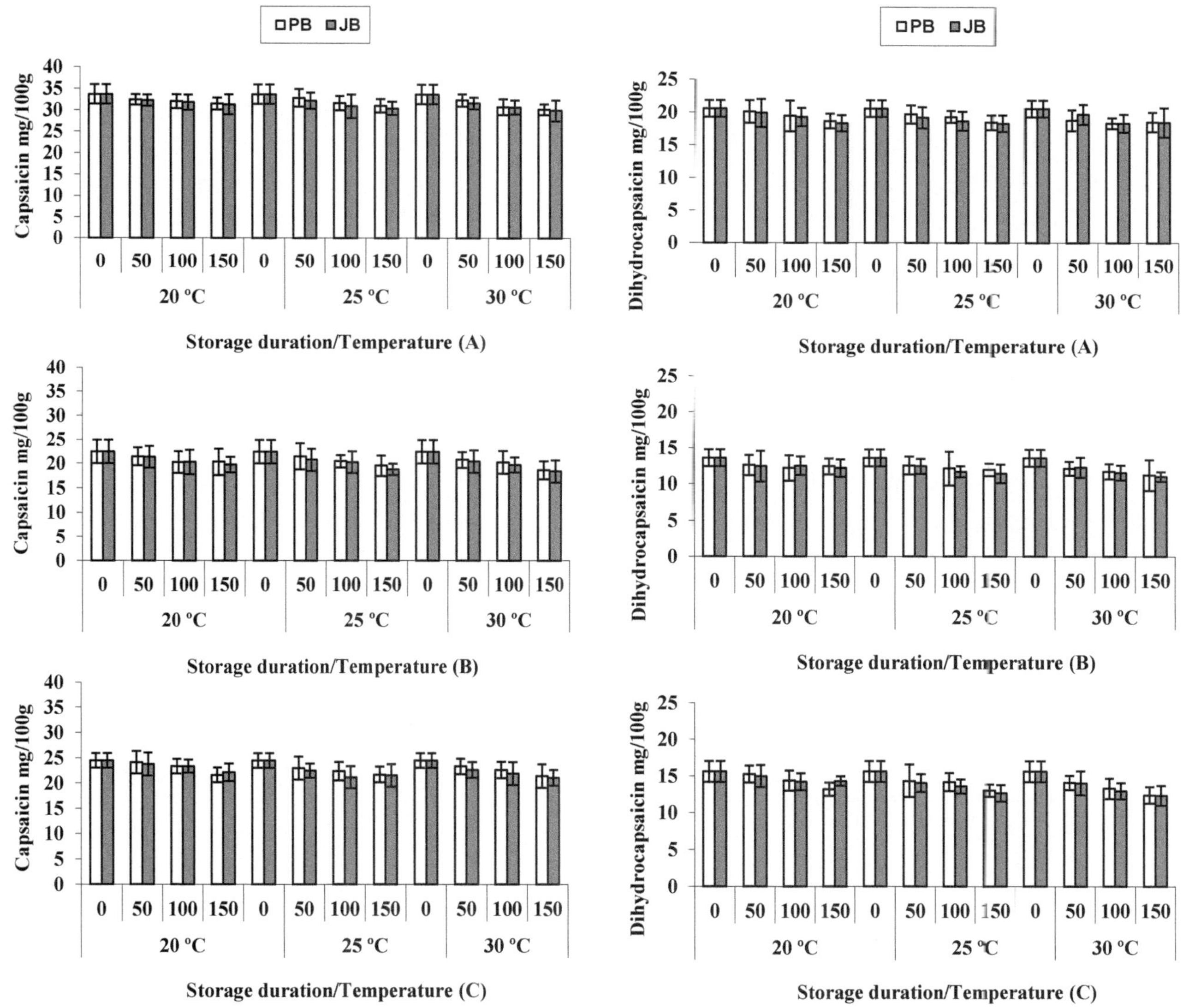

Figure 1. Capsaicin and dihydrocapsaicin concentrations in hot peppers *viz*. Sky Red (**A**), Maha (**B**) and Wonder King (**C**) at different temperatures and packaging materials (PB, polyethylene bag; JB, jute bag) during five months' storage. Vertical bars show ± SD of $n = 3$ replicates.

3.2. Evolution of Total Carotenoids

Increasing interest is being paid to the red hot pepper spice, not only because of its economic importance, but also because of its diversified composition. The intense red color of the ripe peppers and their processed products is due to the presence of carotenoid pigments. When carotenoids are ingested, they show important biological actions, such as being antioxidants, as well as free radical scavengers and reducing cardiovascular diseases and the risk of several types of cancer [19,20]. Hot pepper hybrids differed significantly with respect to total carotenoid concentration (Table 1), and it was highest in Wonder King (111.8 mg/100 g), followed by Sky Red (91.2 mg/100 g), while in case of Maha, it was 83.5 mg/100 g (Figure 2). Trends observed during the storage period showed that total carotenoid concentration in hot pepper hybrids had an inverse relation with storage time (Table 2). Again, carotenoids were significantly reduced during storage for 150 days ($p < 0.05$), and the rate of disappearance increased with storage temperature and tended to be slightly higher in peppers packed in jute bags than in polyethylene. The gradual fading out of total carotenoid concentrations that occurred during storage amounted to 11.5%–20.2% in samples stored in polyethylene bags and from 11.7% to 22.9% in those packed in jute bags. The decreasing rate of total carotenoids was higher between 100 to 150 days at 30 °C, especially in the Maha hybrid (Figure 2). However, the percent reductions of carotenoids during three months' storage were far from what could be considered as the shelf life of the product, established herein as 50% of loss. The highest reductions of carotenoids were observed in the Maha hybrid, followed by Sky Red and Wonder King, which is another example of the influence of genotype on the content of bioactive compounds in red peppers [21].

As carotenoids are thermally sensitive in nature, their disappearance rates were higher at 30 °C than at 20 °C or 25 °C. Likewise, Lee *et al.* [22] showed that carotenoid destruction in red peppers was greatly affected by water activity and storage temperature and reported that carotenoid stability in red peppers was improved by lowering the storage temperature. Carotenoids are fairly stable in their natural environment, but become sensitive to light and temperature and are readily decomposed during some food processing operations [13]. Our results are in line with the findings of Schweiggert *et al.* [23] that total carotenoid concentration during four months' storage at ambient temperature dropped by 9.6% and 16.7% in chili and 38.8% and 39.7% in paprika powders with and without illumination, respectively. Similarly variation or loss of carotenoids during foods processing, e.g., peeling, as well as storage, occurs through geometric isomerization and enzymatic or non-enzymatic oxidation [24]. It has been reported that the loss of the carotenoid fraction during storage correlates with the development of prooxidative processes in paprika [13].

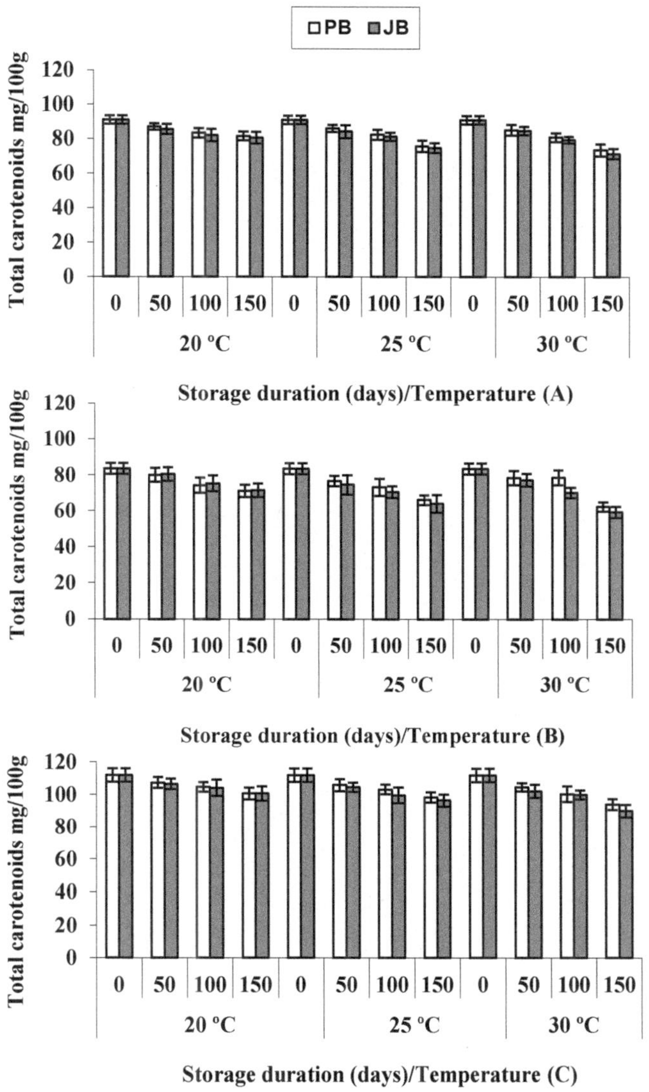

Figure 2. Total carotenoids in hot peppers *viz.* Sky Red (**A**), Maha (**B**) and Wonder King (**C**) at different temperatures and packaging materials during five months' storage. Vertical bars show ± SD of $n = 3$ replicates.

3.3. Evolution of Ascorbic Acid

Ascorbic acid is not only well known as being an antioxidant and biologically-active compound, but also as an important nutritional and functional constituent of hot pepper fruit. Changes in the pattern of ascorbic acid concentration during five months' storage showed that as storage time prolonged, the ascorbic acid concentration in hot peppers decreased significantly ($p \leq 0.05$) (Table 1). The general tendency for the amount of ascorbic acid that was lost was similar for all samples and showed a total retention of 87.8%, 77.9% and 72.4% (20 °C, 25 °C and 30 °C, respectively) in polyethylene bags and 85.3%, 73.8% and 66.8% (20 °C, 25 °C and 30 °C, respectively) in jute bags, after 150 days of storage (Table 2). Then, the ascorbic acid concentration gradually decreased in hot pepper hybrids during storage, and the disappearance rate increased with storage temperature and tended to be higher in jute bags, reaching a maximum of 33.2%. A steep decrease in ascorbic acid concentrations was observed until Day 100, especially in the Wonder King hybrid, but further reduction still occurred between 100 and 150 days of storage (Figure 3).

These results revealed that ascorbic acid is sensitive to high temperatures during storage, and it gradually decreased in all hot pepper hybrids. Similarly, Kalt [12] indicated that ascorbic acid concentration decreased in many major fruits and vegetables during storage. Daood *et al.* [11] reported that ascorbic acid concentration in ground paprika decreased by 10% after 30 days followed by 20% and 35% after 60 and 120 days of storage. It has been reported that modified atmosphere packaging (MAP) can retard the decline of ascorbic acid in intact peppers, as Ornelas-Paz *et al.* [25] observed that its contents remained unchanged in peppers stored under MAP during four weeks.

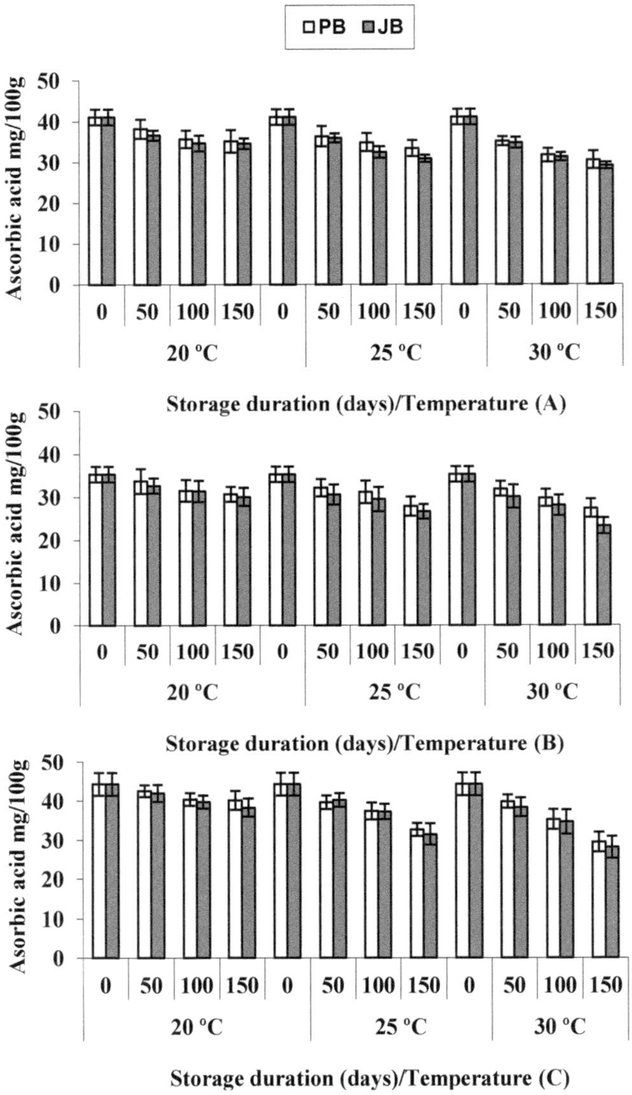

Figure 3. Ascorbic acid in hot peppers *viz.* Sky Red (**A**), Maha (**B**) and Wonder King (**C**) at different temperatures and packaging materials during five months' storage. Vertical bars show ± SD of $n = 3$ replicates.

3.4. Evolutions of Total Phenolic Compounds

Besides the antioxidant activity of phenolic compounds that can protect the human body from free radicals, these compounds also possess antimicrobial properties [26], and they are natural inhibitors of the growth of toxigenic *Aspergillus* and aflatoxin B_1 production [27], which is a significant risk in hot peppers [28]. Table 1 shows that hot pepper hybrids differed significantly ($p \leq 0.05$) with respect to their total phenolic contents, with Maha standing first (67.9 mg/100 g), followed by Sky Red (59.3 mg/100 g) and Wonder King (41.2 mg/100 g) (Figure 4). Total phenolic concentrations decreased continuously in polyethylene-packed hot peppers during 150-day storage by 14.1% at 20 °C, 16.9% at 25 °C and 20.5% at 30 °C, while decreasing amounts in samples packed in jute bags were 15.3%, 19.0% and 22.8%, respectively (Table 2). Then, total phenolic compounds declined significantly during storage for 150 days ($p < 0.05$), while the disappearance rate increased with storage temperature. The trends of phenolics evolution in the three hot pepper hybrids is illustrated in Figure 4. Regarding the stability in the different packing materials, the total phenolic contents followed the same trend in polyethylene and jute-packed peppers, indicating that both packaging materials preserved these compounds to a similar extent.

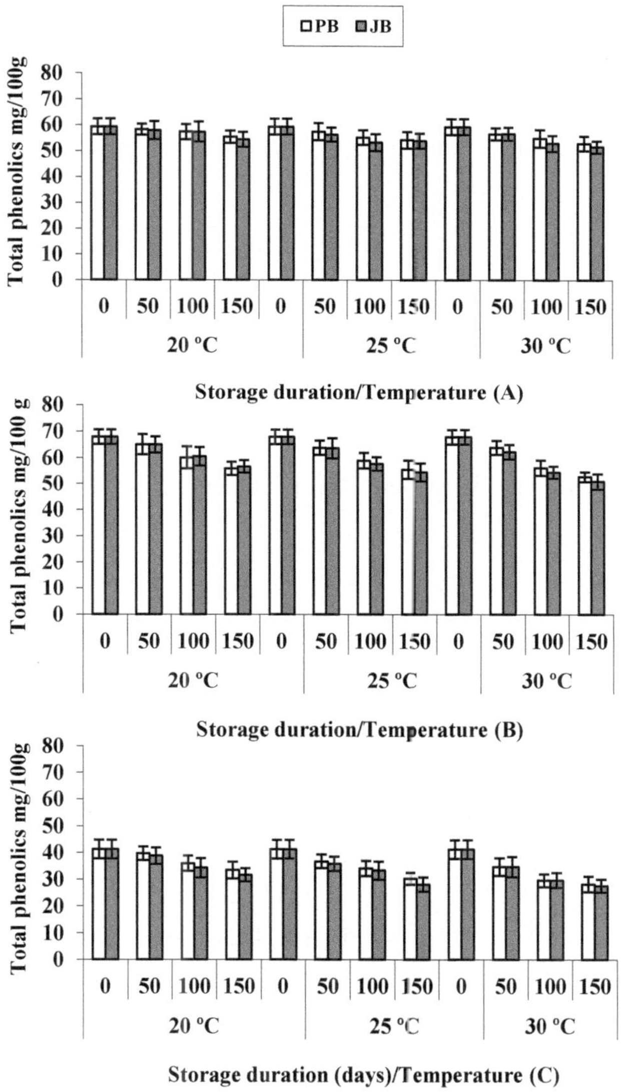

Figure 4. Total phenolic contents in hot peppers *viz.* Sky Red (**A**), Maha (**B**) and Wonder King (**C**) at different temperatures and packaging materials during five months' storage. Vertical bars show + SD of $n = 3$ replicates.

The present studies are in line with the findings of Talcott *et al.* [8] that the fading out of phenolic acids was accentuated by storage at high temperatures and affected the antioxidant activity. Similarly,

Mustapha and Selselet-Attou [9] reported that phenolic compounds in dates decreased gradually during storage for five months, decreasing by 24% at ambient temperature and by 14% at 10 °C.

It has been reported that MAP can retard the ripening process in peppers. Ornelas-Paz *et al.* [25] observed that phenolic contents were not reduced in peppers stored under MAP during four weeks.

4. Conclusions

In light of the present results, it can be concluded that with the increase in storage temperature and time, a significant, but gradual decrease in capsaicinoids and antioxidant levels occurred in dry whole peppers. However, despite the sensitivity of capsaicinoids and antioxidants to oxygen, light and moisture, the packaging in natural jute bags or synthetic LDPE plastic bags, as well as the storage at ambient temperature preserved between 77.4% and 87.3% of the initial amounts of these health and nutrition-promoting compounds during five months' storage. Therefore, the traditional sun-drying of pepper fruits and the application of commercially-controlled conditions of storage is able to maintain the overall quality of peppers in terms of pungency, color and antioxidants.

Acknowledgments

The authors are grateful for being provided laboratory facilities at the Nuclear Institute for Agriculture and Biology, Faisalabad, Pakistan, and financial support by the Higher Education Commission, Islamabad, Pakistan, to complete the research work. The corresponding author is also highly grateful for the support of the University of Agriculture, Faisalabad, Pakistan. Funds from Grupo Consolidado DGA A01 and the European Social Fund of the EC are also acknowledged.

Conflicts of Interest

The authors declare no conflict of interest.

References

1. Hervert-Hernandez, D.; Sayago-Ayerdi, S.G.; Goni, I. Bioactive compounds of four hot pepper varieties (*Capsicum annuum* L.), antioxidant capacity, and intestinal bioaccessibility. *J. Agric. Food Chem.* **2010**, *58*, 3399–3406.
2. Kim, G.D.; Lee, Y.S.; Cho, J.Y.; Lee, Y.H.; Choi, K.J.; Lee, Y.; Han, T.H.; Lee, S.H.; Park, K.H.; Moon, J.H. Comparison of the content of bioactive substances and the inhibitory effects against rat plasma oxidation of conventional and organic hot peppers (*Capsicum annuum* L.). *J. Agric. Food Chem.* **2010**, *58*, 12300–12306
3. Liu, Y.; Nair, M.G. Non-pungent functional food components in the water extracts of hot peppers. *Food Chem.* **2010**, *122*, 731–736
4. Daood, H.G.; Kapitany, J.; Biacs, P.; Albrecht, K. Drying temperature, endogenous antioxidants and capsaicinoids affect carotenoid stability in paprika (red pepper spice) *J. Sci. Food Agric.* **2006**, *86*, 2450–2457.

5. Iqbal, Q.; Amjad, M.; Asi, M.R.; Ariño, A. Characterization of capsaicinoids and antioxidants in hot peppers as influenced by hybrid and harvesting stage. *Plant Foods Hum. Nutr.* **2013**, *68*, 358–363.
6. Munin, M.; Edwards-Levy, F. Encapsulation of natural polyphenolic compounds; a review. *Pharmaceutics* **2011**, *3*, 793–829.
7. Kim, Y.; Welt, B.A.; Talcott, S.T. The impact of packaging materials on the antioxidant phytochemical stability of aqueous infusions of green tea (*Camellia sinensis*) and yaupon holly (*Ilex vomitoria*) during cold storage. *J. Agric. Food Chem.* **2011**, *59*, 4676–4683
8. Talcott, S.T.; Howard, L.R.; Brenes, C.H. Antioxidant changes and sensory properties of carrot puree processed with and without periderm tissues. *J. Agric. Food Chem.* **2000**, *48*, 1315–1321.
9. Mustapha, K.; Selselet-Attou, G. Effect of heat treatment on polyphenol oxidase and peroxidase activities in Algerian stored dates. *Afr. J. Biotechnol.* **2007**, *6*, 790–794.
10. Topuz, A.; Ozdemir, F. Influence of gamma radiation and storage on the capsaicinoids of sun-dried and dehydrated paprika. *Food Chem.* **2004**, *86*, 509–515.
11. Daood, H.G.; Vinkler, M.; Markus, F.; Hebshi, E.A.; Bicas, P.A. Antioxidant vitamin content of spice red pepper (paprika) as affected by technological and varietal factors. *Food Chem.* **1996**, *15*, 365–372.
12. Kalt, W. Effect of production and processing factors on major fruit and vegetable antioxidants. *J. Food Sci.* **2005**, *70*, 11–19.
13. Perez-Galvez, A.; Hornero-Mendez, D.; Minguez-Mosquera, M.I. Stability of paprika without supplementary antioxidants during storage under industrial controlled conditions. *J. Agric. Food Chem.* **2009**, *57*, 4718–4723.
14. AACC (American Association of Cereal Chemists). *Approved Methods of American Association of Cereal Chemists*, 5th ed.; American Association for Cereal Chemistry: St. Paul, MN, USA, 2000.
15. AOAC (Association of Official Analytical Chemists). *Official Methods of Analysis of AOAC International*, 19th ed.; AOAC International: Gaithersburg, MD, USA, 2012.
16. Hill, T.; Lewicki, P. *STATISTICS: Methods And Applications*; StatSoft: Tulsa, OK, USA, 2007.
17. Todd, P.H., Jr.; Bensinger, M.G.; Biftu, T. Determination of pungency due to capsicum by gas-liquid chromatography. *J. Food Sci.* **1977**, *42*, 660–680.
18. Russo, V.M.; Biles, C.L. Fertilizer rate and β-galactosidase and peroxidase activity in pepper fruit at different stages and years of harvest. *Plant Foods Hum. Nutr.* **2004**, *58*, 231–239.
19. Bae, H.; Jayaprakasha, G.K.; Jifon, J.; Patil, B.S. Variation of antioxidant activity and the levels of bioactive compounds in lipophilic and hydrophilic extracts from hot pepper (*Capsicum* spp.) cultivars. *Food Chem.* **2012**, *134*, 1912–1918.
20. Gómez-García, M.R.; Ochoa-Alejo, N. Biochemistry and molecular biology of carotenoid biosynthesis in chili peppers (*Capsicum* spp.). *Int. J. Mol. Sci.* **2013**, *14*, 19025–19053.
21. Gnayfeed, M.H.; Daood, H.G.; Biacs, P.A.; Alcaraz, C.F. Content of bioactive compounds in pungent spice red pepper (paprika) as affected by ripening and genotype. *J. Sci. Food Agric.* **2001**, *81*, 1580–1585
22. Lee, D.S.; Chung, S.K.; Yam, K.L. Carotenoid loss in dried red pepper products. *Int J. Food Sci. Tech.* **1992**, *27*, 179–185.

23. Schweiggert, U.; Kurz, C.; Schieber, A.; Carle, R. Effects of processing and storage on the stability of free and esterified carotenoids of red peppers (*Capsicum annuum* L.) and hot chilli peppers (*Capsicum frutescens* L.). *Eur. Food Res. Tech.* **2007**, *225*, 261–270.
24. Rodriguez-Amaya, D.B. Food carotenoids: Analysis, composition and alterations during storage and processing of foods. *Forum Nutr.* **2003**, *56*, 35–37.
25. Ornelas-Paz, J.de J.; Zamudio-Flores, P.B.; Torres-Cisneros, C.G.; Holguín-Soto, R.; Ramos-Aguilar, O.P.; Ruiz-Cruz, S.; Guevara-Arauza, J.C.; Gonzalez-Aguilar, G.A.; Santana-Rodriguez, V. The barrier properties and potential use of recycled-LDPE films as a packaging material to preserve the quality of Jalapeño peppers by modified atmospheres. *Sci. Hort.* **2012**, *135*, 210–218.
26. Nascimento, P.L.A.; Nascimento, T.C.E.S.; Ramos, N.S.M.; Silva, G.R.; Galindo-Gomes, J.E.; Falcão, R.E.A.; Moreira, K.A.; Porto, A.L.F.; Silva, T.M.S. Quantification, antioxidant and antimicrobial activity of phenolics isolated from different extracts of *Capsicum frutescens* (Pimenta Malagueta). *Molecules* **2014**, *19*, 5434–5447.
27. Nesci, A.; Marin, S.; Etcheverry, M. ; Sanchis, V. Natural maize phytochemicals for control of maize mycoflora and aflatoxigenic fungi. *World Mycotoxin J.* **2009**, *2*, 305–312.
28. Iqbal, Q.; Amjad, M.; Asi, M.R.; Ariño, A. Assessment of hot peppers for aflatoxin and mold proliferation during storage. *J. Food Prot.* **2011**, *74*, 830–835.

10

Purification of Recombinant Peanut Allergen Ara h 1 and Comparison of IgE Binding to the Natural Protein

Barry K. Hurlburt *, Jane K. McBride, Jacqueline B. Nesbit †, Sanbao Ruan ‡ and Soheila J. Maleki

Southern Regional Research Center, Agricultural Research Service, United States Department of Agriculture, 1100 Robert E. Lee, Blvd., New Orleans, LA 70124, USA;
E-Mails: jane.mcbride@ars.usda.gov (J.K.M.); jbnesbit@uno.edu (J.B.N.);
cyoumc@yahoo.com (S.R.); soheila.maleki@ars.usda.gov (S.J.M.)

† Current address: Department of Biology, University of New Orleans, New Orleans, LA 70148, USA.

‡ Current address: Department of Medicine, Louisiana State University Health Sciences Center, New Orleans, LA 70112, USA.

* Author to whom correspondence should be addressed; E-Mail: barry.hurlburt@ars.usda.gov

External Editor: Christopher J. Smith

Abstract: Allergic reactions to food are on the rise worldwide and there is a corresponding increase in interest to understand the molecular mechanisms responsible. Peanut allergies are the most problematic because the reaction often persists into adulthood and can be as severe as anaphylaxis and death. The purpose of the work presented here was to develop a reproducible method to produce large quantities of pure recombinant Ara h 1(rAra h 1) that will enable standardization of immunological tests for patients and allow structural and immunological studies on the wild type and mutagenized forms of the protein. Ara h 1 is initially a pre-pro-protein which, following two endoproteolytic cleavages, becomes the mature form found in peanut. The mature form however has flexible regions that make it refractory to some structural studies including crystallography. Therefore, independent purification of the mature and core regions was desirable. Expression constructs were synthesized cDNA clones for each in a pET plasmid vector without tags. Codons were optimized for expression in *E. coli*. High-level expression was achieved in BL21 strains.

Purification to near homogeneity was achieved by a combination of ammonium sulfate precipitation and ion exchange chromatography. The purified rAra h 1 was then compared with natural Ara h 1 for IgE binding. All patients recognized both the folded natural and rAra h 1, but the IgE binding to the rArah1 was significantly reduced in comparison to the natural allergen, which could potentially make it useful for immunotherapeutic purposes.

Keywords: peanut allergy; immunodominant; Ara h 1; IgE binding

---

1. Introduction

Peanuts (*Arachis hypogaea*) are a tremendously important food with over 30 million metric tons harvested a year worldwide (US Department of Agriculture, [1]). They contain 25% protein comprised of 18 of the 20 naturally occurring amino acids (not asparagine or glutamine) including all of the essential amino acids and are nearly 50% oil. Their high nutritional value and low production cost has made peanut one of the most prevalent foods in the world. They are eaten intact, as a butter, and as additives in a wide variety of processed foods from candy to canned chili. Unfortunately, peanuts are also a frequent cause of food allergy and can cause severe reactions, including fatal anaphylaxis. In the United States more than 1% of the population is peanut allergic [2]. There are currently no good therapies for peanut allergy; avoidance is the only option for patients.

The protein composition of a peanut is almost entirely a small number of seed storage proteins and these are the predominant allergens. Three of these Ara h 1, Ara h 2, and Ara h 3 are immunodominant, in that the majority of peanut allergic patient's sera contain IgE antibodies specific for these proteins [3–5]. Considerable research over many years has resulted in a wealth of knowledge about these antigens. Ara h 1 is a vicilin and member of the 7S globulin family [4] and makes up approximately 15% of the peanut [6]. It is a trimer comprised of ~67 kDa subunits [4,6]. The linear IgE epitopes have been mapped by several groups [4,7–9]. Roasting is the most common processing treatment for peanuts prior to consumption and this procedure heightens the allergenicity of Ara h 1 [6]. During roasting sugars modify the proteins or are attached to the protein via the Maillard reaction. Furthermore, roasting results in an Ara h 1 protein that is less digestible by gastrointestinal enzymes [6].

For several projects that our group has planned having a reproducible system to express recombinant Ara h 1 to high levels and purify it is required. One of those projects is to have a constant source of identical protein to use in development of diagnostic systems to determine peanut allergy. We are also interested in pursuing structural studies with the wild type and mutant forms of Ara h 1. As mentioned above, the linear epitopes bound by IgE were mapped. In addition, individual amino acids within those epitopes were changed and significant reduction in IgE binding could be achieved by one or two substitutions [4]. It is possible that a recombinant mutant Ara h 1 with reduced IgE binding could be developed and may be useful in immunotherapy.

Ara h 1 is translated as a pre-pro-protein (Figure 1). A signal peptide (red in Figure 1) presumably directs the nascent protein to the vacuole prior to its cleavage. There is also a leader sequence (blue in Figure 1) of unknown function that is removed yielding the mature Ara h 1 protein (black and purple in

Figure 1) [10]. Interestingly, three immunodominant epitopes were mapped to this leader sequence [11]. A pET derived construct of the mature Ara h 1 coding region was made. Since it has been reported that the *N*- and *C*-terminal extensions (black in Figure 1) of a highly-conserved core domain are flexible and inhibit crystal formation [12–14], we also generated a pET expression construct of the core domain alone (purple in Figure 1).

Figure 1. Amino acid sequence of Ara h 1. The entire primary translation product is shown. In red is the sequence of a leader sequence that is proteolytically removed. In blue is the signal sequence that is also proteolytically removed. In black and purple is the mature Ara h 1 found in peanut. In purple is the Ara h 1 core sequence. The linear epitopes identified by others are underlined. The immunodominant epitopes are also shown in bold typeface.

```
MRGRVSPLMLLLGILVLASVSATHAKSSPYQKK
TENPCAQRCLQSCQQEPDDLKQKACESRCTKLE
YDPRCVYDPRGHTGTTNQRSPPGERTRGRQPGD
YDDDRRQPRREEGGRWGPAGPREREREEDWRQP
REDWRRPSHQQPRKIRPEGREGEQEWGTPGSHV
REETSRNNPFYFPSRRFSTRYGNQNGRIRVLQR
FDQRSRQFQNLQNHRIVQIEAKPNTLVLPKHAD
ADNILVIQQGQATVTVANGNNRKSFNLDEGHAL
RIPSGFISYILNRHDNQNLRVAKISMPVNTPGQ
FEDFFPASSRDQSSYLQGFSRNTLEAAFNAEFN
EIRRVLLEENAGGEQEERGQRRWSTRSSENNEG
VIVKVSKEHVEELTKHAKSVSKKGSEEEGDITN
PINLREGEPDLSNNFGKLFEVKPDKKNPQLQDL
DMMLTCVEIKEGALMLPHFNSKAMVIVVVNKGT
GNLELVAVRKEQQQRGRREEEEDEDEEEEGSNR
EVRRYTARLKEGDVFIMPAAHPVAINASSELHL
LGFGINAENNHRIFLAGDKDNVIDQIEKQAKDL
AFPGSGEQVEKLIKNQKESHFVSARPQSQSQSP
SSPEKESPEKEDQEEENQGGKGPLLSILKAFN
```

2. Experimental Section

2.1. Plasmid Constructions and Protein Expression

The initial translation product of Ara h 1 is shown in Figure 1. The mature protein found in peanut starts at the arginine at position 85 and ends with the asparagine at position 626 (shown in black and purple in Figure 1). The mature protein sequence was back translated into codons preferred for expression in *E. coli* using programs resident in DNASTAR's Lasergene software (Madison, WI, USA). EZBiolab (Carmel, IN, USA) synthesized the gene and added CATATG to the 5′ end for an Nde I restriction site and start methionine and TAAGGATCC to the 3′ end for a Bam HI restriction site and stop codon. The synthetic gene was excised with Nde I and Bam HI and cloned into equally cut pET9a yielding pET-rAra h 1. A shorter core region (codons 170–586) was subsequently amplified using the synthetic gene as a template in PCR using primers: 5′-GGAGCCCATATGTCTCGTAACAAC CCGTTCTACTT-3′ and 5′-GGAGCCGGATCCTTAACGAGCAGAAACGAAGTGA-3′. The PCR product was then digested with Nde I and Bam HI and cloned into pET9a yielding pET-rsAra h 1. The fidelity of pET-rsAra h 1 was confirmed by DNA sequencing. To express these recombinant proteins, the plasmids were transformed into BL21 (DE3) cells. Individual colonies were inoculated and grown overnight at 37 °C in 5 mL of Luria-Bertani (LB) broth containing 25 μg/mL kanamycin. Five hundred milliliters cultures in 2 L baffle flasks were inoculated with 500 μL of the overnight culture. Growth was monitored by absorbance at 600 nm. IPTG (isopropyl-β-D-thiogalactopyranoside) was added to 1 mM when the culture reached 0.6 O.D. to induce protein expression. Cells that were induced for 3 h were harvested by centrifugation at 5200× *g* at 4 °C for 10 min.

2.2. Purification of Mature rAra h 1

Wet cell paste from 2 L of culture was stored at −20 °C until needed. Cells were thawed on ice then resuspended in 100 mL of SJM-400 buffer (50 mM Tris-Cl, pH 8.4, 1 mM EDTA, 1 mM PMSF, 400 mM NaCl). Two hundred microliters of lysozyme solution (50 mg/mL) was added and held on ice until the solution was viscous indicating cell lysis. The solution was treated with ultrasound on ice until all viscosity was removed. Insoluble material was removed by centrifugation at 17,500× *g*, 4 °C, 30 min. Solid ammonium sulfate was added slowly to the supernatant with stirring at 4 °C to 25% of saturation and stirred further for 30 min. Precipitated material was removed by centrifugation at 17,500× *g*, 4 °C, 30 min. Ammonium sulfate was added to the supernatant to 50% of saturation, stirred and centrifuged as above. The pellet was dissolved in 50 mL of Q column buffer (50 mM Tris-Cl, pH 8.4, 1 mM EDTA) and loaded onto a pre-equilibrated High-Q column (2.5 × 10 cm) (Bio-Rad, Hercules, CA, USA). The column was washed with the same buffer until the optical density (280 nm) of the effluent was zero. The column was further washed with the same buffer containing 100 mM NaCl. The bound rAra h 1 was eluted with a linear gradient of 100–250 mM NaCl in Q column buffer.

2.3. Purification of Core rAra h 1

The cells were lysed exactly as for the mature rAra h 1, except that 0.4% Triton X-100 was present in the lysis buffer. To the cleared lysate solid ammonium sulfate was added to 75% of saturation and cleared as described above. Ammonium sulfate was then added to the supernatant to 100%, stirred 30 min and the precipitate collected by centrifugation. The precipitate was dissolved in 50 mL of Q column buffer and loaded onto a High-Q column. The column was washed with the same buffer until the optical density (280 nm) of the effluent was zero. rsAra h 1was eluted with Q column buffer containing 100 mM NaCl. The eluate was adjusted to pH 7.0 with HCl and NaCl to a final concentration of 280 mM. This was loaded onto a High-S column (Bio-Rad) that was equilibrated with S column buffer (50 mM Tris-Cl, pH 7.0, 1 mM EDTA) with 280 mM NaCl. The column was washed with that buffer until the absorbance at 280 nm was zero. rsAra h 1 was eluted with 450 mM NaCl in column buffer. The High-S column was washed with buffer containing 2 M NaCl, then re-equilibrated with S column buffer with 300 mM NaCl. The eluate was diluted with S column buffer with no salt to give a final NaCl concentration of 300 mM and re-loaded on the column. Short rAra h 1 was eluted with a linear gradient of 300 mM–2 M NaCl in column buffer.

2.4. Gel Electrophoresis and Western Blots

For gel analysis protein samples were mixed with 6× sample buffer and heated to 100 °C for 10 min. Proteins were resolved on commercial 4%–20% gradient gels (Bio-Rad). The molecular weight markers for stained gels were SeeBlue+2 (Invitrogen, Carlsbad, CA, USA) and for western blots Magic Mark 2 (Invitrogen). To visualize the proteins the gels were stained with Gel-Code Blue (Pierce, Rockford, IL, USA). For western blotting, proteins were electrophoretically transferred to PVDF membranes. Membranes were blocked for 1 h at room temperature with 5% non-fat dry milk in PBST (phosphate-buffered saline, 0.05% Tween-20) followed by chicken anti-Ara h 1 (1:5000). After washing, antibody binding was detected with horseradish peroxidase-conjugated anti-chicken IgG (1:100,000) in 2% non-fat dry milk in PBST and ECL Plus chemiluminesence following the manufacturer's instructions (GE Healthcare, Piscataway, NJ, USA). A FujiFilm LAS-1000 recorded the signal.

2.5. Secondary Structure Determination

Far UV (185–250 nm) circular dichroism (CD) spectra of purified rAra h 1, rsAra h 1 and natural Ara h 1 (purified from peanut; [15]) were obtained. Samples were desalted using disposable gel filtration columns (G-25, PD-10, GE Healthcare, Uppsala, Sweden) into Milli-Q water and immediately used in CD measurements. A CD spectrum of Milli-Q water was obtained for background purposes and subtracted from each spectra. Protein concentrations were 0.1 mg/mL and spectra were obtained at room temperature with a JASCO Model J-710 spectropolarimeter (JASCO, Easton, MD, USA). Secondary structural modes were estimated from ellipticities by multiple protein secondary structure prediction and calculation programs such as K2d using a Kohonen neural network with a 2-dimensional output layer [16], Sspro [16], PSIPRED (http://bioinf.cs.ucl.ac.uk/cgi-bin/psipred), PROFsec [17,18], CDPro (http://lamar.colostate.edu/~sreeram/CDPro/main.html) [19,20]. The data from K2d are presented here.

2.6. Patient Sera

Sera were obtained from the blood of peanut allergic individuals, which were collected after informed consent at Tulane Health Science Center (New Orleans, LA, USA) in accordance with the rules and regulations of the institutional review board.

2.7. Western and Spot Blot Analysis with Patient Sera

Purified natural Ara h 1 (nAra h 1) [15] and rAra h 1 proteins (300 ngs) were either electrophoretically transferred to a nitrocellulose membrane following SDS-PAGE (western blot) using standard western blot procedure or spotted directly onto a PVDF membrane and allowed to dry. For IgE western blots and spot blots, membranes were blocked in 2% Blotto (2% dry milk dissolved into phosphate buffered saline (PBS) containing 0.5% TWEEN (PBST) for 30 min and incubated overnight with 1:10 dilution in PBST of patient sera from convincing history/documented peanut allergic individuals with positive skin prick test results. After the incubation patient sera, the membranes were washed 3 times with PBST and incubated with anti-human IgE conjugated to horseradish peroxidase

(HRP)-labeled secondary antibody at 1:10,000, diluted in 2% blotto for 30 min. The membrane was then washed 3 times with PBST and 2 times with PBS and incubated with ECL-Plus Western substrate (Amersham Bioscience Corp., Piscataway, NJ, USA). The signal was then visualized using a CCD camera system (Fuji Photo Film Co., Ltd., Duluth, GA, USA).

3. Results and Discussion

3.1. Expression and Purification of Mature rAra h 1

In Figure 1 is shown the full-length translation product for the Ara h 1 mRNA. The signal and leader peptides (red and blue) are removed to yield the mature, natural Ara h 1 protein (black and purple). The mature protein sequence was back translated using preferred codons for expression in *E. coli*, the gene synthesized and subsequently cloned into pET9a. Expression was tested on two clones (A and B) in strain BL21(DE3) pLysS. Following growth to mid-log phase and three-hour treatment with IPTG, virtually no induction was observed on SDS-PAGE gels of solublized whole cells (data not shown). However, reasonable induction was achieved with the same clones in BL21 (DE3) cells (Figure 2A). The arrow indicates the position of inducibly-expressed Ara h 1. Solubility of the protein was examined following lysis and centrifugation. The results are shown in Figure 2B. It appears that a little more than half of the expressed protein is in the soluble fraction (sup). In Figure 2C the identity of the expressed protein as Ara h 1 is confirmed by western blot analysis. There appears to be a few proteolytic cleavage products also in both the soluble and insoluble fractions.

As a first step in purification, ammonium sulfate precipitation was tested. In Figure 3A is shown the soluble fraction and the results of bringing the soluble fraction to 25%, 50%, 75%, and 100% of saturation in a stepwise manner. It is clear that the vast majority of the Ara h 1 precipitates with 50% ammonium sulfate, with only a small amount at 25%. Therefore, a two-step ammonium sulfate procedure with an undercut of 25% (discarded) and cut of 50% was incorporated into the purification protocol. Various ion exchange resins were tested for binding and elution using the solublized 50% ammonium sulfate precipitate as the load. High-Q was found to have the best binding and elution characteristics. In Figure 3B,C are shown the SDS-PAGE and absorbance analysis of a gradient elution of rAra h 1. The shallow gradient was 100–250 mM NaCl. Over-loading SDS-PAGE gels with peak fraction samples showed small amounts of contaminating protein (not shown). The resultant preparation is estimated to be at least 95% Ara h 1. Peak fractions were pooled and stored at −20 °C.

Figure 2. (**A**) SDS-PAGE of induced (+) and non-induced (−) cultures for clones A and B. The position of the induced mature rAra h 1 is indicated with an arrow. (**B**) SDS-PAGE of the soluble and insoluble fractions of induced cells (A and B clones) following lysis. "sup" is the soluble supernatant. "pell" is the insoluble cellular debris. (**C**) Western blot analysis with anti-Ara h 1 antibody of the same samples shown in B.

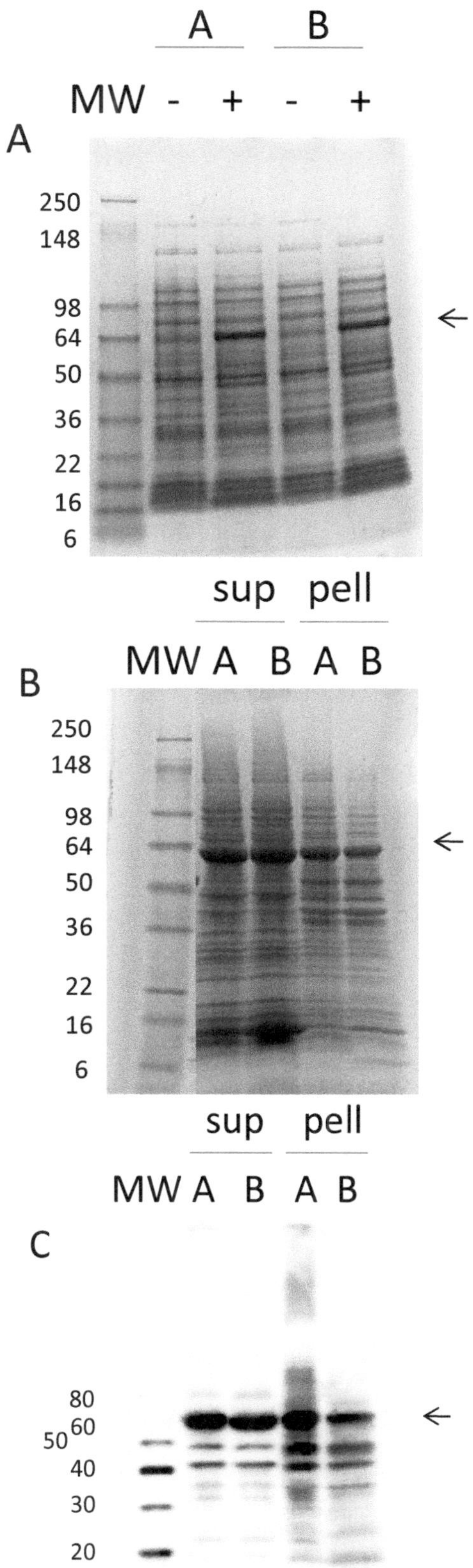

Figure 3. (**A**) SDS-PAGE of ammonium sulfate precipitation tests. "S" indicates the soluble fraction of lysed cells. (**B**) SDS-PAGE of column fractions from High-Q chromatography. (**C**) Elution profile of fractions from High-Q chromatography.

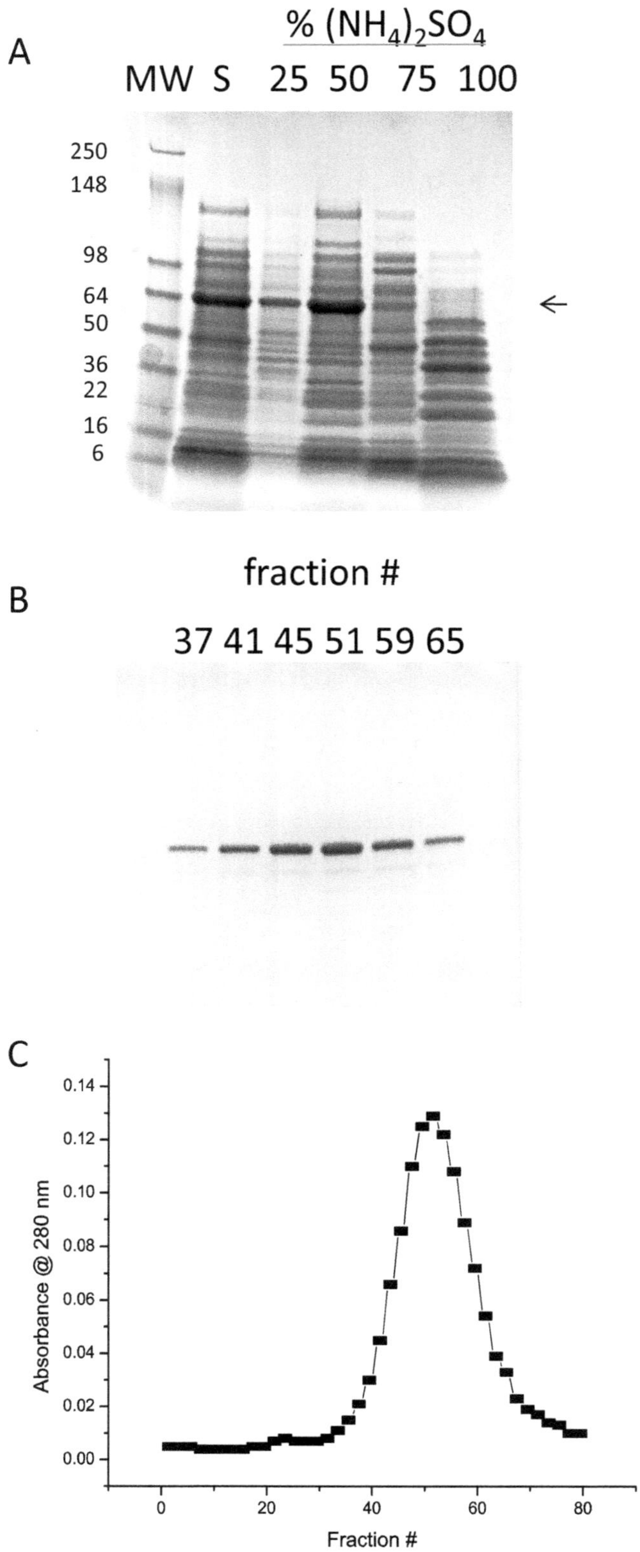

3.2. Expression and Purification of Core rAra h 1

The mature rAra h 1 purified by the procedure described above has been used for structural studies [13,14]. Neither group was able to obtain high-resolution crystals for structure determination.

Therefore, we decided to make a second expression construct containing only the core region of Ara h 1 (black and purple in Figure 1). PCR was used to amplify codons 170–586 and the product was cloned into pET9a. In Figure 4A is shown the IPTG induction test samples and the stepwise ammonium sulfate precipitation tests. Expression was seen after three hours of induction. Western blot analysis (Figure 4B) confirmed the induced protein as rsAra h 1. Ammonium sulfate precipitation properties of this shorter protein were also tested. In Figure 4A, SDS-PAGE analysis shows a band of the appropriate size in the 50%, 75%, and 100% concentrations, with the majority in the 100%. Western blot analysis (Figure 4B) revealed that the co-migrating bands in the 50% and 75% samples were not Ara h 1 and nearly all of it is in the 100% ammonium sulfate sample. As a result, the purification protocol incorporated a 75% undercut and a 100% precipitation to collect the protein.

Figure 4. (**A**) SDS-PAGE of induced (+) and uninduced (−) cells and ammonium sulfate precipitations tests. (**B**) Western blot analysis of the same samples shown in (A).

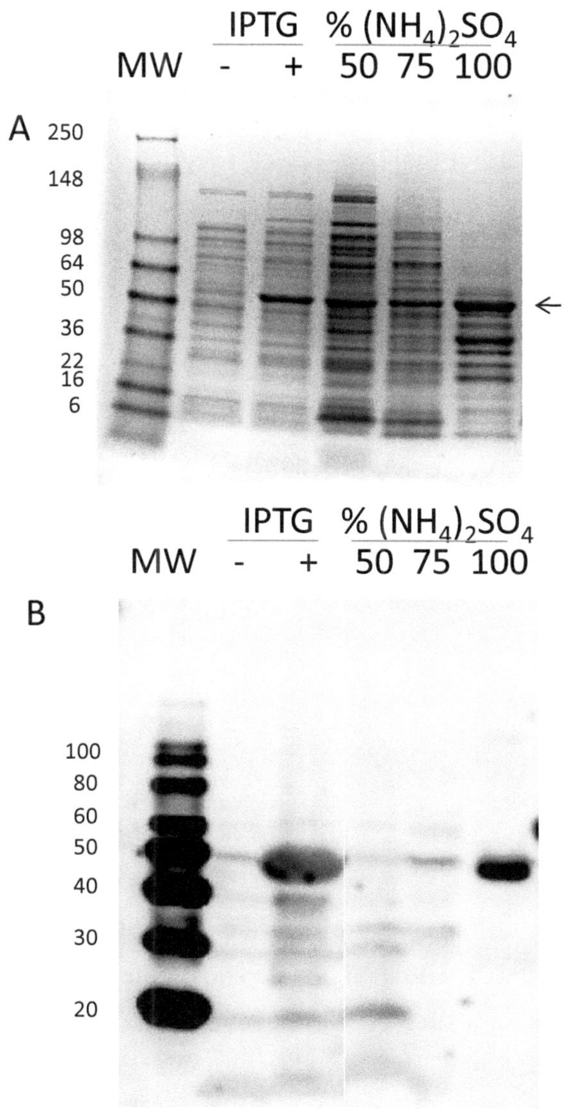

Various ion exchange resins were tested for binding and elution characteristics. In Figure 5A is shown the testing with High-Q resin. Core Ara h 1 would only bind this resin at a relatively high pH (8.4) and very low ionic strength (no salt). To take advantage of this, the purification protocol includes binding to High-Q in Q column buffer without salt and one-step elution with the same buffer with 100 mM NaCl. rsAra h 1 bound much more tightly to High-S than High-Q. However, to achieve binding the pH was reduced to 7.0. In Figure 5B is shown the step-wise elution of the protein. Little or no core protein eluted at 280 mM NaCl, but excellent elution occurred at 450 mM. Therefore, the next step in purification protocol became a load of the Q column eluate onto High-S after the pH was reduced and salt increased to 280 mM. A one-step elution with column buffer with 450 mM NaCl was applied.

Figure 5. (**A**) SDS-PAGE of step gradient fractions from chromatography of rsAra h 1 on High-Q resin. "L" indicates the material loaded on the column. "WT" indicates the wash through material that didn't bind the resin. (**B**) SDS-PAGE of step gradient fractions from chromatography of rsAra h 1 on High-S resin. "L" indicates the material loaded on the column. "WT" indicates the wash through material that didn't bind the resin.

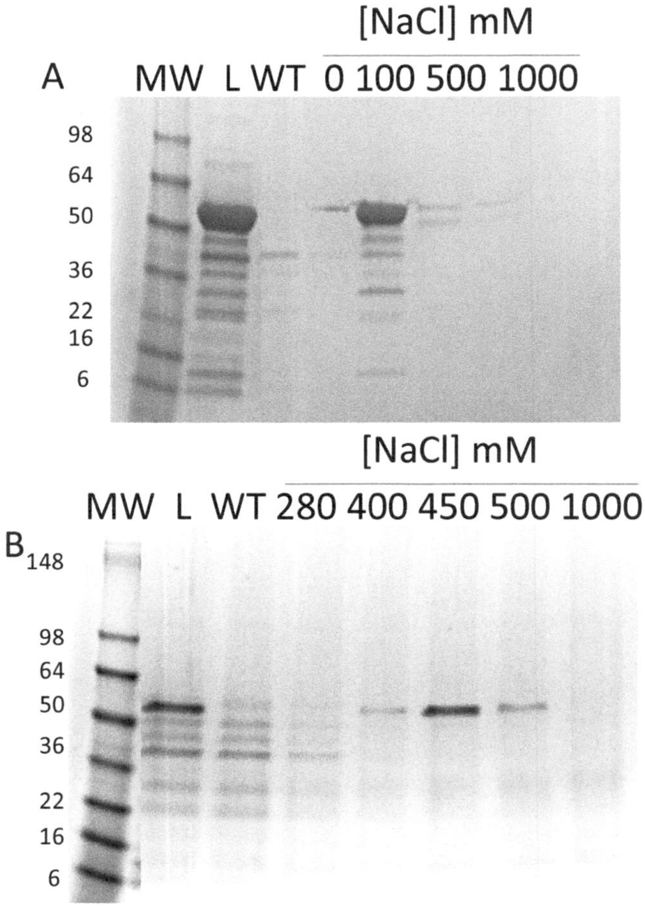

The eluted protein was fairly dilute. To concentrate it, the High-S column was regenerated and equilibrated with S column buffer with 300 mM NaCl. The eluate from the previous column was diluted with buffer containing no salt to get a final NaCl concentration of 300 mM and loaded on the column. A steep gradient of 300 mM to 2 M NaCl was used to elute the core protein. In Figure 6A,B are shown the SDS-PAGE analysis and elution profile. Small amounts of higher molecular weight material can be seen in the highly-concentrated peak fractions. Peak fractions were pooled and stored at −20 °C.

The rAra h 1 and rsAra h 1 behaved very differently in the ammonium sulfate precipitations and on the ion exchange resins. This is likely due to a dramatic change in the charged amino acids. The core protein is missing 31 positively charge residue and 27 negatively charged residues. In general, larger proteins of complexes precipitate at lower concentrations of ammonium sulfate. The core Ara h 1 is 9 kDa smaller than the natural Ara h 1. We know that the core protein forms trimers because it is the protein used for structure determination [14]. However, it has also be reported that natural Ara h 1 can form even higher oligomeric structures [21].

Figure 6. (**A**) SDS-PAGE of gradient fractions from chromatography of rsAra h 1 on High-S resin. "L" indicates the material loaded on the column. "WT" indicates the wash through material that didn't bind the resin. (**B**) Elution profile of fractions from High-S chromatography.

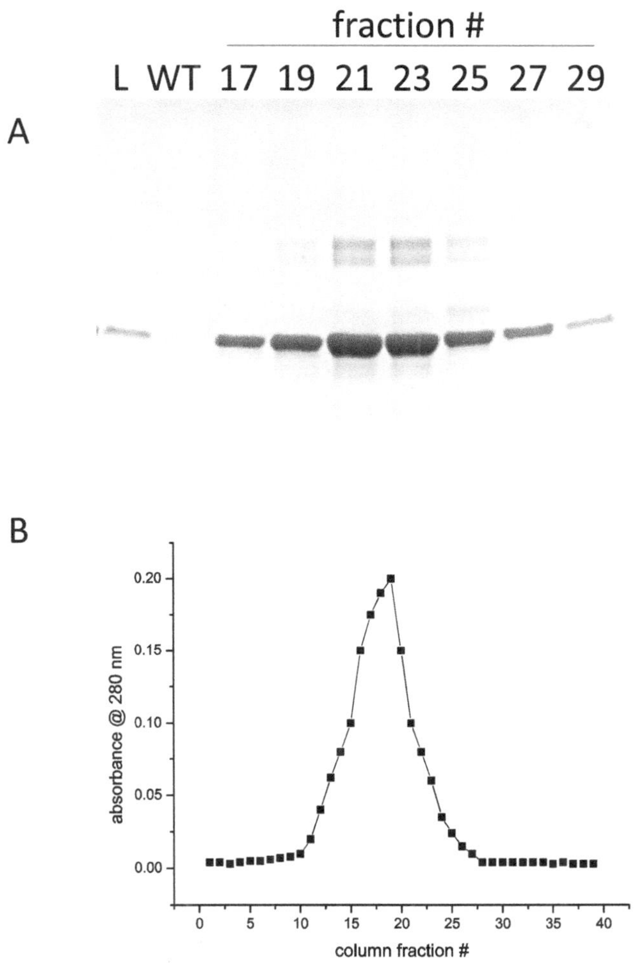

3.3. Secondary Structure Analysis of Recombinant Mature and Core Ara h 1

Far-UV CD (185–260 nm) was used to compare the secondary structure content of mature rAra h 1, core rAra h 1 and natural Ara h 1 (Figure 7). The CD spectra show that the secondary structures of all three Ara h 1 forms are very similar exhibiting minimums at ~208 nm and ~222 nm and a crossover near 200, typical characteristics of proteins with either separate α + β region or combined α/β regions. However, the large 208:222 ratio is indicative of a structure with separate α and β regions, which corroborates the Ara h 1 trimeric models based on phaseolin and β-conglycinin [8] and crystal structures [13,14]. Further analysis of the CD spectra of recombinant and natural Ara h 1 using K2d shows that all three Ara h 1 forms contain similar secondary structure elements (Figure 7, inset) and the values obtained are valid according to normalized root mean square deviation values (NRMSD) that are used for internal software control and confirmed by other software, such as CDPro [22].

Figure 7. Circular dichroism analysis of rAra h 1 (●), rsAra h 1 (■) and natural of Ara h 1 (▲). The percentage of ordered and disordered secondary structure and shown in the inset.

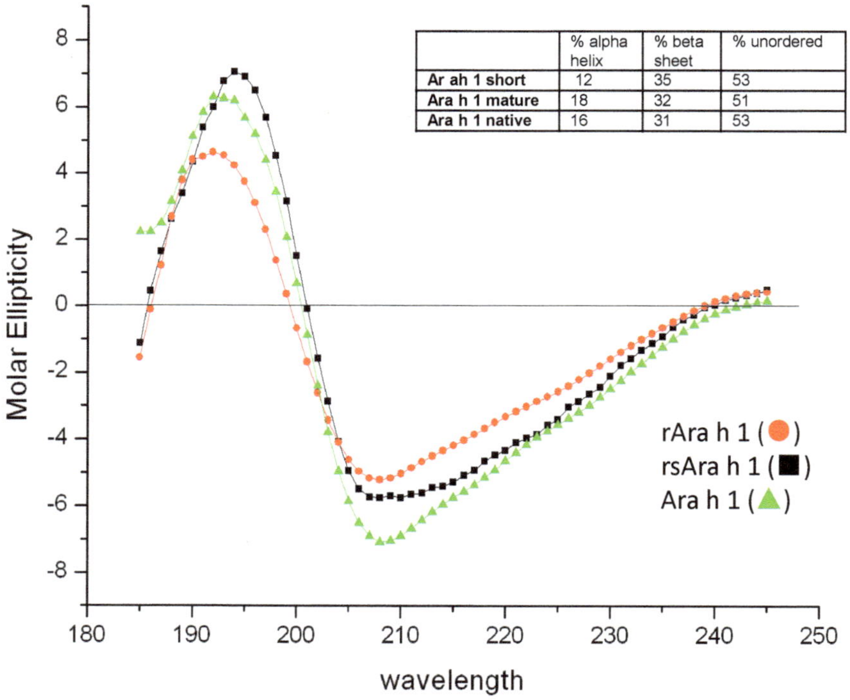

3.4. Western and Spot Blot Analysis

Western blot (on left) and spot blot analysis (on right) were performed to compare IgE binding of 12 peanut allergic sera to natural Ara h 1 (N) *versus* rAra h 1 (R) in linear and folded form (Figure 8). Of the 12 patient sera tested IgE of all patients recognized the natural Ara h 1 in both folded and unfolded forms. In the western blots, five out of 12 (41%) of the sera (6, 7, 9, 10, and 11) did not recognize the linear rAra h 1 at all or at very low levels. The IgE binding was significantly reduced for five (41%) of the sera (3, 4, 5, 8, and 12) and only two (16%) did not show a significant difference in binding (1 and 2). In the spot blot analysis all patient sera recognized both N and R Ara h 1 in folded form, but the IgE binding was significantly reduced for all patient sera with the exception of serum 12, which had increased binding to R compared to N.

Figure 8. IgE western blot and spot blot comparison of natural and recombinant Ara h 1. Natural (N) and rAra h 1 (R) were subjected to western blot (left panels) or spot blot (right panels) with sera from 12 peanut allergic individuals (1 and 2).

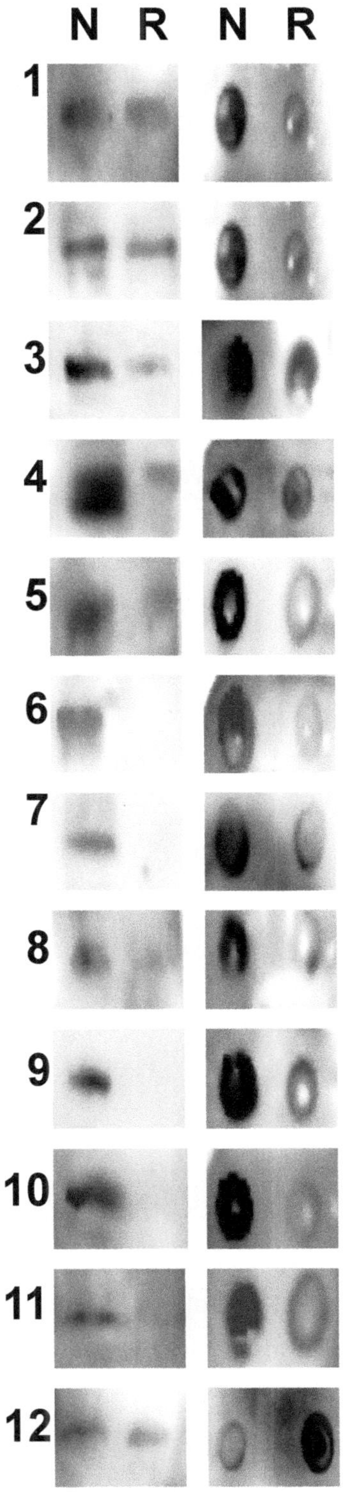

4. Conclusions

While production of recombinant Ara h 1 has been briefly described in other publications [12–14], the methods were laborious and inefficient. Here, we've optimized the DNA clone, the expression and purification for a highly efficient large-scale production of these proteins and mutants thereof without the need for any costly instrument requirements. Purification procedures described here will allow production of sufficient quantities of pure proteins with which to undertake further structural and immunological studies to understand the allergic response better. Here the folded and unfolded forms of the recombinant Ara h 1 showed lower IgE binding by the majority of patient sera tested, which indicates that while it might not be as useful as the natural Ara h 1 for diagnostics, it may provide a safer immunotherapeutic tool. The loss of reactivity with rAra h 1 *versus* natural Ara h 1 in western blot inplicates reactivity with post-translational modifications, namely advanced glycation end products [23]. Importantly, we intend to begin modifying amino acids identified from studies that have mapped and mutated the linear IgE epitopes. Those modified proteins will be tested for IgE binding. Hypoallergenic Ara h 1 could be a very important tool in our overall goal of reducing allergy to peanut.

Acknowledgements

We thank Jay Shockey and P-K Chang for critical review of the manuscript. This work was supported by intramural funding from the Agricultural Research Service, United States Department of Agriculture. Mention of a trade name, proprietary product, or specific equipment does not constitute a guarantee or warranty by the US Department of Agriculture and does not imply approval of the product to the exclusion of others that may be available.

Author Contributions

SJM conceived of the study and contributed to writing the paper. BKH designed the clones, did the cloning, worked on the purifications and wrote the paper. JKM did the western and spot blots. SR contributed to the purifications. JBN performed the CD analysis.

Conflicts of Interest

The authors declare no conflicts of interest.

References

1. Table 13 Peanut Area, Yield, and Production. Available online: http://apps.fas.usda.gov/psdonline/psdreport.aspx?hidReportRetrievalName=BVS&hidReportRetrievalID=918&hidReportRetrievalTemplateID=1#ancor (accessed on 10 December 2014).
2. Sicherer, S.H.; Munoz-Furlong, A.; Godbold, J.H.; Sampson, H.A. US prevalence of self-reported peanut, tree nut, and sesame allergy: 11-year follow-up. *J. Allergy Clin. Immunol.* **2010**, *125*, 1322–1326.
3. Rabjohn, P.; Burks, A.W.; Sampson, H.A.; Bannon, G.A. Mutational analysis of the IgE-binding epitopes of the peanut allergen, Ara h 3: A member of the glycinin family of seed-storage proteins. *J. Allergy Clin. Immun.* **1999**, *103*, S101–S101.
4. Shin, D.S.; Compadre, C.M.; Maleki, S.J.; Kopper, R.A.; Sampson, H.; Huang, S.K.; Burks, A.W.; Bannon, G.A. Biochemical and structural analysis of the IgE binding sites on Ara h1, an abundant and highly allergenic peanut protein. *J. Biol. Chem.* **1998**, *273*, 13753–13759.

5. Stanley, J.S.; King, N.; Burks, A.W.; Huang, S.K.; Sampson, H.; Cockrell, G.; Helm, R.M.; West, C.M.; Bannon, G.A. Identification and mutational analysis of the immunodominant IgE binding epitopes of the major peanut allergen Ara h 2. *Arch. Biochem. Biophys.* **1997**, *342*, 244–253.
6. Maleki, S.J.; Chung, S.Y.; Champagne, E.T.; Raufman, J.P. The effects of roasting on the allergenic properties of peanut proteins. *J. Allergy Clin. Immunol.* **2000**, *106*, 763–768.
7. Cong, Y.J.; Lou, F.; Xue, W.T.; Li, L.F.; Chen, M.H. Characterisation of the IgE-binding immunodominant epitopes on Ara h1. *Food Agric. Immunol.* **2008**, *19*, 175–185.
8. Shreffler, W.G.; Beyer, K.; Chu, T.H.; Burks, A.W.; Sampson, H.A. Microarray immunoassay: Association of clinical history, *in vitro* IgE function, and heterogeneity of allergenic peanut epitopes. *J. Allergy Clin. Immunol.* **2004**, *113*, 776–782.
9. Stanley, J.S.; Helm, R.M.; Cockrell, G.; Burks, A.W.; Bannon, G.A. Peanut hypersensitivity. IgE binding characteristics of a recombinant Ara h I protein. *Adv. Exp. Med. Biol.* **1996**, *409*, 213–216.
10. Wichers, H.J.; de Beijer, T.; Savalkoul, H.F.J.; van Amerongen, A. The major peanut allergen Ara h 1 and its cleaved-off *N*-terminal peptide; possible implications for peanut allergen detection. *J. Agric. Food Chem.* **2004**, *52*, 4903–4907.
11. Burks, A.W.; Shin, D.; Cockrell, G.; Stanley, J.S.; Helm, R.M.; Bannon, G.A. Mapping and mutational analysis of the IgE-binding epitopes on Ara h 1, a legume vicilin proteins and a major allergen in peanut hypersensitivity. *Eur. J. Biochem.* **1997**, *245*, 334–339.
12. Cabanos, C.; Urabe, H.; Masuda, T.; Tandang-Silvas, M.R.; Utsumi, S.; Mikami, B.; Maruyama, N. Crystallization and preliminary X-ray analysis of the major peanut allergen Ara h 1 core region. *Acta Crystallogr. Sect. F Struct. Biol. Cryst. Commun.* **2010**, *66*, 1071–1073.
13. Cabanos, C.; Urabe, H.; Tandang-Silvas, M.R.; Utsumi, S.; Mikami, B.; Maruyama, N. Crystal structure of the major peanut allergen Ara h 1. *Mol. Immunol.* **2011**, *49*, 115–123.
14. Chruszcz, M.; Maleki, S.J.; Majorek, K.A.; Demas, M.; Bublin, M.; Solberg, R.; Hurlburt, B.K.; Ruan, S.; Mattison, C.P.; Breiteneder, H.; Minor, W. Structural and immunologic characterization of Ara h 1, a major peanut allergen. *J. Biol. Chem.* **2011**, *286*, 39318–39327.
15. Maleki, S.J.; Kopper, R.A.; Shin, D.S.; Park, C.W.; Compadre, C.M.; Sampson, H.; Burks, A.W.; Bannon, G.A. Structure of the major peanut allergen Ara h 1 may protect IgE-binding epitopes from degradation. *J. Immunol.* **2000**, *164*, 5844–5849.
16. Pollastri, G.; Przybylski, D.; Rost, B.; Baldi, P. Improving the prediction of protein secondary structure in three and eight classes using recurrent neural networks and profiles. *Proteins* **2002**, *47*, 228–235.
17. Rost, B.; Sander, C. Improved prediction of protein secondary structure by use of sequence profiles and neural networks. *Proc. Natl. Acad. Sci. USA* **1993**, *90*, 7558–7562.
18. Rost, B.; Sander, C. Secondary structure prediction of all-helical proteins in two states. *Protein Eng.* **1993**, *6*, 831–836.
19. Sreerama, N.; Woody, R.W. A self-consistent method for the analysis of protein secondary structure from circular dichroism. *Anal. Biochem.* **1993**, *209*, 32–44.
20. Sreerama, N.; Woody, R.W. Protein secondary structure from circular dichroism spectroscopy. Combining variable selection principle and cluster analysis with neural network, ridge regression and self-consistent methods. *J. Mol. Biol.* **1994**, *242*, 497–507.

21. Van Boxtel, E.L.; Koppelman, S.J.; van den Broek, L.A.M.; Gruppen, H. Determination of pepsin-susceptible and pepsin-resistant epitopes in native and heat-treated peanut allergen Ara h 1. *J. Agric. Food Chem.* **2008**, *56*, 2223–2230.
22. Andrade, M.A.; Chacon, P.; Merelo, J.J.; Moran, F. Evaluation of secondary structure of proteins from UV circular dichroism spectra using an unsupervised learning neural network. *Protein Eng.* **1993**, *6*, 383–390.
23. Mueller, G.A.; Maleki, S.J.; Johnson, K.; Hurlburt, B.K.; Cheng, H.; Ruan, S.; Nesbit, J.B.; Pomes, A.; Edwards, L.L.; Schorzman, A. Identification of Maillard reaction products on peanut allergens that influence binding to the receptor for advanced glycation end products. *Allergy* **2013**, *68*, 1546–1554.

11

Effect of Radio Frequency Heating on Yoghurt, I: Technological Applicability, Shelf-Life and Sensorial Quality

Caroline Siefarth [1,2], Thi Bich Thao Tran [2], Peter Mittermaier [2], Thomas Pfeiffer [2] and Andrea Buettner [1,2,*]

[1] Department of Chemistry and Pharmacy, Emil Fischer Centre, Friedrich-Alexander Universität Erlangen-Nürnberg, Schuhstr. 19, Erlangen 91052, Germany; E-Mail: caroline.siefarth@fau.de

[2] Fraunhofer Institute for Process Engineering and Packaging (IVV), Giggenhauser Str. 35, Freising 85354, Germany; E-Mails: thi_bich_thao_tran@cargill.com (T.B.T.T.); peter.mittermaier@ivv.fraunhofer.de (P.M.); thomas.pfeiffer@ivv.fraunhofer.de (T.P.)

* Author to whom correspondence should be addressed; E-Mail: andrea.buettner@ivv.fraunhofer.de

---

Abstract: This first part of a two-part study focuses on the technical feasibility of applying radio frequency (RF) heating at different temperatures (58, 65 and 72 °C) to a stirred yoghurt gel after culturing. For comparison, a convectional (CV) heating process was also applied. The aim was to increase the yoghurt shelf-life, by preventing post-acidification and the growth of yeasts and molds. At the same time, the viability of lactic acid bacteria (LAB) was investigated in view of existing legal regulations for yoghurts. Additionally, the yoghurt color, aroma and taste profiles were evaluated. It was found that the application of RF heating was effective for the rapid attainment of homogenous temperatures of 58 and 65 °C, respectively. For RF heating at 72 °C, it was not possible to establish a stable heating regime, since in some cases, there was significant overheating followed by strong contraction of the yoghurt curd and whey separation. Hence, it was decided not to continue with the RF heating series at 72 °C. In the case of CV heating, heat transfer limitations were observed, and prolonged heating was required. Nevertheless, we showed that yeasts and molds survived neither the RF nor CV heat treatment. LAB were found not to survive the CV treatment, but these beneficial microorganisms were still present in reduced numbers after RF heating to 58 and 65 °C. This important observation is most likely related to the mildness of RF treatment. While post-acidification was not observed on yoghurt storage, slight color changes occurred after heat treatment. The flavor and taste profiles were shown to be similar to the reference product. Furthermore, a trained sensory panel was not able to distinguish between, for example, the reference yoghurt and the RF 65 °C sample by triangular testing ($\alpha = 5\%$), showing the potential of novel strategies for further improvements of heat-treated yoghurt.

Keywords: heating; pH; radio frequency; sensory; shelf-life; storage; yoghurt

---

Abbreviations

ANOVA, analysis of variance; APA, aroma profile analysis; CFU, colony forming units; CV, convectional heating; ε_r, relative permittivity/relative dielectric constant; LAB, lactic acid bacteria; n.s., not significant; RF, radio frequency heating; SD, standard deviation; sig., significant.

1. Introduction

Thermal treatment is a common and important strategy in the dairy industry for inactivating microorganisms and enzymes and, thus, guaranteeing safe products throughout the predicted shelf-life. However, traditional thermal treatments rely on heat transfer by conduction and convection, resulting in relatively long heating-up times, depending on the respective food matrix. These limitations can lead to strong physicochemical changes within the product, resulting in sensorial and textural modifications, as well as potentially decreasing the nutritive value. Hence, the dairy industry is always searching for new technologies. Over the last few decades, new technologies have been described in scientific publications, but many have not been broadly transferred to manufacturing processes in the food industry. One of these techniques is the radio frequency (RF) heating of foods with common frequencies of 13.56 MHz and 27.12 MHz. RF heating was first described in the middle of the last century in the context of thawing and curing meat [1–3]. The advantage of electromagnetic heating is its ability to generate heat inside the food material by orientation polarization of dipoles, such as water, or the forced movement of ions [4]. In this way, the limitations of conventional heat transfer and heat diffusion are overcome and very rapid heating becomes possible, at least in principle. However, RF heating installations in most cases use very high voltages and are prone to electric flashovers. More recently, RF heating was applied to bottled or packaged food in a water bath equipped with electrodes, as described by Bach [5] and Felke *et al.* [6]. By using water, with its high relative permittivity or so-called relative dielectric constant (ε_r), instead of air as the dielectric field transfer medium between electrodes and food packages, the electrode voltage can be much reduced without reducing the heating rate. This fact minimizes the risk of electric flashovers. Moreover, the ε_r of water is closer to that of food materials than the ε_r of air. Exposing food to the electric field in a dielectric environment with similar ε_r, field concentrations at the edges and corners of packages, which cause local overheating, can be avoided and heating is more uniform. In addition, the preheated water acts as a thermal buffer and provides additional temperature uniformity to the surface of the food package. In order to avoid the absorption of RF energy by the water, de-ionized water is used.

Microwave heating has also been used to heat milk products, and this has resulted in improved sensorial characteristics compared to conventional heating on a laboratory scale [7]. However, microwave fields with the commonly used frequency of 2.45 GHz have a limited ability to penetrate larger food volumes. The limited penetration is described by the penetration depth parameter. This is the distance from the food surface into the food volume at which 63% of the electromagnetic power has already been absorbed. The penetration depth of microwave fields of 2.45 GHz into milk or yoghurt is in the order of 1 cm, while, according to Felke *et al.* [6], in the case of 27.12 MHz RF fields, the penetration depths into food materials is about 20 cm and, thus, leads to the more uniform heating of products of a larger diameter. A further disadvantage of microwave fields compared to 27.12 MHz RF fields is the presence of patterns of constructive and destructive wave interference inside the food, which leads to patterns of hot and cold spots. In the case of RF fields, the waves are so long, that interference is not relevant on the scale of a food item or a food package.

Due to the demand of consumers for minimally processed, but safe foods, fuelling the need for constant technological progress, heating using radio frequencies was chosen as a promising technological approach for the present study. The aim was to perform a feasibility study on the technological applicability of RF heating on stirred yoghurt gels after culturing.

Generally, yoghurt is consumed not only because of its high nutritive value and appealing organoleptic properties, but also due to its health-promoting effects. Especially, the living microflora of lactic acid bacteria has a positive influence on human digestion (probiotic effect). Moreover, bioactive peptides that are released, e.g., during fermentation as a result of the enzymatic cleavage of milk proteins, are known to have various effects on human health (biogenic effect) [8]. According to the Code of Federal Regulations [9], yoghurt may be heat-treated to destroy viable microorganisms to extend shelf-life. Nevertheless, if dairy ingredients are heat-treated after culturing, then the name of the food must be followed by the parenthetical phrase "heat treated after culturing". To increase the shelf-life of yoghurt products beyond three to four weeks at refrigeration temperatures, Kessler [10] describes several temperature ranges that are sufficient for the pasteurization of the yoghurt curds: 65 to 75 °C and holding times of 30 to 60 s. Based on this, a comparable approach was chosen in the present study for post-heating stirred yoghurt filled in glass jars. Three temperature regimes were tested: 58, 65 and 72 °C. In detail, we aimed to prolong the shelf-life by mild heat treatment using an adapted heating technology, reducing the microbial numbers, while, at the same time, maintaining the yoghurt's sensorial and textural profiles as close as possible to those of the stirred yoghurt reference. RF heating was expected to result in shorter heating-up times compared to convectional (CV) heating and, thus, to impart less damage or deterioration with respect to favorable aspects, such as the sensory properties of the products. A further goal of our study was to determine whether RF heating can achieve the same required homogeneous temperature distribution in the yoghurt curd, as expected from CV heating. Accordingly, for comparison, the yoghurts were also heated via CV treatment using steam as the heat carrier medium. The properties of the treated products (changes in pH, color and sensory properties) were then compared to the stirred yoghurt reference without any additional heat treatment. Additionally, in view of the microbial aspects and the related effects on yoghurt shelf-life, potential changes in microbial numbers (lactic acid bacteria (LAB), yeasts and molds) were investigated.

A second, separate part of this feasibility study will focus on the changes in microstructure and texture caused by a post-fermentative heat treatment [11].

2. Experimental Section

2.1. Yoghurt

Commercially available plain stirred yoghurt (4.4% protein, 5.4% carbohydrates, 3.8% fat) was purchased from a local supermarket. The yoghurt was filled in 500 mL glass jars, which were closed with a metal screw cap (twist-off, 70 mm) and were filled up to about 100 mm. The jars had a diameter of 87 mm. All yoghurts were purchased at the same time and had the same best-before date, which was four weeks after purchase. According to the manufacturer disclosure, the yoghurts had a shelf-life of four weeks and, thus, were delivered immediately after manufacturing.

For further investigations, a set-style yoghurt reference (3.8% fat, 3.4% protein, 4.4% carbohydrates), matured in a cup (150 mL), was purchased from a local supermarket and stored with the other yoghurt samples.

2.2. Heat Treatment

Post-fermentative heat-treatments were performed within three days after the samples were purchased. Prior to heating, the yoghurt samples were pre-tempered to a controlled starting temperature of 40 ± 1 °C. The following target temperatures were applied: 58, 65 and 72 °C. Immediately after heating, the glass jars were placed in an ice water bath for rapid cooling. Before and after heat treatment (Week 0), the samples were placed into a cooling chamber at refrigeration temperatures of 8 ± 1 °C up to a storage period of five weeks. Thus, the storage period exceeded the yoghurt best-before date by one week. In weekly cycles, various quality parameters were investigated, as explained in the respective sections (Sections 2.3–2.6). Samples without any additional heating step were held as references for comparative investigations.

2.2.1. Radio Frequency (RF) Treatment

RF heating of yoghurt was performed in an RF water bath on pilot scale, as displayed in Figure 1. RF power was provided by a tube generator with 27.12 MHz operating frequency and 16 kW rated power (Type 16000 K, Kiefel AG, Freilassing, Germany) together with an impedance matching network by the same manufacturer. The power flow to the products was controlled by automatically adjusting the electrode voltage, which was measured directly at the electrodes. Applied electrode voltages during the yoghurt heating trials were between 2.2 and 2.4 kV; with resulting field intensities in the water bath between 17.8 $kV \cdot m^{-1}$ and 19.2 $kV \cdot m^{-1}$. By the additional control of exposure time, the average temperature in the product at the end of RF heating could be determined with a tolerance of ±2 °C.

Figure 1. Schematic set-up of the RF water bath, modified from Felke *et al.* [6].

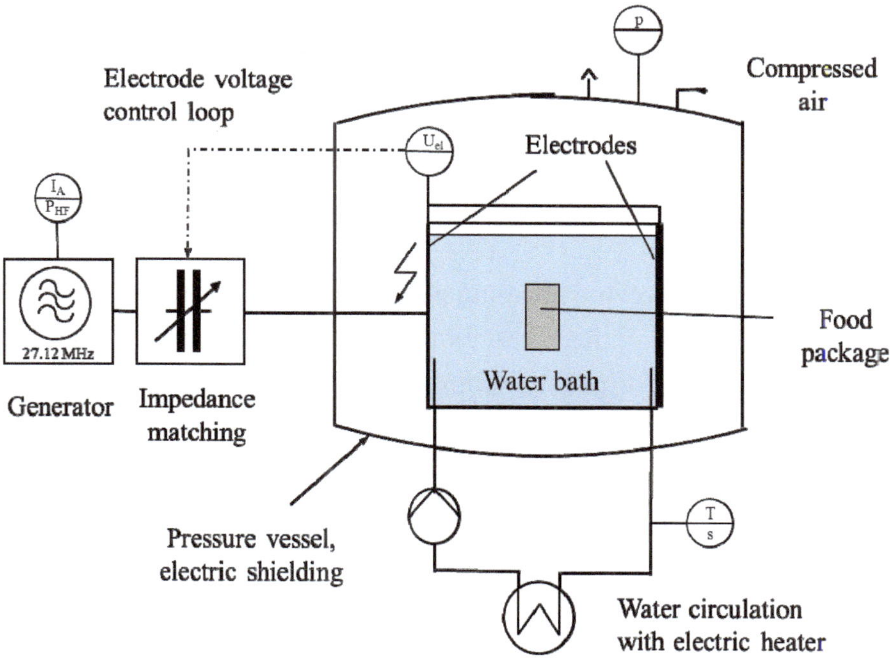

Heating procedure: Two glass jars with tightly closed screw caps placed on a support were immersed in the water bath under atmospheric pressure (Figure 2). The water bath was preheated to a temperature slightly above the process target temperature. RF exposure started immediately after immersion and continued for a preset time. The electrode voltages and the exposure times had been determined in previous heating experiments according to the respective target temperatures. Samples for further microbiological, structural, textural and sensory evaluation were exposed to an additional temperature holding time and were not subjected to inline-temperature measurements. However, to guarantee temperature control, three to four spare samples were opened for supervisory purposes during a sample heating series. All details on the specific heating parameters used to produce the product samples are compiled in Table 1.

Figure 2. Experimental set-up: Yoghurt jars on a support immersed between the electrodes of the RF water bath (here: inline-temperature measurement with fiber-optic temperature sensors).

Table 1. Heating procedure of RF heating in an RF water bath.

Process target temperature	Electrode voltage	RF exposure time	Holding time [1] (RF switched off)	Water bath temperature
58 °C	2.3 ± 0.1 kV	60 s	60 s	63 ± 1 °C
65 °C	2.3 ± 0.1 kV	90 s	60 s	70 ± 1 °C
72 °C	2.3 ± 0.1 kV	120 s	60 s	77 ± 2 °C

[1] At the process target temperature.

Temperature measurements: In previous heating experiments, inline-measurements with fiber-optic temperature sensors (Fotemp 4 fiber-optical thermometer with TS5 0.5-mm diameter sensors, Optocon AG, Dresden, Germany) were performed to adjust the electrode voltage and the exposure times according to the respective target temperatures. The fibers were introduced through septa glued onto holes, which were drilled into the metal screw caps of the jars. Temperatures were measured in the center of the jars, as well as near the walls, at mid-height of the filled yoghurt mass.

During the main series of the heating experiments, manual temperature measurements were performed directly after heating with a fast response thermocouple (0.5 mm diameter, type K, Thermocoax GmbH, Stapelfeld, Germany) connected to an electronic thermometer (Ebro TFN 520-SMP, Ebro GmbH, Ingolstadt, Germany). Temperatures were measured at different height levels within the yoghurt mass in the jars. In addition, the average temperature of the yoghurt filling was measured after mixing with a plastic spoon, because of its low thermal capacity.

The electric conductivity ($mS \cdot cm^{-1}$) of the stirred reference yoghurt was measured with a laboratory conductometer (Innolab Cond level 2, WTW GmbH, Weilheim, Germany) within a temperature range of 25 to 55 °C.

2.2.2. Convectional (CV) Treatment

CV heating of yoghurts was performed in a convection oven (Rational white efficiency SCC WE61, Rational AG, Landsberg am Lech, Germany) in steam mode with 100% saturated steam.

Heating procedure: Five glass jars with tightly closed screw caps were placed on a fence that was positioned in the center of the convection oven. The oven temperature was set to slightly above the process target temperatures. An inline-thermocouple was used to control the products temperature, with its tip placed in the core of one additional yoghurt jar. Therefore, the lid of the jar was perforated to yield a center hole; this temperature control sample was discarded afterwards. The samples were exposed to CV heating until the process target temperature was reached and an additional temperature holding time was maintained. All details on the respective CV heating parameters used to produce the product samples are compiled in Table 2.

Temperature measurements: In addition to the inline-thermocouple, the entire temperature profile was recorded once by the use of a fast response thermocouple (0.5 mm diameter, type K, Thermocoax GmbH, Stapelfeld, Germany), placed in the core of the product and connected to a data logging instrument (ALMEMO 2890-9, Ahlborn Mess- und Regelungstechnik GmbH, Holzkirchen, Germany).

Table 2. Heating procedure of convectional (CV) heating in a convection oven (100% saturated steam).

Process target temperature	CV exposure time	Holding time [1]	Oven/ steam Temperature
58 °C	60.0 ± 0.5 min	60 s	63 °C
65 °C	61.5 ± 4.5 min	60 s	70 °C
72 °C	58.0 ± 0.5 min	60 s	77 °C

[1] At process target temperature.

2.3. Microbiological Investigations

The microbial numbers of yeasts and molds, as well as lactic acid bacteria (LAB) were investigated directly after (post-)manufacturing of the different yoghurt samples (Week 0) and after a storage period of five weeks. No further inoculation of the samples with LAB or other microorganisms was performed, and the jars were opened just immediately before the samples were taken for further microbial investigations. For the preparation of sample dilutions, 10 g of yoghurt were dissolved in 90 mL of Ringer's solution (25%, OXOID LTD., Hampshire, U.K.). Ringer's solution was also used for any further dilution steps. In detail, for each treatment and temperature regime, three independent yoghurt jars were randomly selected and three dilution steps were prepared in threefold repetition to analyze microbial numbers.

Yeasts and molds: yeast extract glucose chloramphenicol (YGC) agar (Merck KGaA) was selected as the medium for the viability investigation. The incubation was performed at 25 °C aerobically for three to five days. LAB: de Man, Rogosa and Sharpe (MRS) agar (Merck KGaA) was used for enumeration, and aerobic incubation was performed at 37 °C for 72 h. Microbial numbers were expressed as colony forming units (CFU) g^{-1}.

2.4. pH and Color Measurement

The determination of the pH of the yoghurt samples was performed over the entire storage period (Weeks 0, 2, 4 and 5) using a pH meter (pH 538, WTW GmbH, Weilheim, Germany) together with a pH electrode (BlueLine 11pH, SI Analytics GmbH, Mainz, Germany) and temperature electrode (TFK 325, WTW GmbH). Instrumental color analysis was carried out directly after manufacturing (Week 0) by the use of a Chroma-meter CR-300 (Konika Minolta Inc., Marunouchi, Japan) with a DP-301 data processor. Calibration was performed on a white standard (CR-A43, Konika Minolta Inc.). The stirred samples were filled in Petri dishes, and the surface was sleeked. Each sample was analyzed at ten different points above the surface, and the color was expressed in $L^*a^*b^*$ mode, in which L^* represents the lightness value and a^* and b^* values the chromaticity coordinates. For color and pH measurements, the samples were analyzed in triplicate.

2.5. Aroma and Taste Profile Analysis

Stirred yoghurt samples (20 mL) were filled into sensory glass beakers (140 mL, J. Weck GmbH u. Co. KG, Wehr, Germany) and closed with a lid. Sensory analyses were performed in a sensory panel room at 21 ± 1°C. Trained panelists (n = 12, male/female, age 24 to 45) with normal olfactory and gustatory function participated in the sensory sessions and exhibited no known illness at the time of examination. Prior to this study, the assessors were recruited in weekly training sessions in the recognition of about 100 selected odor-active compounds according to their odor qualities by means of an in-house developed flavor language. The order of the presentation of the different yoghurts was randomized, and no information on the purpose of the experiment or the composition of the samples was given to the panelists. The results were averaged for each attribute. Sensory analyses were performed up to Week 4 of storage (best-before date) in intervals of two weeks (Weeks 0, 2 and 4).

Aroma profile analysis (APA): in the first session, the panelists were asked to evaluate the yoghurt gels (retronasal), and the named odor attributes of the different products were collected. Attributes that were detected by more than 50% of the panelists were selected for subsequent evaluations. In subsequent sessions, the panelists were asked to score the perceived retronasal intensities of the selected attributes on a seven-point-scale from 0 (*no perception*) to 3 (*strong perception*) in increments of 0.5.

Taste Profile Analysis: in addition to the APA, the panelists were requested to evaluate the following taste attributes: sweet, sour, salty and bitter. The panelists had to score the attributes' intensities on a visual analogue scale from 0 (*not perceivable*) to 10 (*strongly perceivable*).

2.6. Triangle Test

A triangle test was performed according to DIN EN ISO 4120:2007 [12]. Two triangles of the following yoghurt gels were tested: reference *vs.* RF 65 °C and RF 58 °C *vs.* CV 58 °C. During the tests, the panelists had their eyes bandaged to avoid any influence of the possible differences in the yoghurts' appearance. Twelve panelists evaluated each triad in duplicate, leading to 24 evaluations in total. The test was performed after a storage period of four weeks (best-before date).

2.7. Statistical Analysis

Statistical analyses were performed by the use of the software OriginPro 9G (OriginLab Co., Northampton, MA, USA) and Statistica 10 (StatSoft Europe GmbH, Hamburg, Germany), respectively. For all groups of data, one-way analysis of variances (ANOVA) and Fisher LSD post-hoc testing were carried out to elaborate differences between the differently treated yoghurts and during storage (repeated measures one-way ANOVA). The level of statistical significance was set at 5%.

3. Results and Discussion

This section presents the results of the RF and CV heating of yoghurt products after culturing and describes the technological applicability, shelf-life, pH, color and sensorial changes. The details of the textural and microstructural changes are given elsewhere in the second part of this two-part study [11].

3.1. Technological Applicability of the RF Heating of Yoghurt Gels

To get initial information about the electrical properties of the plain stirred reference yoghurt, its electric conductivity was measured. As can be seen in Figure 3, the electric conductivity of the yoghurt was directly proportional to the temperature. Generally, the measured electric conductivity was slightly higher than the literature values for milk, most likely due to the contribution of lactic acid to conductivity [11].

Figure 3. Electric conductivity of the plain stirred reference yoghurt.

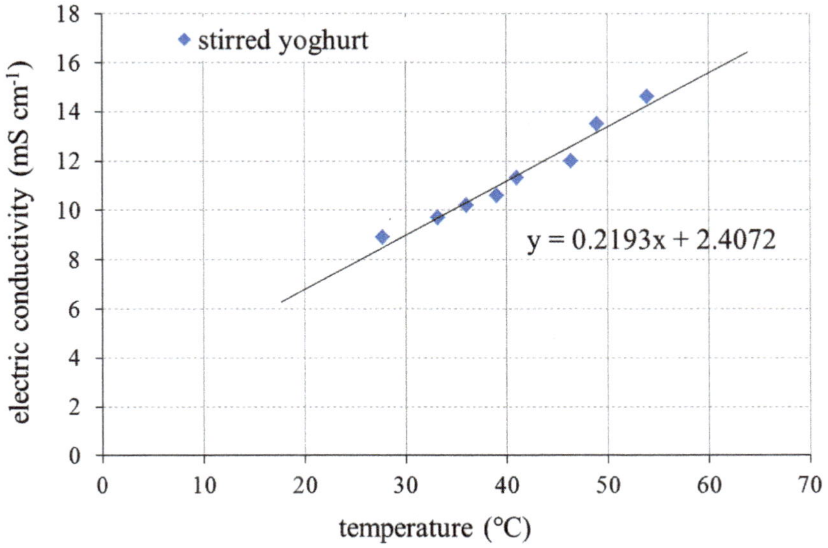

RF heating was carried out as described in Section 2.2.1, and three different temperature regimes were applied, which were chosen based on the death line of vegetative cells [10]. Starting at a temperature of 40 °C, 60 s were necessary to reach 58 °C, 90 s to obtain 65 °C and 120 s to achieve 72 °C. Thus, the heating rate for all temperature regimes was 0.28 ± 0.02 $K \cdot s^{-1}$ (*cf.* Figure 4).

Figure 4. RF heat treatment: the temperature-time profile for heating a stirred yoghurt (in glass jars) to 65 °C in an RF water bath (T = 70 °C).

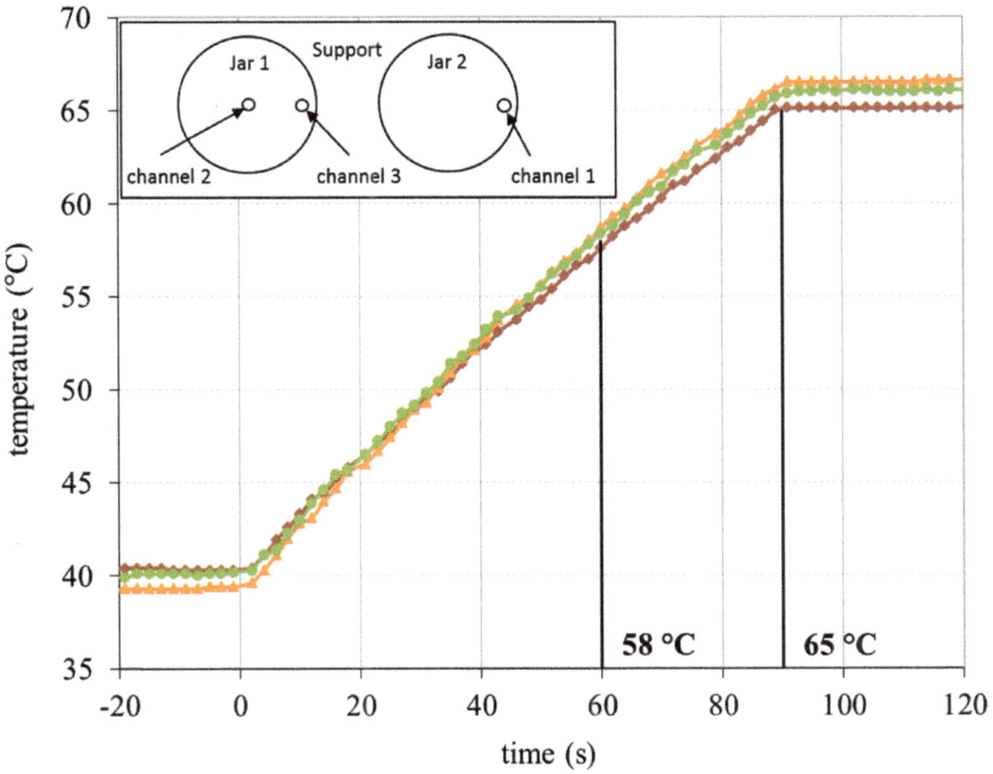

Besides a fast heating rate, temperatures of 58 and 65 °C could also be applied very homogeneously to stirred yoghurt in an RF water bath. After heating yoghurt jars to 72 °C, in some cases, there was significant overheating followed by the strong contraction of the yoghurt curd and whey separation. It was not possible to establish a stable heating regime for a 72 °C process. The reason for the heating instability is not clear. A possible explanation could be a sudden change in the dielectric properties above 70 °C, which provoked faster heating. It was therefore decided not to undertake an RF heating series at 72 °C, and this is the reason why no temperature-time profile at 72 °C is shown in Figure 4. Further details concerning syneresis and the texture analysis of the few yogurt samples that were heated to 72 °C are described in the second part of this two-part study [11].

In the frame of this feasibility study, experiments were performed on a small pilot scale. However, RF technology with power ratings of up to several 100 kW are currently implemented on the production scale for manufacturing processes in general industries, including the food industry. Furthermore, large pilot scale equipment of the RF water bath heating process used in this study exists, and its economic feasibility has been estimated [13]. Although investment costs are higher for the RF heater compared to conventional technologies, the energy costs are comparable, even when taking into account the electricity usage compared to the fossil fuel consumption of conventional heaters. Notably, energy costs with emerging RF power generators constructed with semi-conductor technology are expected to fall, due to their higher efficiency, and they are more robust than traditional tube generators.

3.2. Technological Applicability of CV Heating of Yoghurt Gels

Temperatures of 58, 65 and 72 °C were also applied to stirred yoghurt in a convection oven. Unlike with RF heating, heat transfer limitations were encountered in the convection oven. This was even so when the heat energy was transferred using steam as the heat carrier medium rather than air. The result was that the heating rate was comparatively low, with the heat curve showing a slowly ascending sigmoidal behavior. Unlike with RF heating, different heating rates for each temperature regime were observed: 0.30, 0.41 and 0.55 $K \cdot min^{-1}$ (58, 65 and 72 °C), respectively (Figure 5).

Figure 5. CV heat treatment: temperature-time profile for heating a stirred yoghurt (in glass jars) to 65 °C in a convection oven in steam mode (100%, T = 70 °C).

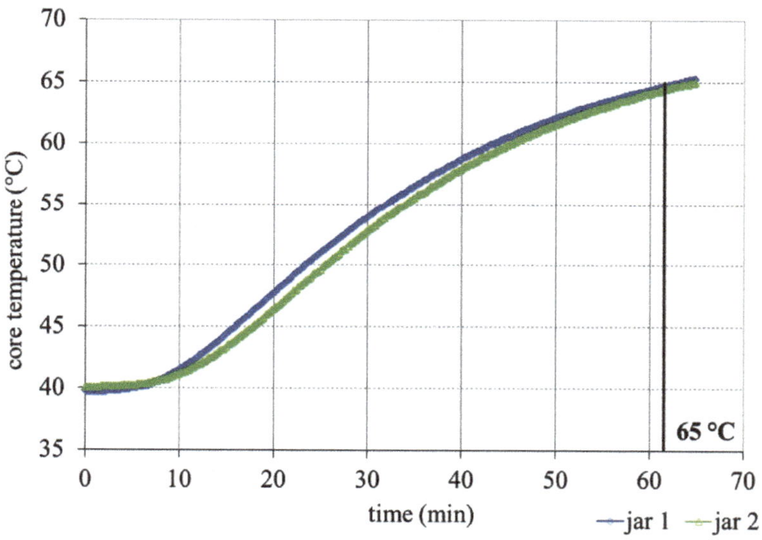

While difficulties were reported for the dielectric heating of yoghurt gels to 72 °C, CV heating was successfully applied to all temperature regimes. However, the aim of the present study was to improve yoghurt shelf-life, while maintaining its high quality with regard to living LAB and the expected texture and sensorial properties. As significant textural changes occurred in the CV 72 °C samples, as reported in the second part of this study [11], these products were produced in the same small batch numbers as the RF 72 °C samples.

3.3. Microbiological Investigations

Bach [5] was the first and, to the best of our knowledge, the only scientist who has applied mild temperatures to yoghurts filled in plastic containers via electromagnetic fields. Nevertheless, no detailed information was provided about microbial numbers.

Thus, LAB, as well as yeasts and molds were characterized in the heat-treated yoghurt products of the present study. Microbiological investigations were performed directly after manufacture and also after a storage period of five weeks. The results are shown in Figure 6.

Figure 6. Colony forming units (CFU) g^{-1} at Week 0 (grey) and Week 5 (black): (**A**) yeasts and molds, (**B**) lactic acid bacteria (LAB). CFU g^{-1} are expressed as mean values ± standard deviation (SD), with ">" denoting "more than" or "<" denoting "less than" the proposed CFU g^{-1}.

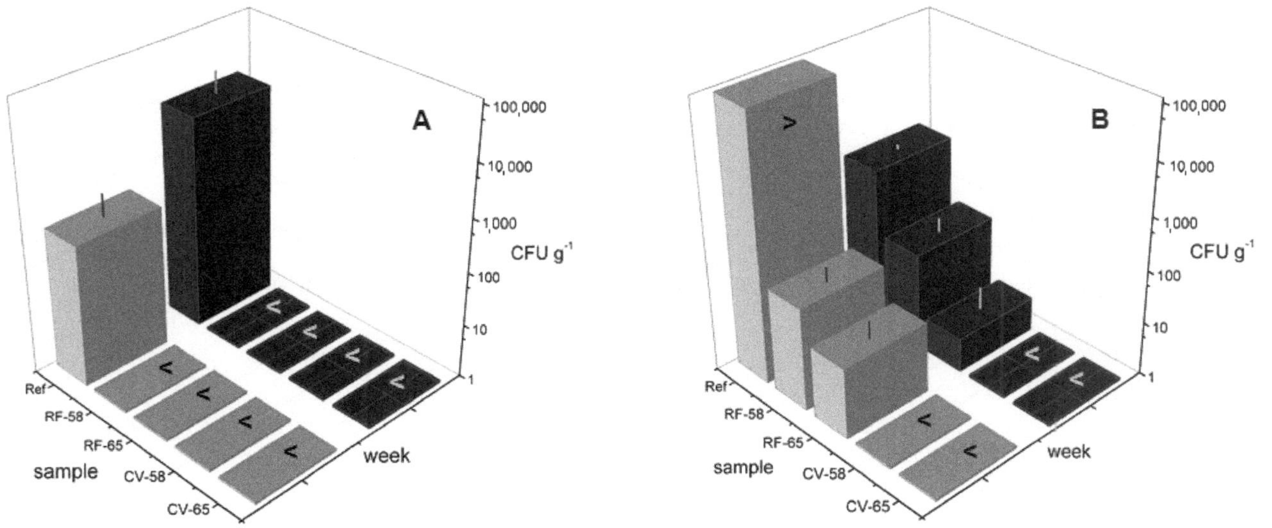

In Figure 6, the reference yoghurt without any additional heat treatment (Ref) demonstrates the initial population (Week 0) of yeasts and molds (A) and LAB (B). Generally, a reduction in microbial numbers was observed after any additional heat treatment of the yoghurt (Figure 6). In detail, yeasts and molds did not survive post-fermentative RF and CV heat treatments, even at 58 °C. On the other hand, LAB were found to partially survive RF heating to 58 and 65 °C, while they were inactivated by CV heating at the same temperatures. The decrease in LAB numbers was in good agreement with the temperature and time of heat treatment (thermal stress).

Regarding RF and microwave processing, there is an ongoing discussion about the possible non-thermal effects on microorganisms due to exposure to an electrical alternating field. It was shown by Heddleson and Doores [14] that the inactivation of microorganisms can be attributed to the heat energy developed within the product, which was also described by Datta and Davidson [4] on comparing the inactivation curves of microorganisms from RF, microwave and conventional heating. These observations are further supported by our data on LAB numbers, which are in good agreement with the degree of thermal stress applied to the yoghurts. Accordingly, our work shows that RF is a mild heating method.

Generally, these results are in agreement with the literature findings on the thermal lability of these microorganisms (*cf.* Section 1). Kessler [10] described that mild heat treatment at 50 to 55 °C for several minutes might be sufficient for fermented milk products in view of the targeted destruction of yeasts, molds and rods that produce much acid and whose proteases are able to break down proteins, yielding unwanted bitter peptides. It might be that even milder temperatures suffice to prolong yoghurt shelf-life. However, this should be investigated in more detail in a further study.

Within the framework of this feasibility study, microbial numbers were also investigated after five weeks of storage. In the case of the stirred reference yoghurt, an increase in the number of yeasts and molds was found in storage (Figure 6A), while LAB numbers were generally found to decrease in storage (Figure 6B); an observation that is in accordance with previous reports on LAB numbers [15,16]. Summarizing our results, one can conclude that the RF heating of yoghurt showed promise: an increased shelf-life can be predicted, since a reduction in microbial numbers was achieved. Hence, the potential to prolong yoghurt shelf-life is related not only to the inactivation of yeasts and molds, but also to the prevention of post-acidification due to reduced LAB numbers. This assumption follows from the work of Chandan and O'Rell [17], who described that heating to 60–65 °C stabilizes the product, so that the yoghurt shelf-life is prolonged to eight to 12 weeks (after manufacture) at a storage temperature of 12 °C or lower. In a follow-up study, the experimental design should therefore include a storage time up to at least three months.

3.4. pH and Color Measurement

The pH value was measured over the entire storage period in order to reveal possible post-acidification due to the activity of living LAB. A storage period of up to five weeks did not reveal any changes in pH. The pH stayed very constant at a value of 4.3 ± 0.1 (10.8 ± 0.6 °C) for all samples. Thus, any ongoing post-acidification can be excluded, also for the reference yoghurt, without any further heat treatment and, thus, high LAB numbers.

Additionally, color changes in the yoghurt gels due to post-fermentative heat treatment were investigated directly after manufacture (Table 3). It was found that the color between the different yoghurt matrices differed slightly. The L^* value (whiteness), negative a^* value (green fluorescence) and positive b^* value (yellow fluorescence) significantly increased after an additional heat treatment at 58 and 65°C. A direct correlation to the final temperature applied was shown for the positive b^* value and, thereby, a change towards a yellow color. Typically, yoghurt shows a blue fluorescence. This fluorescence occurs due to the conversion of riboflavin (vitamin B2), which has a characteristic yellow to green fluorescence, to colorless leucoriboflavin in the presence of LAB [18]. This explains the lower b^* value of the reference yoghurt. An increase in the yellow character of the yoghurts with the increasing temperature of the heat treatment might be due to a slight caramelization processes; Maillard reactions are less probable, due to the low pH of yoghurt and the mild temperatures applied.

Another reason might be the conversion of riboflavin to lumiflavin, which also shows a yellow-to-green fluorescence, but again, such reactions are favored in alkaline solutions [18].

To the best of our knowledge, no other literature describes the color changes in yoghurt due to an additional heating step after culturing.

Table 3. Color changes in yoghurt due to post-fermentative heat treatment and the results of ANOVA (one-way).

Color analysis	Reference	RF-treated yoghurts			CV-treated yoghurts			ANOVA results	
($L^*a^*b^*$)	yoghurt	58 °C	65 °C	72 °C	58 °C	65 °C	72 °C	F value	p value
L^*	90.36 ± 0.08 b	90.71 ± 0.05 d	90.63 ± 0.02 c,d	89.51 ± 0.10 a	90.67 ± 0.04 c,d	90.59 ± 0.09 c	90.45 ± 0 04 b	119.27	3.33×10^{-11}
a^*	−3.34 ± 0.06 a	−3.47 ± 0.04 b,c	−3.45 ± 0.04 b,c	−3.73 ± 0.02 d	−3.49 ± 0.01 c	−3.48 ± 0.01 c	−3.42 ± 0.03 b	37.99	7.05×10^{-8}
b^*	9.75 ± 0.17a	10.31 ± 0.05 b	10.27 ± 0.06 b	11.12 ± 0.12 d	10.41 ± 0.10 b	10.33 ± 0.09 b	10.66 ± 0 10 c	47.11	1.72×10^{-8}

Intensity values with different letters indicate significant differences between products ($p < 0.05$, Fisher LSD post-hoc).

3.5. Aroma and Taste Profile Analysis

The aroma profile of yoghurt is a mixture of various odor-active compounds [19–21]. Further characteristics of yoghurt flavor are, for example, the sour impressions and the specific textural properties that are jointly evaluated when tasting yoghurt [22,23]. In the present study, the heat-treated yoghurt samples and the stirred reference yoghurt were evaluated retronasally by means of an APA and taste profile analysis in order to reveal possible profile changes due to an additional heating step (Table 4).

Regarding the APA, the following descriptors were detected in the yoghurts by more than 50% of the panelists and were hence selected for subsequent evaluations: yoghurt-like, fatty, buttermilk-like, cream-like, sour cream-like and cream cheese-like. At Week 0, the odor attribute "yoghurt-like" was described for all samples as being clearly to strongly perceivable. All the other odor attributes were perceived rather weakly in the yoghurts, with just a few exceptions: the attributes, fatty, cream-like and cream cheese-like, were described for the RF 58 °C yoghurt as being slightly more strongly perceivable than in the other yoghurts. The attributes, buttermilk-like and sour cream-like, were evaluated highest in the case of the set-style reference yoghurt. A closer look at Table 4 reveals that only three attributes were evaluated as being significantly different between the products (Week 0): yoghurt-like, cream-like and cream cheese-like. In detail, those variances arose either between the references and the heat-treated samples (yoghurt-like, cream-like) or between the yoghurts heated to 58 °C and the other products (cream cheese-like). After two and four weeks of storage, these differences decreased and were no longer significant. Hence, during storage, the aroma profiles of the reference and heated yoghurts were found to be very similar to each other.

Table 4. Retronasal intensity rating of specific flavor attributes by aroma profile analysis (APA) of yoghurt gels with/without additional heat treatment after culturing and the results of ANOVA (one-way): Weeks 0, 2 and 4.

Aroma profile analysis (APA) (scale: 0–3; mean values)		Reference yoghurt	RF-treated yoghurts		CV-treated yoghurts		Set reference	ANOVA results	
			58 °C	65 °C	58 °C	65 °C		*F* value	*p* value
Week 0	yoghurt-like	2.6 a	1.6 c	2.0 b,c	2.1 b	2.2 a,b	2.3 a,b	4.27	1.86×10^{-3}
	fatty	1.3	1.6	1.2	1.1	1.1	0.9	1.22	3.10×10^{-1} *
	buttermilk-like	0.9	1.2	1.3	0.9	1.2	1.4	0.78	5.70×10^{-1} *
	cream-like	0.6 c	1.3 a	0.7 b,c	1.2 a,b	0.7 b,c	0.3 c	3.53	6.53×10^{-3}
	sour cream-like	0.6	1.1	1.1	0.8	1.1	1.4	1.28	2.80×10^{-1} *
	cream cheese-like	0.6 b	1.4 a	0.6 b	1.0 a,b	0.8 b	0.8 b	2.60	3.24×10^{-2}
Week 2	yoghurt-like	2.3	1.8	2.1	2.0	1.7	-	1.01	4.12×10^{-1} *
	fatty	1.2	0.9	1.1	0.7	0.9	-	0.74	5.71×10^{-1} *
	buttermilk-like	0.9	0.8	1.0	1.0	1.3	-	0.51	7.26×10^{-1} *
	cream-like	0.8	0.9	1.1	0.9	1.0	-	0.29	8.82×10^{-1} *
	sour cream-like	1.3	1.0	1.0	1.1	1.1	-	0.40	8.11×10^{-1} *
	cream cheese-like	0.7	1.2	1.0	1.3	1.3	-	1.08	3.73×10^{-1} *
Week 4	yoghurt-like	2.2	2.0	2.2	2.3	2.3	-	0.27	8.98×10^{-1} *
	fatty	1.0	0.8	0.7	1.0	0.8	-	0.13	9.70×10^{-1} *
	buttermilk-like	1.1	0.8	1.5	1.1	1.1	-	0.87	4.90×10^{-1} *
	cream-like	0.5	0.6	0.5	0.4	0.3	-	0.74	5.68×10^{-1} *
	sour cream-like	0.9	1.0	1.0	0.8	1.2	-	0.36	8.34×10^{-1} *
	cream cheese-like	0.7	1.0	0.9	1.2	1.1	-	0.59	6.72×10^{-1} *

* No post-hoc test necessary ($p \geq 0.10$, n.s.). Intensity values with different letters indicate significant differences between products ($p < 0.05$, Fisher LSD post-hoc).

A taste profile analysis was additionally performed with the results being shown in Table 5. At Week 0, all yoghurt products were described as being sour with medium intensity (4.5–6.3); thereby, the set-style reference was evaluated as being the sourest (6.3). The attributes sweet, salty and bitter were just slightly perceivable and, in general, no significant differences between the samples could be detected. Thus, no formation of bitter peptides in the course of the additional heating step was observed. Additionally, no significant differences between the products became evident. In storage, the taste profiles of the samples remained quite close to each other, and no significant differences arose. Accordingly, it can be concluded that the trained sensory panel not only evaluated the aroma, but also the taste profiles of all heat-treated samples as being very close to the stirred reference product.

Table 5. The intensity rating of specific taste attributes by taste profile analysis of yoghurt gels with/without additional heat treatment after culturing and the results of ANOVA (one-way): Weeks 0, 2 and 4.

Taste profile analysis (visual analogue scale, mean values)		Reference yoghurt	RF-treated yoghurts 58 °C	RF-treated yoghurts 65 °C	CV-treated yoghurts 58 °C	CV-treated yoghurts 65 °C	Set reference	ANOVA results *F* value	ANOVA results *p* value
Week 0	sweet	1.8	2.0	2.2	2.3	2.1	1.3	0.55	7.37×10^{-1} *
	sour	5.3	5.6	4.9	4.5	5.2	6.3	1.07	3.85×10^{-1} *
	salty	1.2	0.8	1.0	1.0	0.9	1.1	0.09	9.93×10^{-1} *
	bitter	0.7	0.9	1.0	0.5	0.5	0.9	0.34	8.84×10^{-1} *
Week 2	sweet	1.5	0.9	1.8	1.9	1.2	-	0.76	5.55×10^{-1} *
	sour	5.9	5.8	5.8	5.3	6.5	-	0.47	7.56×10^{-1} *
	salty	1.2	1.4	1.3	1.4	1.5	-	0.04	9.97×10^{-1} *
	bitter	0.4	0.2	0.2	0.2	0.2	-	0.26	9.05×10^{-1} *
Week 4	sweet	0.9	1.7	2.0	1.3	1.2	-	1.15	3.45×10^{-1} *
	sour	5.1	5.6	5.5	5.2	5.4	-	0.12	9.76×10^{-1} *
	salty	0.9	1.1	0.9	1.5	0.7	-	0.51	7.29×10^{-1} *
	bitter	0.5	0.5	0.6	0.5	0.7	-	0.12	9.74×10^{-1} *

* No post-hoc test necessary ($p \geq 0.10$, n.s.). Intensity values with different letters indicate significant differences between products ($p < 0.05$, Fisher LSD post-hoc).

Moreover, with the storage of the individual yoghurts, no significant changes in the flavor or taste profiles were detected. The only exceptions were the 58 °C samples (RF, CV) in the case of the attribute cream-like: its intensity was evaluated as significantly higher in the samples at Week 0, but a subsequent decrease occurred thereafter. Generally, the observed lack of changes in flavor (intensity) in storage is in good agreement with previous finding [16,24]. Martin *et al.* [24] reported very low differences in the typical retronasal and orthonasal odor profiles of yoghurt (e.g., yoghurt-like, cream-like, butter-like) during 21 days of storage. Only the attribute, cream-like, was rated significantly lower in storage, and the same was observed for the 58°C heated products in the present study. Overall, Martin *et al.* [24] reported only minor olfactory differences in the aroma profiles of such plain yoghurt products fermented using different starter cultures. In that work, only the odor attribute, yoghurt-like, was ranked with medium intensity, and all other attributes were found to be only weakly perceivable, showing, again, the overall agreement with the findings of the present study.

In the present study, triangular testing was additionally performed to determine whether general differences in the sensory profiles of the differently treated yoghurts were statistically significant at the end of their predicted shelf-life (Week 4). To avoid any influence of the yoghurts appearance on the results of the triangular evaluation (e.g., due to slight color differences, as reported in Section 3.4), the testing was performed blindfolded by the panelists. In detail, two triads were tested: reference *vs.* RF 65 °C and RF 58 °C *vs.* CV 58 °C. The first triad was chosen to compare the reference yoghurt with an RF heated yoghurt, since this treatment was generally found to be very mild in the case of 58 and 65 °C (Section 3.3). The second triad was chosen to reveal the differences between RF and CV heated products for the mildest temperature of 58 °C, because huge varieties in heating rates were found (Sections 3.1 and 3.2). In accordance with the results of the APA and the taste profile analysis, the tests confirmed that all of the samples were very close to each other with respect to their overall sensory properties: for none of the triads was the divergent sample identified with statistical significance. Out of 24 evaluations, 13 correct answers would have been necessary to indicate a significant difference between the samples at a level of significance of 5% (alpha and beta error, respectively). In the case of the triad, reference *vs.* RF 65 °C, 12 out of 24 evaluations led to the correct identification of the divergent sample; for the triad, RF 58 °C *vs.* CV 58 °C, only seven evaluations identified the divergent sample correctly. Accordingly, the samples were not distinguishable based on their overall sensory properties.

4. Conclusions

The present study evaluated the effect of additional heat treatment (CV and RF) after the fermentation process in yoghurt production. Such additional heat treatment has been little studied previously. Both heating methods showed promise concerning the homogeneity of the temperature distribution. However, RF appeared to be superior, due to a very fast heating rate of 0.3 $K \cdot s^{-1}$. This study also focused on the yoghurt shelf-life, pH, color and sensorial quality. Microbial characterization demonstrated the mildness of the RF treatment, because LAB partially survived. Nevertheless, a reduction in LAB highlighted the potential of this method for prolonging the yoghurt shelf-life, not only by inactivating the yeasts and molds, but also by preventing strong post-acidification. Whereas generally, no significant changes in pH and sensorial quality were observed, slight color changes were found to occur after heat treatment, possibly due to the onset of the caramelization processes. Based on these observations, we believe RF heating is a promising technological tool for future applications in the dairy industry (e.g., milk pasteurization, sterilization or post-heating of fermented products). However, further improvements with regard to batch-wise manufacturing are necessary in order to open up this technique to industry, such as the application of radio frequencies in a continuous process.

Acknowledgements

The present study was awarded and financially supported by the Danone Innovation Prize 2012. The authors would like to thank Danone GmbH (Haar, Germany) for this support. Furthermore, the authors would like to thank Angela López Pinar from the Friedrich-Alexander University of Erlangen-Nürnberg, Department of Chemistry and Pharmacy, as well as Monika Steinbuechl and Sigrid Regnet from the Fraunhofer IVV, Department Retention of Food Quality, for their technical support. We acknowledge support by Deutsche Forschungsgemeinschaft and Friedrich-Alexander-Universität Erlangen-Nürnberg (FAU) within the funding programme Open Access Publishing.

Conflicts of Interest

The authors declare no conflict of interest.

References

1. Sanders, H.R. Dielectric thawing of meat and meat products. *Int. J. Food Sci. Technol.* **1966**, *1*, 183–192.

2. Cathcart, W.H.; Parker, J.J. Defrosting frozen foods by high-frequency heat. *J. Food Sci.* **1946**, *11*, 341–344.
3. Pircon, L.J.; Loquercio, P.; Doty, D.M. High-frequency cooking. High-frequency heating as a unit operation in meat processing. *J. Agric. Food Chem.* **1953**, *1*, 844–847.
4. Datta, A.K.; Davidson, P.M. Microwave and radio frequency processing. *J. Food Sci.* **2000**, *65*, 32–41.
5. Bach, J. Improvements in or Relating to Methods for Effecting Uniform Heating of Organic Products. *British Patent 1 507 069*, 1978.
6. Felke, K.; Pfeiffer, T.; Eisner, P. Neues Verfahren zur schnellen und schonenden Erhitzung von verpackten Lebensmitteln: Hochfrequenzerhitzung im Wasserbad. *Chem. Ing. Tech.* **2009**, *81*, 1815–1821.
7. Villamiel, M.; Schutyser, M.A.I.; de Jong, P. Novel methods of milk processing. In *Milk Processing and Quality Management*, 1st ed.; Tamime, A.Y., Ed.; Blackwell Publishing: Oxford, UK, 2009; pp. 205–236.
8. Shiby, V.K.; Mishra, H.N. Fermented milks and milk products as functional foods—A review. *Crit. Rev. Food Sci. Nutr.* **2013**, *53*, 482–496.
9. Food and Drug Administration (FDA). CFR—Code of Federal Regulations, Title 21—Foods and Drugs, Subchapter B—Food for Human Consuption, Part 131—Milk and Cream, Subpart B—Requirements for Specific Standardized Milk and Cream, Section 131.200—Yogurt. Available online: http://www.gpo.gov/fdsys/pkg/CFR-1999-title21-vol2/pdf/CFR-1999-title21-vol2-sec131-200.pdf (accessed on 27 February 2014).
10. Kessler, H.G. *Food and Bio Process Engineering: Dairy Technology*, 5th ed.; Verlag A. Kessler: München, Germany, 2002.
11. Siefarth, C.; Tran, T.B.T.; Mittermaier, P.; Pfeiffer, T.; Buettner, A. Effect of radio frequency heating on yoghurt—II. Microstructure and texture. *Foods* 2014, submitted for publication.
12. Deutsches Institut für Normung (DIN). *DIN-EN-ISO-4120:2007-10. Sensory Analysis—Methodology—Triangle Test (German version)*; DIN: Berlin, Germany, 2007.
13. *Cost Comparison Retort* versus *RF Heater, Internal Report: 24.1.2014*; Fraunhofer Institute for Process Engineering and Packaging (IVV): Freising, Germany, 2014.
14. Heddleson, R.A.; Doores, S. Factors affecting microwave heating of foods and microwave induced destruction of foodborne pathogens—A review. *J. Food Prot.* **1994**, *57*, 1025–1037.
15. Damin, M.R.; Minowa, E.; Alcântara, M.R.; Oliveira, M.N. Effect of cold storage on culture viability and some rheological properties of fermented milk prepared with yogurt and probiotic bacteria. *J. Texture Stud.* **2008**, *39*, 40–55.
16. Salvador, A.; Fiszman, S.M. Textural and sensory characteristics of whole and skimmed flavored set-type yogurt during long storage. *J. Dairy Sci.* **2004**, *87*, 4033–4041.
17. Chandan, R.C.; O'Rell, K.R. Manufacture of various types of yogurt. In *Manufacturing Yogurt and Fermented Milks*, 1st ed.; Chandan, R.C., White, C.H., Kilara, A., Hui, Y.H., Eds.; Blackwell Publishing: Oxford, UK, 2006; pp. 211–236.
18. Töpel, A. *Chemie und Physik der Milch: Naturstoff, Rohstoff, Lebensmittel*, 1st ed.; Behr's Verlag: Hamburg, Germany, 2004; pp. 350–351.

19. Ott, A.; Fay, L.B.; Chaintreau, A. Determination and origin of the aroma impact compounds of yogurt flavor. *J. Agric. Food Chem.* **1997**, *45*, 850–858.
20. Ott, A.; Germond, J.-E.; Baumgartner, M.; Chaintreau, A. Aroma comparisons of traditional and mild yogurts: Headspace gas chromatography quantification of volatiles and origin of α-diketones. *J. Agric. Food Chem.* **1999**, *47*, 2379–2385.
21. Pollien, P.; Ott, A.; Montigon, F.; Baumgartner, M.; Muñoz-Box, R.; Chaintreau, A. Hyphenated headspace-gas chromatography-sniffing technique: Screening of impact odorants and quantitative aromagram comparisons. *J. Agric. Food Chem.* **1997**, *45*, 2630–2637.
22. Belitz, H.-D.; Grosch, W.; Schieberle, P. *Lehrbuch der Lebensmittelchemie*, 6th ed.; Springer: Berlin Heidelberg, Germany, 2008.
23. Buettner, A. *Aroma Release and Perception during Consumption of Food Taking into Account Physiological Aspects*; Verlag Dr. Hut: Munich, Germany, 2007.
24. Martin, N.C.; Skokanova, J.; Latrille, E.; Beal, C.; Corrieu, G. Influence of fermentation and storage conditions on the sensory properties of plain low fat stirred yogurts. *J. Sens. Stud.* **1999**, *14*, 139–160.

12

Effect of Radio Frequency Heating on Yoghurt, II: Microstructure and Texture

Caroline Siefarth [1,2], Thi Bich Thao Tran [2], Peter Mittermaier [2], Thomas Pfeiffer [2] and Andrea Buettner [1,2,*]

[1] Department of Chemistry and Pharmacy, Emil Fischer Centre, Friedrich-Alexander Universität Erlangen-Nürnberg, Schuhstr. 19, Erlangen 91052, Germany; E-Mail: caroline.siefarth@fau.de

[2] Fraunhofer Institute for Process Engineering and Packaging (IVV), Giggenhauser Str. 35, Freising 85354, Germany; E-Mails: thi_bich_thao_tran@cargill.com (T.B.T.T.); peter.mittermaier@ivv.fraunhofer.de (P.M.); thomas.pfeiffer@ivv.fraunhofer.de (T.P.)

* Author to whom correspondence should be addressed; E-Mail: andrea.buettner@ivv.fraunhofer.de

---

Abstract: Radio frequency (RF) heating was applied to stirred yoghurt after culturing in order to enhance the shelf-life and thereby meet industrial demands in countries where the distribution cold chain cannot be implicitly guaranteed. In parallel, a convectional (CV) heating process was also tested. In order to meet consumers' expectations with regard to texture and sensory properties, the yoghurts were heated to different temperatures (58, 65 and 72 °C). This second part of our feasibility study focused on the changes in microstructure and texture caused by post-fermentative heat treatment. It was shown that there were always microstructural changes with additional heat treatment. Compared to the dense and compact casein network in the stirred reference yoghurt, network contractions and further protein aggregation were observed after heat treatment, while at the same time larger pore geometries were detected. The changes in microstructure as well as other physical and sensorial texture properties (syneresis, hardness, cohesiveness, gumminess, apparent viscosity, G', G'', homogeneity) were in good agreement with the temperature and time of the heat treatment (thermal stress). The RF heated products were found to be very similar to the stirred reference yoghurt, showing potential for further industrial development such as novel heating strategies to obtain products with prolonged shelf-life.

Keywords: heating; microstructure; radio frequency; cryo-SEM; texture; yoghurt

---

Abbreviations

A: area under curve; AD: adhesiveness (g·s); ANOVA: analysis of variance; CO: cohesiveness; cryo-SEM: cryo-scanning electron microscopy; CV: convectional heating; F: force (g); GU: gumminess (g); HD: hardness (g); ND: not detected; n.s.: not significant; RF: radio frequency heating; sig.: significant; STPA: sensory texture profile analysis; TPA: texture profile analysis.

1. Introduction

Yoghurt is a cultured milk product, typically obtained by fermentation with *Lactobacillus delbrueckii* subsp. *bulgaricus* and *Streptococcus salivarius* subsp. *thermophilus*. These bacterial strains live symbiotically and thus activate each other and promote each other's growth. Their primary function is to generate lactic acid, amongst others, by fermentation of lactose. In the course of the associated changes in the pH of yoghurt milk, casein micelles are affected, further physico-chemical interactions with other milk components occur, and an acidic milk gel is formed in which the aqueous phase and milk fat are entrapped [1,2].

Depending on the mechanical forces, for example due to stirring or pumping the yoghurt mass following gelation, the casein network can be disrupted, irreversibly leading to stirred yoghurt products with a smooth and viscoelastic texture. According to Walstra [3], the most interesting physical characteristics of either set-style or stirred yoghurt gels include their heterogeneity/homogeneity of structure, fineness, fragility, stiffness (G'), tendency to exhibit syneresis, and gel strength (breaking stress). In any case, the texture of acidic milk gels is an important quality characteristic and the presence of any hard or irregularly formed particles may affect the sensory perception and, consequently, consumer acceptance.

An additional heat treatment after fermentation can further affect the texture of yoghurt [4]. In the literature there is hardly any information about the microstructural and textural changes caused by further heat treatment of yoghurt after fermentation. Chandan & Kilara [5] mentioned a post-ripening heat treatment and claimed that it was possible, together with the use of appropriate stabilizers and a further homogenization process, to redevelop the texture and body of the yoghurt. However, a detailed study comparing such products, for example with and without stabilizer(s), has not yet appeared.

A few patents were published in the late 1970s and 1980s on the production of sterilized yoghurt products (for example, reference [6–10]). Barua & Hampton [6] and Keefer & Murray [7] aimed to produce sterile natural yoghurt having the taste and texture of natural yoghurt. In these studies, and in contrast to the approach described in a patent of Egli & Egli [8], a mix of additives was added in a single step to the product after fermentation, resulting in less complex and less time-consuming processing to obtain sterile yoghurt. Also Amrein & Widmer [9] and, more than ten years later, Güldas & Atamer [10] improved the processing step by adding just a single stabilizer to the yoghurt product after fermentation and before an additional pasteurization or sterilization. The effect of post-fermentative heat treatment without any addition of stabilizers was reported by Bach [11], who applied small temperature changes to yoghurts filled in plastic containers via electromagnetic fields. Nevertheless, besides mentioning a smooth and non-gritty texture, the authors of the above cited patents and publications generally provided no further information about the resulting yoghurt texture or about textural changes caused by the respective additional heat treatment.

In the first part of this two-part study, the applicability of post-fermentative radio frequency (RF) heating and its effect on the shelf-life, pH, color, aroma, and taste of yoghurt were discussed [12]. It was found that living lactic acid bacteria were reduced in number but still present in the RF heated yoghurts, while yeasts and moulds were no longer detected. Also, the aroma and taste profiles were found to be very similar to the reference yoghurt, overall supporting the potential to prolong the yoghurts' shelf-life.

Based on these observations, this second part of the study investigates the effect of gentle post-fermentative heat treatment on yoghurt microstructure and texture. Following Bach [11], temperatures of 58, 65 and 72 °C, respectively, were applied to a stirred reference yoghurt, either via RF heating or convectional (CV) heating in a convection oven. Microstructural changes were investigated via cryo-scanning electron microscopy (cryo-SEM), and texture changes were measured by texture profile analysis (TPA), rheometry (rotation and oscillation analysis), and measurements on whey separation (syneresis). Textural changes in the yoghurt curd on storage were analyzed over several weeks. Additionally, a trained sensory panel performed sensory texture profile analysis (STPA) during the whole storage period.

2. Experimental Section

2.1. Yoghurt

Commercially available creamed yoghurt (3.8% fat, 4.4% protein, 5.4% carbohydrates) was purchased from a local supermarket. The yoghurt was filled in 500 mL glass jars, which were closed with a metal screw cap (twist-off, 70 mm). The jars had a diameter of 87 mm and were filled up to about 100 mm. All yoghurts were purchased at the same time and had the same best-before date, which was four weeks after start of the storage period of the present study.

For further investigations, a set-style yoghurt reference (3.8% fat, 3.4% protein, 4.4% carbohydrates), matured in the cup (150 mL), was purchased from a local supermarket and stored with the other yoghurt samples.

2.2. Heat Treatment

Prior to heat treatment, the yoghurt samples were pre-tempered to 40 ± 1 °C. The target temperatures of 58, 65 and 72 °C were applied. Immediately after heating, the glass jars were placed in an ice water bath for rapid cooling. For a storage period of five weeks, the samples were placed into a cooling chamber at refrigeration temperatures of 8 ± 1 °C. Samples without any additional heating step were held as references for comparative investigations.

2.2.1. Radio Frequency (RF) Treatment

RF heating of yoghurt was performed in a RF water bath [12]. The 27.12 MHz RF generator was a free-running tube oscillator 16,000 K with 16 kW nominal power and an impedance matching network by the manufacturer Kiefel AG (Freilassing, Bavaria, Germany). The power flow to the products was controlled by automatically adjusting the electrode voltage, which was measured directly at the electrodes. Applied electrode voltages during the yoghurt heating trials were between 2.2 and 2.4 kV, the resulting field intensities in the water bath were between 17.8 and 19.2 $kV \cdot m^{-1}$. By additional control of exposition time, the temperature in the product at the end of RF heating could be determined with narrow tolerance [12].

Heating procedure: Two glass jars with tightly closed screw caps standing on a support were immersed in the water bath under atmospheric pressure. The water bath was preheated to a temperature slightly above the process target temperature (+5 K). RF exposition started immediately after immersion and continued for a pre-set duration of time. The electrode voltage and the exposition times had been determined in previous heating experiments according to the respective target temperatures and were 2.3 ± 0.1 kV for 60 s (58 °C), 90 s (65 °C), and 120 s (72 °C), respectively. Samples for further microbiological, structural, textural, and sensory evaluation were exposed to an additional temperature holding time of 60 s and were not subjected to inline-temperature measurements. Further details on the experimental set-up and temperature measurements are described in the first part of this two-part study [12].

2.2.2. Convectional (CV) Treatment

CV heating of yoghurts was performed in a convection oven (Rational white-efficiency SCC 61, Rational AG, Landsberg am Lech, Germany) in steam mode.

Heating procedure: Five glass jars with tightly closed screw caps were placed on a fence in the centre of the convection oven. The steam temperature was set to a temperature slightly above the process target temperatures (+5 K). An inline-thermocouple controlled the product temperature with its tip placed in the core of one additional yoghurt jar. Therefore, the lid of the jar was provided with a center hole, and the sample was discarded afterwards. The samples were exposed to CV heating until the process target temperature and an additional temperature holding time of 60 s were reached. For all three temperature regimes, a CV exposition time of *ca.* 60 min was necessary [12].

2.3. Structural Analysis via Cryo-Scanning Electron Microscopy (Cryo-SEM)

Directly after heat treatment, about 50 mg samples of each yoghurt and the stirred reference were collected using an ice cream scoop. These samples were flash frozen in liquid nitrogen (−196 °C) and stored at −80 °C to maintain the microstructure of the products. For cryo-preparation, a small piece (about 5 mg) was cut out of the center part of the flash-frozen yoghurts using a cryo-frozen knife. This small fraction was mounted onto a sample carrier using a cryo-glue consisting of agar-agar. Immediately after preparation, the specimen was plunged into liquid nitrogen at atmospheric pressure (−196 °C) and further cooled down to −210 °C. The frozen specimen was transferred under vacuum to a Polaron PP2000T preparation stage (Quorum Technologies Ltd., Lewes, UK) and freeze-fractured at

−150 °C with a cold scalpel blade. Subsequently, the fractured specimen was freeze-etched at −80 °C for 15 min and sputter coating was performed at −150 °C with 5 nm of platinum. The platinum-coated specimen was transferred under vacuum into an ABT-55 scanning electron microscope (ISI ABT, Akashi Beam Technology, Tokyo, Japan) using a transportable cryo-stage. On that cryo-stage, the specimen was maintained at a minimum of −130 °C and imaged at an accelerating voltage of 3 kV.

2.4. Whey Separation (Syneresis)

Modified by Kessler [13], the whey drip-off volume of intact yoghurt gels was estimated using a drainage test for set-style yoghurts. The test for syneresis was performed for 2 h at 8 ± 1 °C and over the entire storage period (Weeks 0, 2, 4, and 5). The yoghurt jars were placed upside down on a mesh with a mesh size of 1 × 1 mm. The mesh was tightly stretched on a funnel, which was placed on top of a graduated cylinder (100 mL). The whey drip-off volume was recorded in quadruplicate on independent yoghurt jars.

2.5. Textural Analysis via Texture Analyzer and Rheometer

Textural analyses were performed on the intact and stirred yoghurt gels. The texture was analyzed throughout the entire storage period of five weeks in the following intervals: Weeks 0, 2, 4, and 5. At Week 0, a set-style reference was additionally investigated. For all texture analyses, the samples were kept in the refrigerator (8 ± 1 °C) until analysis. The texture of each sample was analyzed in quadruplicate on independent yoghurt jars.

2.5.1. Intact Yoghurt Gels

The texture profiles of the intact yoghurt gels were analyzed directly in the yoghurt jars to avoid any damage or modifications of structure. Texture profile analysis (TPA) was performed with a TA.XTplus Texture Analyzer (Stable Microsystems Ltd., Surrey, UK) equipped with a 50 kg load cell. Two sequential compression events were performed with an aluminum cylinder probe (25 mm, P/25, Stable Microsystems) with a diameter that was about a third of the jar diameter to minimize side-wall effects [14]. The TPA settings modified from Kaur *et al.* [15] were 12 mm penetration at 0.5 $mm \cdot s^{-1}$ during compression as well as pre-test and post-test speed. During the measurements, the yoghurt jars were held manually on the base plate. The texture profile was analyzed via Exponent software (Stable Microsystems). Typical force-time curves, imitating the conditions in mouth after two subsequent product compressions, were obtained as shown in Figure 1.

Data on forces (F) and areas under curve (A) of the force-time curves were used to calculate the following TPA parameters as described previously by Bourne [14]: *hardness* (HD, (g)), as maximum force of first compression cycle (=F1); *cohesiveness* (CO), as ratio of the positive force areas under first and second compression and thus, parameter representing the strength of internal bonds or so called inner gel strength (=A2/A1); *adhesiveness* (AD, (g·s)), as negative force area under the first bite (=A3) and thus, parameter representing the work to pull the cylinder probe out of the sample which is correlated with removing a sample from the mouth during eating that adheres to tongue and mucosal tissues [15]; and *gumminess* (GU, (g)), as product of HD and CO and thus, the energy required to treat the food for being ready to be swallowed.

Figure 1. Force-time curve of texture profile analysis (TPA) with F—force and A—area under curve, modified from Rosenthal [16].

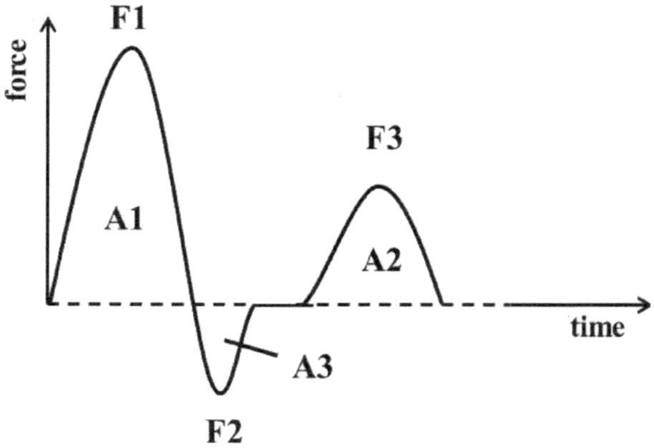

2.5.2. Stirred Yoghurt Gels

The yoghurt gels were stirred manually by use of a spoon, which was rotated very quickly. The procedure was performed until the yoghurts appeared to be very homogenous, with a number of rotations from 20 to 25.

The stirred yoghurt gels were analyzed with a Bohlin CVO50 rheometer (Malvern Instruments Ltd., Worcestershire, UK) using a parallel plate geometry (PP40) with a gap size of 1 mm. Both, rotational measurements as well as oscillatory measurements were carried out to describe the overall flow properties of the viscoelastic samples [17]. The following tests were performed according to Vercet *et al.* [18] and Afonso & Maia [19]: (1) *Flow*. Up-and-down flow curves were performed at shear rates from 10 to 290 s^{-1}. Overall, 26 data points were obtained with a delay time of 5 s. (2) *Viscosity*. Apparent viscosity (η) was determined within the linear viscoelastic section of the flow curve for 2 min. The difference in initial and final apparent viscosity was calculated using the following formula adopted from Vercet *et al.* [18]: % broken structure = [(initial viscosity − final viscosity)/initial viscosity] × 100. (3) *Oscillation*. An amplitude sweep was performed at a frequency of 5 Hz and strain amplitude from 0.1% to 100% to determine the viscoelastic region of the yoghurt samples. Further oscillation tests for the determination of storage modulus (G', representing the elastic behavior of the yoghurt sample) and loss modulus (G'', representing the viscous behavior of the yoghurt sample) were performed at 5 Hz and 2% strain, which was found to be inside the linear viscoelastic range.

2.6. Sensory Texture Profile Analysis (STPA) and Hedonic Rating

Stirred yoghurt samples (20 mL) were filled into sensory glass beakers (140 mL, J. Weck GmbH u. Co. KG, Wehr, Germany) and closed with a lid. STPA were performed in a sensory panel room at 21 ± 1 °C. Trained panelists (*n* = 13, male/female, 23 to 45 years) from the Friedrich-Alexander University of Erlangen-Nürnberg (Erlangen, Germany) and Fraunhofer IVV (Freising, Germany) participated in the STPA sessions and exhibited no known illness at the time of examination. The panelists were asked to score the textural properties on a visual analogue scale (10 cm). Prior to this study, the assessors were recruited in a training session on the evaluation of the following texture attributes adapted from Clark *et al.* [20]: *creaminess*, from watery (0) to creamy (10); *stickiness*, from sticky and thus, adherent to the mouth (0) to soft (10); *homogeneity* (mouthfeel), from grainy (0) to smooth/homogenous (10); and *visual homogeneity* ("v"), from grainy (0) to smooth/homogenous (10).

The order of presentation of the different samples was randomized and no information on the purpose of the experiment or the composition of the samples was given to the panelists. STPA were performed in the following intervals: Weeks 0, 2, and 4. Since the best-before date of the reference yoghurt expired after four weeks of storage, no further sensory testing was performed afterwards. Additionally, a set-style yoghurt reference was evaluated by the panelists at Week 0. The results for each texture attribute and sample were averaged and plotted in a box-plot diagram.

For the hedonic rating, the panelists were asked to rate the pleasantness of the differently treated yoghurt samples on a scale from 1 (dislike extremely) through 5 (neither like nor dislike) to 9 (like extremely).

2.7. Statistical Analysis

Statistical analyses were performed using the software OriginPro 9G (OriginLab Co., Northampton, MA, USA) and Statistica 10 (StatSoft Europe GmbH, Hamburg, Germany), respectively. One-way analysis of variances (ANOVA) and Fisher LSD *post-hoc* testing were carried out to elaborate differences between the differently treated yoghurts and over storage time (repeated measures ANOVA) for all groups of data. The level of statistical significance was set at 5%.

3. Results and Discussion

Gentle radio frequency (RF) heating was applied to stirred yoghurt, a matrix that represents an acidic casein gel with a very sensitive structure. In the following section, the observed microstructural and textural effects associated with this post-fermentative heat treatment are discussed and compared to the effects of the respective convectional (CV) heating process.

Thereby, it has to be noted that it was decided to abstain from the RF 72 °C heating in the heating series. As already described in the first part of this study [12], it was not possible to establish a stable heating regime for RF 72 °C products since in some cases a significant overheating followed by strong contraction of the yoghurt curd and whey separation was observed. The few 72 °C samples of the heating experiments were kept, and textural analyses were performed at Week 0. Further textural analyses, as well as microstructural analyses and sensory texture profile analyses (STPA) were only performed on yoghurts heated to 58 °C and 65 °C.

3.1. Microstructural Analysis via Cryo-SEM

Cryo-SEM is a very valuable technique in dairy research. The latter fact is related to the rapid freezing and sample handling and thus to the formation of vitreous ice with an amorphous, non-crystalline structure which prevents the occurrence of any deformation or distortion of the microstructure [21]. The results of cryo-SEM are shown in Figure 2 at two different magnifications.

Figure 2. Microstructure of heated yoghurt gels (obtained by cryo-scanning electron microscopy (SEM)) at different magnification: (**A**) Reference; (**B**) radio frequency heating (RF) 58 °C; (**C**) convectional heating (CV) 58 °C; (**D**) RF 65 °C; (**E**) CV 65 °C; with **1**—network of micellar caseins linked together in clusters, chains, and strands; **2**—pores originally confining whey; **3**—fat globules; **4**—unknown formations within the network; **5**—frozen water. Microstructural observations are based on similar observations reported by other authors [22,23].

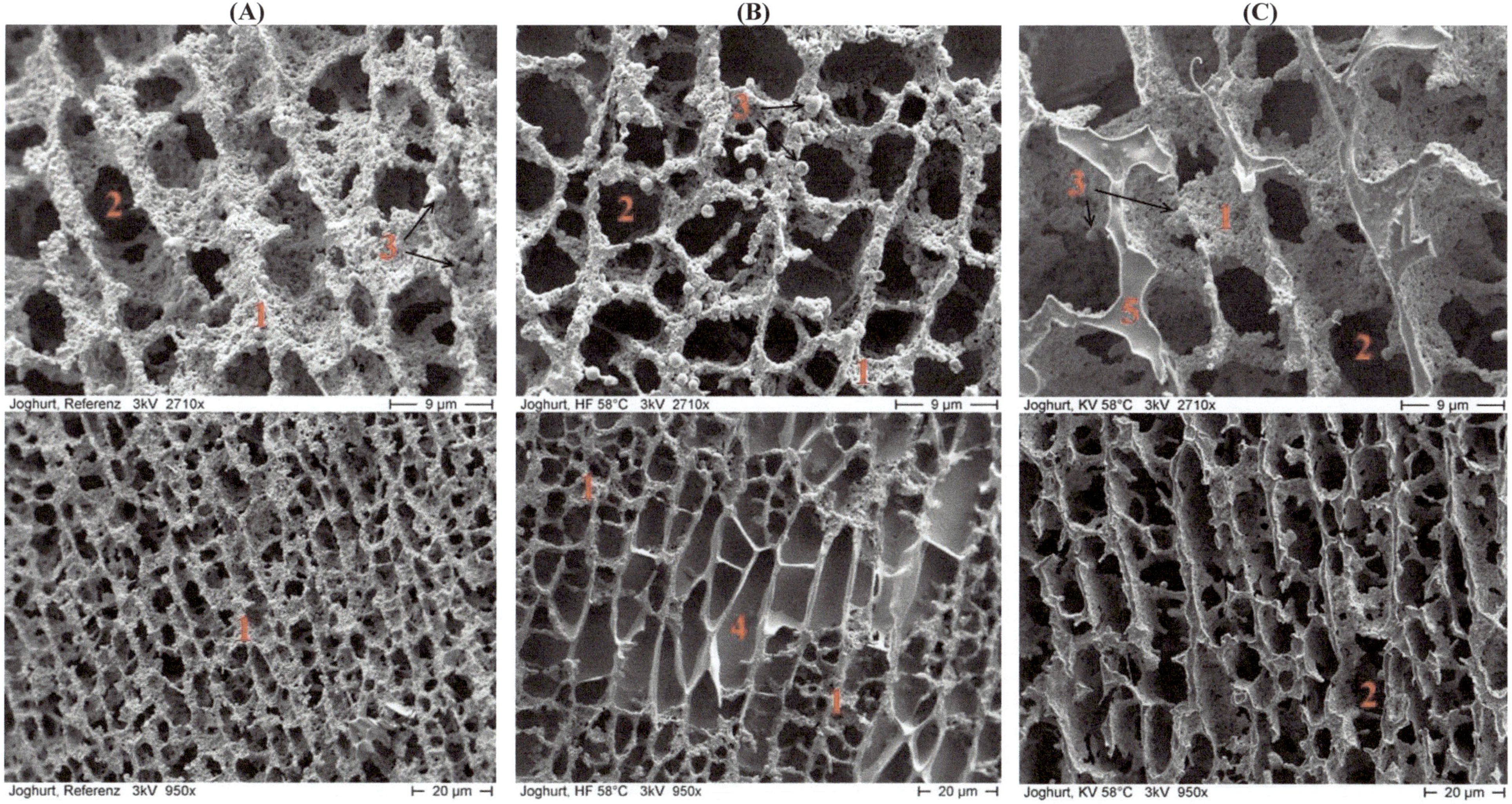

Figure 2. *Cont.*

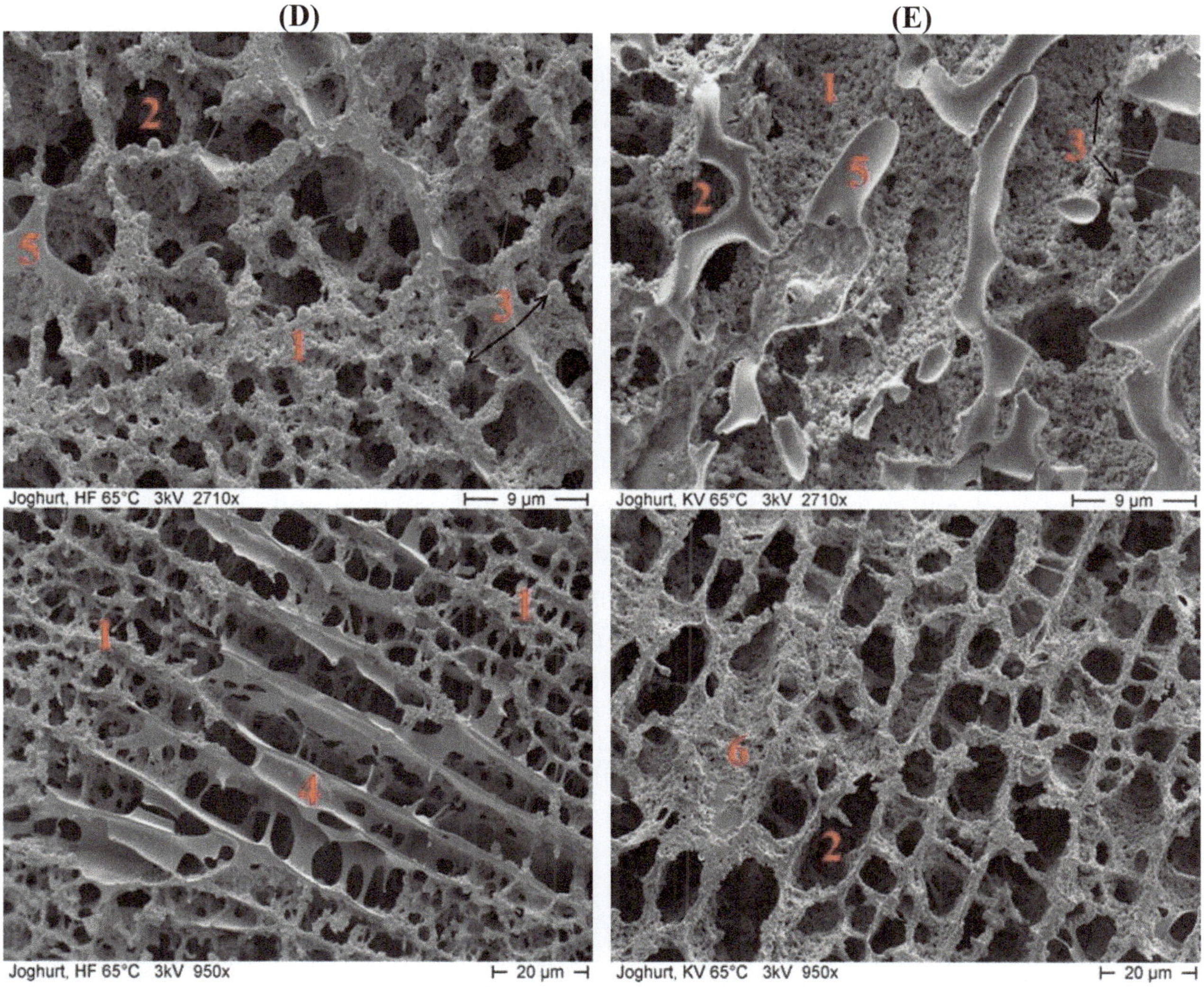

As expected from literature surveys, cryo-SEM was able to provide detailed information about changes to the yoghurt's microstructure caused by post-fermentative heat treatment (Figure 2). The gel structure of yoghurt is a three-dimensional network of micellar caseins, which are linked together in clusters, chains, and strands. According to Lee & Lucey [24], acidic gels made from heated milk are reported to be highly branched and thus possess a fine structure as can be seen in Figure 2A in the cryo-SEM image of the reference yoghurt. The network is very homogeneous, showing a well-defined and compact structure of micellar caseins (**1**) with intact pores inside the casein network originally confining the milk serum phase (whey) (**2**), as reported before by Hassan *et al.* [22] and Sandoval-Castilla *et al.* [23]. Fat globules (**3**) are also entrapped within the protein network. Due to homogenization processes of the yoghurt milk, their size was reduced to about 2 µm [25].

As can be seen in Figure 2B–E, the casein network changed due to post-fermentative heat-treatment. In the case of RF heating (Figure 2B,D), the micellar casein strands appear to undergo a contraction process, resulting in gel microstructures with larger interstitial spaces (**2**). Fat globules (**3**), on the other hand, are less entrapped in the network even if their size remains stable (~2 µm). In addition, some unknown formations (**4**) presumably filled with milk serum phase can be observed after RF heating, which are segregated from the protein network and remain intact after mild heat treatment at 58 °C

(Figure 2B, **4**) but are disrupted after treatment at 65 °C (Figure 2D, **4**). A closer look at the cryo-SEM images at lower magnification shows that these unknown formations are partially distributed across the casein network; however, overall, the network of micellar caseins still appears uniform and very similar to the reference but with larger interstitial spaces. After CV heating, the visual appearance of the casein network is very dense with even larger interstitial spaces (Figure 2C,E, **2**). Moreover, large regions of frozen water are seen on the surface layer (Figure 2C,E, **5**) as observed by Hassan *et al.* [22] who also investigated the structural properties of yoghurts and other dairy products using cryo-SEM. In contrast, small amounts of frozen water were only observed for RF heating at 65 °C (Figure 2D, **5**).

In previous studies, there were hardly any reports of physico-chemical interactions that might occur due to post-fermentative heating, thereby leading to an open structure with large(r) pores. With any network contraction, not only the milk serum in the networks' pores but also the serum entrapped in the casein network in the form of capillary water (between micellar casein chains and strands, ~10%–30% of total milk serum) and hydration water (~40% of total milk serum) [26] might be affected and emanate. Such an effect could lead to increased amounts of free serum phase, the formation of unknown conglomerates within the network after RF heating (Figure 2B,D, **4**), and/or large amounts of frozen water after CV heating (Figure 2C,E, **5**). Indeed, Töpel [26] reported that the amount of capillary-bound water was strongly related to the network's microstructure, with decreasing amounts of enclosed water with opening of the structure, thereby forming a coarse-meshed network (Figure 2B–E). Another aspect that has to be considered is (further) protein aggregation due to an additional heat treatment, as can be suspected to have occurred in Figure 2E, **6**. Certainly, on the macro-scale, the formation of white flakes (lumps) was observed by direct visual inspection in the stirred yoghurts both after RF and CV heating; these flakes increased in number and size with increasing temperature and duration of heat treatment. It is known from the manufacturing process for stirred yoghurts that textural defects, known as lumpiness or graininess, can occur [27]. Töpel [26] reported flocculation in heated milk products in the context of surface modifications of micellar caseins due to low pH and denatured whey proteins at the caseins' surface. Thus, dense aggregates as seen in cryo-SEM image Figure 2E, **6** may reveal the presence of white flakes, which are reported in the range of 1 to 5 mm [28]. Such textural defects are known to be reversible and can be removed by stirring, low-pressure homogenization or pumping the yoghurt mass through special devices such as meshes, filters, smoothening valves or colloid mills [27,29].

3.2. Whey Separation (Syneresis)

In a further part of the study, whey drainage volumes were investigated in order to observe changes within the yoghurt curds due to the post-fermentative heat treatment. All products were analyzed before stirring and thus still had their intact gel network. Data on whey separation over the whole storage period is shown in Table 1.

Whey separation was observed in all the heated yoghurts (Table 1). In general, whey separation is a commonly detected long-term storage effect of yoghurt products, since yoghurt is a dairy product with a very high percentage of water content [24]. Nevertheless, in our case the reference yoghurt showed no syneresis over the entire storage period. Accordingly, the observed syneresis effects were directly linked to the heat treatments that were performed.

As reported in the first part of our study [12], for some cases of RF 72 °C heated products a significant overheating was observed during manufacturing followed by strong contraction of the yoghurt curd and whey separation. In detail, the separated whey volume after 72 °C RF heating was greater than 50 mL and thus about 1/10 of the original yoghurt volume. With around 20 mL, the 72 °C CV samples also showed a significantly higher syneresis volume than yoghurts heated to lower temperatures at Week 0. For CV heating to 65 °C, significant differences in whey drainage volumes were observed compared to RF heating and CV heating to 58 °C (Weeks 0, 4) (Table 1). After a storage period of two weeks, significant differences in whey drainage volumes were found between the RF and CV products in general.

Via cryo-SEM, different pore radii within the casein network became visible, as well as some unknown formations presumably filled with whey (4) and frozen water (5) (Figure 2). This information about the yoghurts' microstructure explains the increasing whey drainage volumes with increasing thermal stress. As mentioned in Section 3.1, a decrease in capillary-bound water or even hydration water could have been responsible for an increase in free serum phase and thereby syneresis volumes [26].

On storage, the whey drainage volumes increased for the RF heated products and the CV heated samples showed the same tendency (F = 2.63 (n.s.), 13.99 (sig.), 11.87 (sig.), 3.57 (n.s.); $p < 0.05$ (RF 58 °C, RF 65 °C, CV 58 °C and CV 65 °C)). These results are in line with the findings of other researchers on the syneresis behavior of yoghurt products on long-term storage [30]. Presumably, structural rearrangements occur within the gel networks during storage, causing a breakage of casein strands and fusion processes of casein particles [24].

3.3. Textural Analysis via Instrumental Assessment

3.3.1. TPA on Intact Yoghurt Gels

The textures of differently treated intact yoghurts gels were evaluated using a texture analyzer by means of the following attributes: hardness (HD), cohesiveness (CO), adhesiveness (AD) and gumminess (GU), respectively. The results for different storage periods are shown in Table 2 (Week 0) and Table 3 (Weeks 2, 4, 5).

Table 1. Whey separation (mL) on second heat treatment (RF, CV) of yoghurt to different temperatures including results of analysis of variance (ANOVA) (one-way).

Whey Separation (mL)	Reference Yoghurt	RF-Treated Yoghurts			CV-Treated Yoghurts			ANOVA Results	
		58 °C	65 °C	72 °C	58 °C	65 °C	72 °C	*F* Value	*p* Value
Week 0	ND (0.0 ± 0.0)	9.5 ± 1.4a,b	6.7 ± 0.7a	56.2 ± 10.0d	7.6 ± 2.0a	13.7 ± 1.0b	19.3 ± 4.2c	74.88	1.59×10^{-10}
Week 2	ND (0.0 ± 0.0)	9.2 ± 0.2a	8.5 ± 2.7a [1]	-	18.5 ± 1.5b	20.0 ± 8.1b [1]	-	6.75	7.57×10^{-3}
Week 4	ND (0.0 ± 0.0)	9.6 ± 7.6a [1]	12.8 ± 2.9a	-	14.0 ± 1.5a	22.9 ± 3.5b	-	6.45	7.56×10^{-3}
Week 5	ND (0.0 ± 0.0)	16.1 ± 2.0b	15.4 ± 1.5a,b	-	11.2 ± 4.3a [1]	15.1 ± 1.0a,b	-	2.92	8.16×10^{-2}

[1] Coefficient of variation >25%. Intensity values with different letters indicate significant differences between products ($p < 0.05$, Fisher LSD *post-hoc*).

Table 2. Rheological parameters of intact yoghurt gels and results of ANOVA (one-way): week 0.

Texture Profile Analysis (TPA)	Reference Yoghurt	RF-Treated Yoghurts			CV-Treated Yoghurts			Set Reference	ANOVA Results	
		58 °C	65 °C	72 °C	58 °C	65 °C	72 °C	Yoghurt	*F* Value	*p* Value
Hardness HD (g)	32.961 ± 2.660a	57.271 ± 1.577b	99.066 ± 12.430d	86.605 ± 6.814c	112.142 ± 4.848e	107.038 ± 3.689de	142.693 ± 11.755f	160.936 ± 5.282g	102.38	2.22×10^{-16}
Cohesiveness CO	0.617 ± 0.016a	0.543 ± 0.040b	0.487 ± 0.007c	0.570 ± 0.011b	0.466 ± 0.019c	0.489 ± 0.015c	0.455 ± 0.025c	0.390 ± 0.031d	33.33	7.28×10^{-11}
Adhesiveness AD (g·s)	73.000 ± 12.268bc	126.000 ± 11.591d	115.522 ± 21.873d	0.003 ± 0.005a [1]	81.566 ± 18.221bc	49.537 ± 10.743b	54.465 ± 38.268b [1]	101.750 ± 20.092cd	13.03	8.25×10^{-7}
Gumminess GU (g)	20.293 ± 1.176a	33.064 ± 1.295b	48.215 ± 5.769c	49.331 ± 3.489c	51.356 ± 0.582c	51.643 ± 2.293c	64.648 ± 2.525d	62.818 ± 5.263d	60.36	1.08×10^{-13}

[1] Coefficient of variation >25%. Intensity values with different letters indicate significant differences between products ($p < 0.05$, Fisher LSD *post-hoc*).

Table 3. Rheological parameters of intact yoghurt gels and results of ANOVA (one-way): Week 2, 4, 5; with hardness (HD), cohesiveness (CO), adhesiveness (AD) and gumminess (GU).

Texture Profile Analysis (TPA)		Reference Yoghurt	RF-Treated Yoghurts		CV-Treated Yoghurts		ANOVA Results	
			58 °C	65 °C	58 °C	65 °C	*F* Value	*p* Value
Week 2	HD (g)	33.379 ± 0.962a	57.646 ± 3.626b	96.335 ± 13.917c	90.643 ± 2.195c	114.678 ± 13.599d	40.19	7.57×10^{-8}
	CO	0.596 ± 0.010a	0.532 ± 0.015b	0.487 ± 0.011c	0.495 ± 0.011c	0.480 ± 0.022c	38.90	9.44×10^{-8}
	AD (g·s)	89.750 ± 20.584a	113.000 ± 14.612b	89.750 ± 19.434a	65.000 ± 4.583a	83.750 ± 21.487a [1]	2.95	5.54×10^{-2}
	GU (g)	19.771 ± 0.357a	30.621 ± 1.274b	46.818 ± 6.050c	44.847 ± 1.453c	58.031 ± 1.445d	80.68	5.87×10^{-10}
Week 4	HD (g)	32.037 ± 0.723a	81.367 ± 2.685b	104.357 ± 17.762c	104.933 ± 7.409c	121.986 ± 11.941c	35.28	1.82×10^{-7}
	CO	0.602 ± 0.014a	0.511 ± 0.016b	0.483 ± 0.015c	0.483 ± 0.012c	0.473 ± 0.014c	46.50	2.80×10^{-8}
	AD (g·s)	69.000 ± 11.247a	120.000 ± 9.721b	77.750 ± 46.885a [1]	82.500 ± 11.715ab	69.250 ± 15.610a	2.38	9.78×10^{-2}
	GU (g)	19.265 ± 0.053a	41.598 ± 1.457b	50.247 ± 7.780c	50.596 ± 2.700c	57.523 ± 4.429c	37.19	1.28×10^{-7}
Week 5	HD (g)	42.390 ± 2.285a	80.797 ± 0.986b	98.642 ± 16.684c	101.107 ± 8.037c	135.789 ± 2.174d	51.64	1.36×10^{-8}
	CO	0.545 ± 0.004a	0.503 ± 0.012bc	0.509 ± 0.047ab	0.486 ± 0.014bc	0.466 ± 0.013c	4.95	9.53×10^{-3}
	AD (g·s)	103.000 ± 6.595bc	104.250 ± 23.626c	62.500 ± 9.206a	65.250 ± 19.227a [1]	77.250 ± 6.942ab	5.50	6.28×10^{-3}
	GU (g)	23.078 ± 1.099a	40.596 ± 0.726b	49.489 ± 4.825c	49.018 ± 2.725c	63.196 ± 0.840d	104.99	8.88×10^{-11}

[1] Coefficient of variation >25%. Intensity values with different letters indicate significant differences between products ($p < 0.05$, Fisher LSD *post-hoc*).

Directly after manufacturing, there was a significant difference between the reference and all the heat-treated yoghurt samples with respect to the TPA attributes HD, CO and GU (Table 2). For these attributes, the RF 58 °C yoghurt was closest to the reference product and differed significantly from the RF 65 °C sample and the yoghurts treated via CV heating.

In detail, the maximum force of first compression (HD) increased with increasing temperature of the post-fermentative heat treatment and with heating time (Reference < RF 58 °C < RF 65 °C < CV 58 °C and 65 °C < CV 72 °C) (Table 2). The highest HD value was obtained for the set-style reference yoghurt. Thus, it appears that an additional heat treatment alters the texture of the yoghurt from stirred towards set-style, an observation that can also be confirmed by the visual appearance of the intact yoghurt samples within the glass jar before stirring as seen in Figure 3 (here: CV 65 °C). The top view of the glass jar shows the yoghurt curd to be contracted, firm, and textured, attributes that are typical for set-style yoghurt. The only "outlier" in Table 2 is the RF 72 °C product. As described above and according to the first part of this study [12], it was not possible to establish a stable heating regime for this product temperature. In some cases, significant overheating was observed followed by strong contraction of the yoghurt curd and whey separation. Hence, unknown (structural) changes within the product are the most likely reason for the statistical deviation.

Figure 3. Visual appearance of intact yoghurt gels after post-fermentative heat treatment and before stirring (here: CV 65 °C).

The inner gel strength, represented by CO, decreased with increasing temperature and duration of heat treatment (Table 2). The set-style reference showed the lowest inner gel strength and differed significantly from the other samples. Thus, the inner gel strength of the RF and CV heated yoghurts also changed more and more towards a set-style yoghurt. The parameter GU, representing the energy required to get the food ready for swallowing again, increased with increasing temperature and heating time. Thus, increasing HD values for the acidic casein gels and decreasing inner gel strengths should impair, to a certain degree, the swallowing of the yoghurt. The parameter AD, representing the effort required to remove a sample from the mouth due to adherence, shows less aligned results in Table 2 for Week 0: with lowest values for the CV heated products, highest values for the RF heated products, and the values of the stirred and set-style references lying in between.

It can thus be concluded that the syneresis volumes increased and the yoghurts' texture contracted with increasing thermal stress of post-fermentative heat treatment, with a rise in HD and GU values and a decrease in the strength of internal bonds (CO). Roefs *et al.* [31] reported that most textural parameters for the characterization of casein gels depend on the number of bonds between the micellar caseins and thus the three-dimensional distribution of the network's chains and strands. These considerations of Roefs *et al.* [31] are in good agreement with the findings of the present study on CO values, and are further supported by the cryo-SEM images, e.g., with respect to the density of casein chains and strands or different pore radii).

Table 3 summarizes the data obtained on monitoring the textural changes of the intact yoghurt gels during storage. Once again, the reference yoghurt shows significant differences to the samples that were additionally heat-treated. The RF 58 °C product was most similar to the reference with respect to the textural attributes HD, CO and GU. With temperature and time of heat treatment, these three TPA parameters increased or decreased in the same manner as observed for the Week 0 samples.

Related to the alteration of the yoghurt gels, the HD and GU values of most samples significantly increased in storage while the CO values slightly decreased (HD: F = 19.94 (sig.), 100.41 (sig.), 0.12 (n.s.), 16.34 (sig.), 4.41 (sig.); CO: F = 32.78 (sig.), 2.34 (n.s.), 0.54 (n.s.), 2.98 (n.s.), 1.33 (n.s.); and GU: F = 11.66 (sig.), 62.95 (sig.), 0.16 (n.s.), 8.69 (sig.), 12.57 (sig.); $p < 0.05$ (Reference, RF 58 °C, RF 65 °C, CV 58 °C and CV 65 °C)). As described above for the data for Week 0, the data for the TPA parameter AD during storage did not reveal any specific trend towards a steady increase or decrease. Hence, it can be deduced that in storage there is again a further rise in the HD and GU values and a decrease in the strength of internal bonds (CO). These data on the textural changes during storage are in good agreement with previous findings reported in the literature. According to Lucey [27] and Lee & Lucey [24], shrinkage of the yoghurt network was observed during storage, accompanied by altered related properties (e.g., increasing yoghurt hardness).

3.3.2. Texture Analysis on Stirred Yoghurt Gels

Rotational measurements as well as oscillatory measurements were carried out by application of rheometry to measure the overall viscoelastic properties of the differently treated yoghurts after stirring. For rotation, up-and-down flow curves were investigated. A shear rate of 112 s^{-1} was found to be within the linear viscoelastic section of the products' different flow curves (Figure 4). Similar shear rates have been described before and these have been reported to be within the linear viscoelastic region of stirred yoghurts and can thus be used to determine their apparent viscosities [18,32]. Characteristic of yoghurt, the flow, hysteresis, and viscosity curves in Figure 4 indicate a non-Newtonian, shear thinning, thixotropic behavior of all samples. For the thixotropic behavior, the hysteresis areas in Figure 4 were observed to increase with increasing thermal stress.

Results on apparent viscosities are shown in Table 4 (Week 0) and Table 5 (Weeks 2, 4, 5). In accordance with previous reports, all yoghurt samples in Table 4 showed a decrease from initial to final apparent viscosity and thus a time-dependent behavior (e.g., reference [18]). The initial and final apparent viscosity and percental broken structure of the yoghurts significantly increased with increasing temperature regime and time of post-fermentative heat treatment. According to the observations described in Section 3.3.1 about the texture properties of intact yoghurt gels, the reference sample showed significantly different results compared to the heat-treated yoghurts, with the lowest initial and final apparent viscosities, the lowest percental broken structure, and thus the highest gel strength. The stirred RF 58 °C product was again most similar to the reference with respect to its textural properties. The set-style reference showed a low initial and final viscosity but, like the yoghurts heated to 72 °C, a high percentage of broken structure and thus comparably low gel strength.

Figure 4. Flow and viscosity curves of yoghurt reference (Ref), RF and CV heated yoghurts to different temperatures (58 °C, 65 °C, 72 °C): shear stress (black) and apparent viscosity (blue).

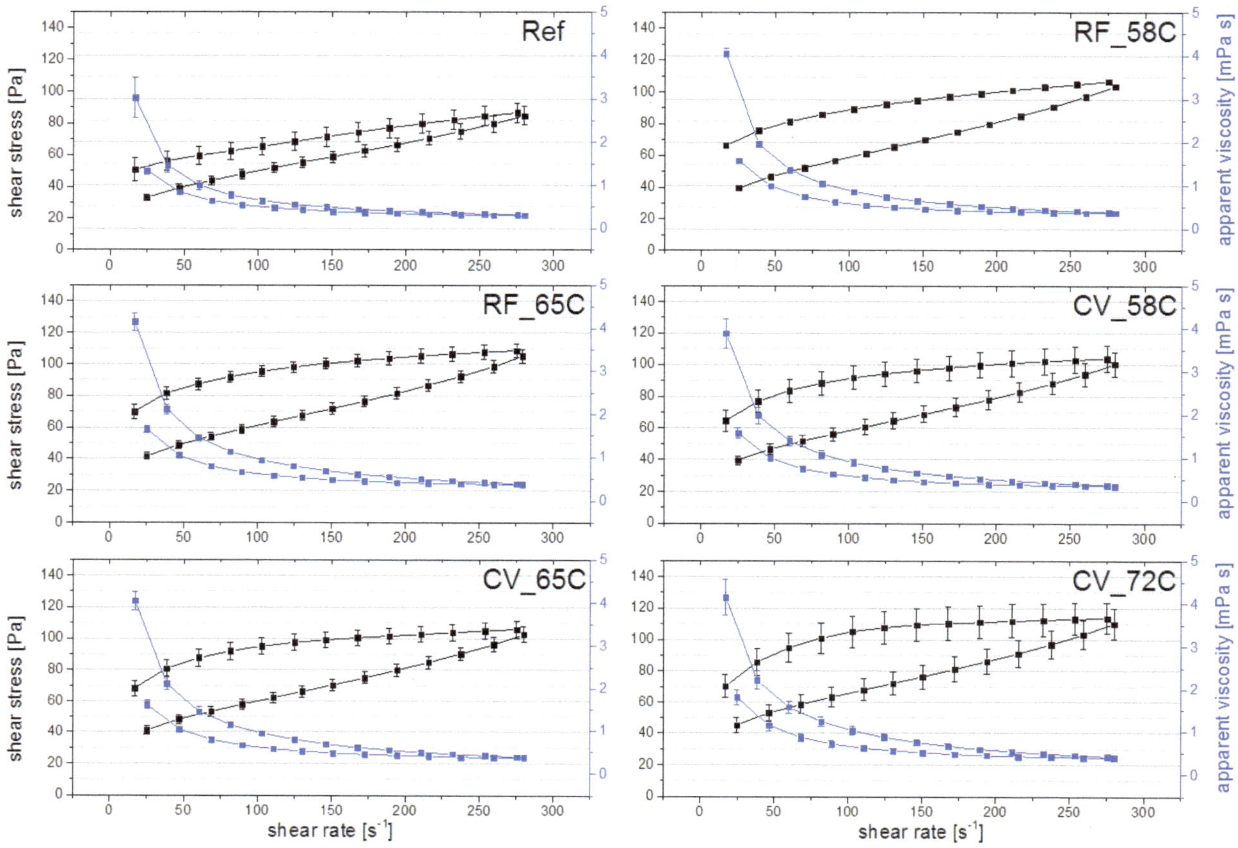

Table 5 reveals similar findings on apparent viscosity and percental broken structure over the entire storage period compared to week 0 (Table 4). The reference was significantly different compared to the heat-treated yoghurts, with the stirred RF 58 °C yoghurt being closest to the stirred reference. With increasing thermal stress, all the values increased significantly.

On storage, the apparent viscosities and percental broken structures significantly increased for most of the yoghurts (apparent viscosity (initial): F = 10.34 (sig.), 19.34 (sig.), 21.54 (sig.), 32.02 (sig.), 1.88 (n.s.); apparent viscosity (final): F = 4.36 (sig.), 29.96 (sig.), 8.37 (sig.), 17.87 (sig.), 2.56 (n.s.); and % broken structure: F = 4.76 (sig.), 5.56 (sig.), 8.31 (sig.), 12.24 (sig.), 7.97 (sig.); $p < 0.05$ (Reference, RF 58 °C, RF 65 °C, CV 58 °C and CV 65 °C)). These results agree quite well with the TPA findings described in Section 3.3.1, where there was an increase in gel firmness on storage and at the same time a slight decrease in the inner gel strength of the yoghurts.

Table 4. Rheological parameters of stirred yoghurt gels and results of ANOVA (one-way): Week 0.

Rotation and Oscillation Analysis		Reference Yoghurt	RF-Treated Yoghurts			CV-Treated Yoghurts			Set Reference Yoghurt	ANOVA Results	
			58 °C	65 °C	72 °C	58 °C	65 °C	72 °C		*F* Value	*p* Value
Apparent viscosity (Pa·s)	Initial	0.852 ± 0.014a	1.079 ± 0.039c	1.141 ± 0.043cd	1.317 ± 0.090e	1.198 ± 0.038d	1.318 ± 0.005e	1.457 ± 0.042f	0.971 ± 0.065b	58.57	1.51×10^{-13}
	Final	0.555 ± 0.021b	0.621 ± 0.017c	0.632 ± 0.021c	0.449 ± 0.057a	0.653 ± 0.018c	0.698 ± 0.003d	0.737 ± 0.025d	0.453 ± 0.025a	55.78	2.61×10^{-13}
	% Broken structure	34.875 ± 1.560a	42.464 ± 1.628b	44.588 ± 0.665bc	65.609 ± 5.440f	45.444 ± 0.730bc	47.028 ± 0.317cd	49.434 ± 0.643d	53.273 ± 0.712e	52.52	5.10×10^{-13}
Oscillatory test (Pa)	G'	385.128 ± 141.595a [1]	400.585 ± 60.124a	561.600 ± 60.186a	1401.226 ± 389.870c [1]	591.024 ± 59.334a	576.934 ± 99.024a	868.278 ± 12.132b	644.001 ± 61.057ab	13.23	7.19×10^{-7}
	G''	144.374 ± 51.547 a [1]	124.086 ± 16.562a	173.494 ± 13.801a	436.935 ± 112.452c [1]	178.469 ± 16.495a	175.900 ± 29.154a	268.420 ± 6.590b	201.644 ± 18.500ab	13.96	4.40×10^{-7}

[1] Coefficient of variation >25%. Intensity values with different letters indicate significant differences between products ($p < 0.05$, Fisher LSD *post-hoc*).

Table 5. Rheological parameters of stirred yoghurt gels and results of ANOVA (one-way): Weeks 2, 4, 5.

Rotation and Oscillation Analysis			Reference Yoghurt	RF-Treated Yoghurts		CV-Treated Yoghurts		ANOVA Results	
				58 °C	65 °C	58 °C	65 °C	*F* Value	*p* Value
Week 2	Apparent viscosity (mPa·s)	Initial	0.807 ± 0.043a	1.180 ± 0.034b	1.291 ± 0.035b	1.472 ± 0.062c	1.427 ± 0.137c	42.23	5.41×10^{-8}
		Final	0.523 ± 0.024a	0.615 ± 0.010b	0.657 ± 0.033b	0.723 ± 0.027c	0.713 ± 0.055c	19.18	9.19×10^{-6}
		% Broken structure	35.139 ± 0.813a	47.792 ± 1.568b	49.123 ± 1.389bc	50.816 ±1.676c	50.032 ± 1.083bc	69.79	1.64×10^{-9}
	Oscillatory test (Pa)	G'	177.643 ± 22.163a	464.479 ± 264.951b [1]	829.497 ± 144.263c	671.827 ± 37.555bc	614.750 ± 168.148bc [1]	7.55	1.53×10^{-3}
		G''	58.985 ± 6.280a	178.843 ± 11.699b	252.599 ± 52.660c	201.149 ± 10.039bc	189.612 ± 46.672b	14.72	4.47×10^{-5}
Week 4	Apparent viscosity (mPa·s)	Initial	0.909 ± 0.005a	1.304 ± 0.038b	1.369 ± 0.046c	1.389 ± 0.023c	1.452 ± 0.049d	120.27	3.32×10^{-11}
		Final	0.568 ± 0.004a	0.711 ± 0.009bc	0.705 ± 0.005b	0.726 ± 0.011c	0.763 ± 0.021d	129.52	1.94×10^{-11}
		% Broken structure	37.541 ± 0.463a	45.474 ± 1.531b	48.499 ± 1.218c	47.719 ± 0.255c	47.441 ± 0.717c	65.69	2.52×10^{-9}
	Oscillatory test (Pa)	G'	311.313 ± 121.421a [1]	461.110 ± 89.371ab	696.739 ± 6.195c	652.366 ± 68.599bc	526.338 ± 187.800bc [1]	5.70	5.39×10^{-3}
		G''	95.718 ± 33.897a [1]	142.018 ± 23.555ab	213.894 ± 2.990c	194.065 ± 19.226bc	162.922 ± 52.584bc [1]	6.59	2.87×10^{-3}
Week 5	Apparent viscosity (mPa·s)	Initial	0.922 ± 0.031a	1.327 ± 0.059b	1.332 ± 0.022b	1.391 ± 0.035bc	1.428 ± 0.045c	80.28	6.08×10^{-10}
		Final	0.580 ± 0.023a	0.712 ± 0.030b	0.706 ± 0.004b	0.731 ± 0.003b	0.736 ± 0.014b	38.04	1.10×10^{-7}
		% Broken structure	37.088 ± 0.963a	46.318 ± 1.656b	46.966 ± 0.778bc	47.415 ± 0.987bc	48.440 ± 0.678c	56.19	7.54×10^{-9}
	oscillatory test (Pa)	G'	173.049± 1.544a	564.869 ± 151.042bc [1]	534.679 ± 163.877b [1]	723.439 ± 31.596cd	830.820 ± 27.264d	18.41	1.18×10^{-5}
		G''	59.458± 2.488a	169.716 ± 41.109bc	164.178 ± 45.164b [1]	213.616 ± 8.514cd	245.669 ± 8.373d	19.40	8.58×10^{-6}

[1] Coefficient of variation >25%. Intensity values with different letters indicate significant differences between products ($p < 0.05$, Fisher LSD post-hoc).

As a decrease from initial to final apparent viscosity was reported for all yoghurts, the storage and loss moduli (G', G'') were also investigated in the present study and are shown in Tables 4 and 5. According to Vercet *et al.* [18], a strain amplitude of 2% was found to be within the linear viscoelastic region of the yoghurt samples. With the exception of the 72 °C samples, the storage and loss moduli of the samples were not significantly different directly after manufacturing (Table 4). Accordingly, their viscoelastic properties were quite similar directly after heat treatment. Thereby, the loss tangent (=G''/G') was <1.0, as is typical for (yoghurt) gels; thus all yoghurts behaved more like a solid with the storage modulus G' showing higher values compared to G'' over the entire linear viscoelastic region [33].

On storage, significant differences in the viscoelastic parameters developed in relation to the applied temperature and time of post-fermentative heat treatment. Again, the stirred reference was significantly different from all the heat-treated samples and RF 58 °C yoghurts were closest to the reference, with the sample of Week 5 being the only exception. More specifically, the lowest G' and G'' values were observed for the stirred reference which is in accordance with previous findings of other groups [34,35]. Smith & Hui [36] reported that low G' and G'' values were typical for very fine and homogenous yoghurt networks. This in turn correlates with the findings obtained from our cryo-SEM investigations. Lucey *et al.* [37] stated that cross-linking, presumably by heat-denatured whey proteins, could be responsible for increased rigidity and G' of yoghurt networks.

Regarding changes in the individual yoghurt samples, it was observed that the storage and loss moduli G' and G'' of the CV treated samples (slightly) increased with storage time. A significant increase was only found for the CV 65 °C sample: G', $F = 4.16$ (sig.); G'', $F = 3.86$ (sig.); $p < 0.05$. Similar observations were made by Damin *et al.* [35] who investigated the effect of cold storage (35 days) on various rheological properties of yoghurt such as firmness, G' and G''. Those authors observed increasing values for G' and G'' towards the end of the storage period. Serra *et al.* [34] also reported an increase in G' when investigating the texture of set and stirred yoghurts during the first 14 days of storage. In the same context, Lucey *et al.* [37] hypothesized a presumably ongoing fusion of casein particles leading to an increase in viscoelastic parameters during storage.

3.4. Texture Profile Analysis by Sensory Evaluation

The heating of yoghurt gels after culturing was shown in the first part of this study to have just a minor effect on the aroma profile and no influence on the taste profile of the samples [12]. Since this second part of the study focuses on the effect of RF heating on the textural and microstructural properties of yoghurt, STPA (Sensory Texture Profile Analysis) was performed on the stirred samples as a complementary non-instrumental, human-sensory evaluation. The results are shown in Table 6. A trained sensory panel evaluated the creaminess, stickiness, homogeneity by mouthfeel and the visual homogeneity ("v") over the entire storage period. For direct comparison of textural changes, a set-style reference yoghurt was also investigated (Week 0).

Table 6. Intensity rating of specific texture attributes by sensory texture profile analysis (STPA) of yoghurt gels with/without additional heat treatment after culturing and results of ANOVA (one-way): Weeks 0, 2, 4.

Sensory Texture Profile Analysis (STPA) (Visual Analogue Scale, Mean Values)		Reference Yoghurt	RF-Treated Yoghurts		CV-Treated Yoghurts		Set Reference	ANOVA Results	
			58 °C	65 °C	58 °C	65 °C		*F* Value	*p* Value
Week 0	Creaminess	6.6b	7.7ab	7.5ab	8.3a	7.7ab	4.8c	5.49	2.47×10^{-4}
	Stickiness	7.6	7.5	7.1	6.8	6.5	6.9	0.41	8.42×10^{-1} *
	Homogeneity	8.8a	6.8b	5.3b	6.9b	5.9b	1.7c	14.80	5.69×10^{-10}
	Homogeneity (v)	9.2a	8.3abc	7.3c	8.8ab	7.6bc	4.0d	16.30	1.00×10^{-10}
Week 2	Creaminess	7.7ab	7.3abc	5.7c	8.2a	5.9bc	-	2.94	2.76×10^{-2}
	Stickiness	7.7a	5.6b	4.8b	5.8ab	5.4b	-	2.43	5.76×10^{-2}
	Homogeneity	8.2a	4.2bc	2.5c	5.2b	3.5bc	-	7.60	5.05×10^{-5}
	Homogeneity (v)	9.5a	7.2b	3.8d	6.8bc	5.2cd	-	8.86	1.09×10^{-5}
Week 4	Creaminess	8.1	7.0	8.2	7.4	6.5	-	1.43	0.24 *
	Stickiness	7.2	6.0	7.3	5.7	6.4	-	0.75	0.56 *
	Homogeneity	10.0a	7.1b	6.2bc	4.7cd	4.2d	-	15.31	3.07×10^{-8}
	Homogeneity (v)	9.7a	8.5ab	8.5a	6.8bc	6.5c	-	5.23	1.34×10^{-3}

* No *post-hoc* test necessary ($p \geq 0.10$, n.s.). Intensity values with different letters indicate significant differences between products ($p < 0.05$, Fisher LSD *post-hoc*).

Besides the set-style reference, all the investigated yoghurt products had a creamy, smooth, and homogenous texture directly after manufacturing with only minor differences (Table 6). Regarding the creaminess, homogeneity, and visual homogeneity attributes, there was a significant difference between the set-style reference and the other samples, with the set-style reference being the least creamy and grainiest sample. These findings can be related to the different microstructure of set-style yoghurts [24]. A closer look at the STPA results of the heat-treated samples directly after manufacturing reveals only minor differences to the stirred reference with regard to the perceived creaminess. In detail, the heat-treated yoghurts were evaluated as being even slightly creamier than the reference (Table 6). In view of the visual homogeneity, only the yoghurts heated to 65 °C (RF, CV) were evaluated to be significantly different from the reference. In Section 3.1, the presence of white flakes/lumps was reported after post-fermentative heat treatment. In agreement with our STPA results, these flakes increased in number and size with increasing temperature and time of heat treatment.

A difference in the homogeneity of the overall mouthfeel was more pronounced after an additional heat treatment. These differences in the perceived homogeneity of the samples were presumably dependent on the microstructural changes of the samples described in Section 3.1: with increasing thermal stress, the structure of the network opened while instrumental texture analysis revealed increasing differences in the samples' inner gel strength (Section 3.3: CO, % broken structure). As the perceived viscosity and consistency is reported to decrease with a loss of inner gel strength, this is also in good agreement with our results on homogeneity as rated by mouthfeel [38].

The stickiness attribute was rated quite differently. No significant differences between the heat-treated yoghurts and the stirred and set-style references were obtained. Moreover, even in storage the evaluation of this textural attribute remained very similar between all samples, with only small differences perceived by the panelists after two weeks of storage (Table 6). Likewise, the creaminess of all samples was also evaluated as being very similar between the samples after two weeks of storage, and no significant sensory differences of any kind were found after four weeks of storage.

Greater differences between the stirred reference and the other yoghurt samples were observed with storage time for the homogeneity and visual homogeneity attributes. However, this difference was not significant after four weeks of storage for the RF heat-treated yoghurts. It is noteworthy that the latter findings from the sensory texture evaluation agree quite well with the experimental results described in Section 3.3 where textural analysis confirmed that RF heated products were closer to the reference than were yoghurts after CV heating.

Figure 5 summarizes in more detail the results given in Table 6, comparing Week 0 and Week 4 and highlighting the panelists' inter-individual variation. The data are shown as median values (instead of the arithmetic mean) in box-plot diagrams due to their robustness towards outliers. Figure 5 clearly shows that with regard to the homogeneity and visual homogeneity attributes the samples were evaluated as grainier by the panelists with increasing temperature and time of the heat treatment. These differences were even more pronounced after four weeks of storage and on highlighting the median values and the 25%–75% quartiles of the box-plots.

Differences in the homogeneity and visual homogeneity on longer storage were perceived by the panelists in the case of the CV heated yoghurts (Table 6, Figure 5). Panelists reported a tendency towards a grainier texture/mouthfeel; nevertheless, the ratings for the CV 58 °C yoghurt were not statistically significant (homogeneity: $F = 2.26$ (n.s.), 3.72 (sig.); homogeneity (visual): $F = 3.23$ (n.s.), 3.46 (sig.); $p < 0.05$ (CV 58 °C and CV 65 °C)).

Figure 5. Box-plot of STPA data from a trained sensory panel, comparing different textural attributes of yoghurts directly after heat treatment and after four weeks' storage. Median: •; whiskers: ±minimum-maximum ratings (without outliers); box: percentiles 25%–75%.

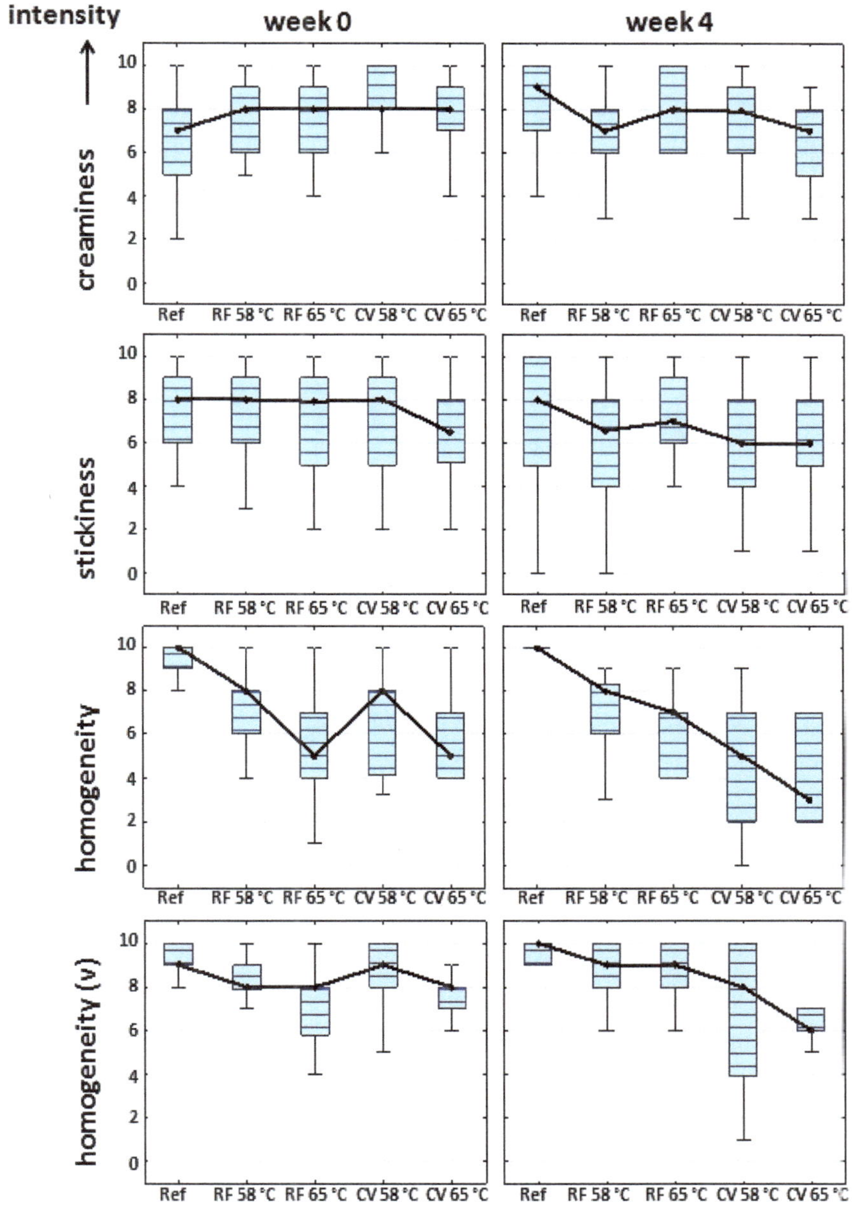

3.5. Hedonic Rating of Stirred Yoghurt Gels

Besides the evaluation of the textural properties of the samples, the flavor and taste of the reference(s) and heat-treated samples were evaluated by a trained sensory panel according to the protocol previously described [12]. After completion of all the profile analyses, the panelists were additionally asked to rank the products' popularity. The hedonic rating revealed the following ranking after four weeks of storage: reference (8, like very much) > RF 58 °C, RF 65 °C and CV 58 °C (7, like moderately) > CV 65 °C (6, like slightly). The highest discrepancies in popularity were observed between the reference product and the CV 65 °C product.

In the first part of our study [12], we also reported on a triangular testing that was additionally performed. The aim was to reveal if the differences in the yoghurts' sensory profile ratings would make it more or less complicated for the panelists to distinguish the differently treated yoghurts. The testing was performed blindfolded so that visual properties such as differences in color and visual graininess did not influence the outcome of the triangle evaluation. The CV 65 °C yoghurt was excluded from the triangle test since its flavor and taste were found to differ significantly from the other samples and, as mentioned above, was the least liked in the hedonic ranking. When performing the triangular testing, it became evident that the panel could not perceive any statistically significant differences between the stirred reference sample and the RF and CV heated yoghurts with respect to the their overall sensory properties (flavor, taste, texture). Further details are described in our previous publication [12].

4. Conclusions

This second part of our study aimed to characterize potential microstructural and textural effects in stirred yoghurt caused by post-fermentative heat treatment to prolong yoghurts' shelf-life, a procedure that is currently rarely used. In a comparative evaluation, the focus was on the effects of CV and RF heating. Microstructural changes, such as changes in pore radii and increases in the free serum phase were observed which increased with increasing thermal stress. Syneresis, textural changes, and popularity of the products were shown to be in agreement with cryo-SEM images: RF treatment produced yoghurts were more similar to the stirred reference product than after CV heat treatment. One major drawback of even gentle additional heat treatment by RF was found to be the formation of white flakes and a decrease in the inner gel strength. Nevertheless, textural defects such as white flakes are known to be reversible. A further improvement could possibly be achieved by an additional homogenization step after heat treatment, for example by pumping the curd through continuous flow coolers or special devices.

Overall, this two-part feasibility study was able to characterize the microbial, sensorial, textural, and microstructural changes caused by post-fermentative heat treatment, and such data has not hitherto been reported in the literature. A quick, homogeneous, and gentle RF treatment up to 65 °C was shown to prolong the shelf-life of yoghurts by reduced microbial numbers while only minor sensorial and presumably reversible textural changes were observed [12]. Such findings on less susceptible yoghurt products with satisfactory sensorial and textural properties open up opportunities for improved "heat treated" yoghurt sales, especially in countries where the distribution cold chain cannot be guaranteed at all times. In a follow-up step, radio frequencies could be applied to stirred yoghurt in an RF tube. The application of radio frequencies in a continuous manufacturing process would open up this technique to industry, while the yoghurts' textural and sensorial properties would be expected to be further improved.

Acknowledgements

The present study was awarded and financially supported by the Danone Innovation Prize 2012. The authors would like to thank Danone GmbH (Haar, Germany) for this support. Furthermore, the authors would like to thank Isabel Muranyi from the Fraunhofer IVV, Department of Process Engineering, for her technical support. We acknowledge support by Deutsche Forschungsgemeinschaft and Friedrich-Alexander-Universität Erlangen-Nürnberg (FAU) within the funding programme Open Access Publishing.

Author Contributions

The present manuscript was carefully written by Caroline Siefarth within the frame of her PhD thesis, who carried out the experimental work together with her diploma student Thi Bich Thao Tran. RF experiments were guided by Peter Mittermaier and Thomas Pfeiffer. The study and diploma thesis were supervised by Prof. Dr. Andrea Buettner.

Conflicts of Interest

The authors declare no conflict of interest.

References

1. Walstra, P. On the stability of casein micelles. *J. Dairy Sci.* **1990**, *73*, 1965–1979.
2. Walstra, P.; Wouters, J.T.M.; Geurts, T.J. *Dairy Science and Technology*, 2nd ed.; Taylor & Francis: Boca Raton, FL, USA, 2006.
3. Walstra, P. Relation between structure and texture of cultured milk products. In *Texture of Fermented Milk Products and Dairy Desserts, Proceedings IDF Symposium 1997*; International Dairy Federation: Brussels, Belgium, 1998; pp. 9–15.
4. Chandan, R.C.; O'Rell, K.R. Manufacture of various types of yogurt. In *Manufacturing Yogurt and Fermented Milks*, 1st ed.; Chandan, R.C., White, C.H., Kilara, A., Hui, Y.H., Eds.; Blackwell Publishing: Oxford, UK, 2006; pp. 211–236.
5. Chandan, R.C.; Kilara, A. *Manufacturing Yogurt and Fermented Milks*, 2nd ed.; Wiley-Blackwell: Hoboken, NJ, USA, 2013; pp. 294–295.
6. Barua, N.N.; Hampton, R.J. Aseptic Yoghurt. US Patent 4609554 A, 2 September 1986.
7. Keefer, D.K.; Murray, M.E. Process for Production of a No-Starch Shelf Stable Yogurt Product. US Patent 4748026 A, 31 May 1988.
8. Egli, F.; Egli, F. Production of Sterile Yoghurt. US Patent 4235934-A, 25 November 1980.
9. Amrein, R.; Widmer, W. Sterile Yoghurt Production. Swiss Patent 634464 A, 1983.
10. Güldas, M.; Atamer, M. Effect of pasteurization norm and storage temperature on the quality of long-life yoghurt. *GIDA* **1995**, *20*, 313–319.
11. Bach, J. Improvements in or relating to methods for effecting uniform heating of organic products. British Patent 1507069-A, 1978.

12. Siefarth, C.; Tran, T.B.T.; Mittermaier, P.; Pfeiffer, T.; Buettner, A. Effect of radio frequency heating on yoghurt, I: Technological applicability, shelf-life and sensorial quality. *Foods* **2014**, *3*, 318–335.
13. Kessler, H.G. The structure of fermented milk products as influenced by technology and composition. In *Texture of Fermented Milk Products and Dairy Desserts, Proceedings IDF Symposium 1997*; International Dairy Federation: Brussels, Belgium, 1998; pp. 93–105.
14. Bourne, M.C. *Food Texture and Viscosity: Concept and Measurement*, 2nd ed.; Academic Press: San Diego, CA, USA, 2002.
15. Kaur, H.; Mishra, H.N.; Kumar, P. Textural properties of mango soy fortified probiotic yoghurt: Optimisation of inoculum level of yoghurt and probiotic culture. *Int. J. Food Sci. Technol.* **2009**, *44*, 415–424.
16. Rosenthal, A.J. *Food Texture: Measurement and Perception*; Springer: New York, NY, USA, 1999.
17. Everett, D.W.; McLeod, R.E. Interactions of polysaccharide stabilisers with casein aggregates in stirred skim-milk yoghurt. *Int. Dairy J.* **2005**, *15*, 1175–1183.
18. Vercet, A.; Oria, R.; Marquina, P.; Crelier, S.; Lopez-Buesa, P. Rheological properties of yoghurt made with milk submitted to manothermosonication. *J. Agric. Food Chem.* **2002**, *50*, 6165–6171.
19. Afonso, I.M.; Maia, J.M. Rheological monitoring of structure evolution and development in stirred yoghurt. *J. Food Eng.* **1999**, *42*, 183–190.
20. Tribby, D. Yogurt. In *The Sensory Evaluation of Dairy Products*, 2nd ed.; Clark, S., Costello, M., Bodyfelt, F., Drake, M.A, Eds.; Springer: New York, NY, USA, 2009; pp. 205–218.
21. Kalab, M. Practical aspects of electron-microscopy in dairy research. *Food Struct.* **1993**, *12*, 95–114.
22. Hassan, A.N.; Frank, J.F.; Elsoda, M. Observation of bacterial exopolysaccharide in dairy products using cryo-scanning electron microscopy. *Int. Dairy J.* **2003**, *13*, 755–762.
23. Sandoval-Castilla, O.; Lobato-Calleros, C.; Aguirre-Mandujano, E.; Vernon-Carter, E.J. Microstructure and texture of yogurt as influenced by fat replacers. *Int. Dairy J.* **2004**, *14*, 151–159.
24. Lee, W.J.; Lucey, J.A. Formation and physical properties of yogurt. *Aust. J. Anim. Sci.* **2010**, *23*, 1127–1136.
25. Soukoulis, C.; Panagiotidis, P.; Koureli, R.; Tzia, C. Industrial yogurt manufacture: Monitoring of fermentation process and improvement of final product quality. *J. Dairy Sci.* **2007**, *90*, 2641–2654.
26. Töpel, A. *Chemie und Physik der Milch: Naturstoff, Rohstoff, Lebensmittel*, 1st ed.; Behr's Verlag: Hamburg, Germany, 2004; pp. 402–430 (in German).
27. Lucey, J.A. Cultured dairy products: An overview of their gelation and texture properties. *Int. J. Dairy Technol.* **2004**, *57*, 77–84.
28. Lucey, J.A.; Singh, H. Formation and physical properties of acid milk gels: A review. *Food Res. Int.* **1997**, *30*, 529–542.
29. Weidendorfer, K.; Bienias, A.; Hinrichs, J. Investigation of the effects of mechanical post-processing with a colloid mill on the texture properties of stirred yogurt. *Int. J. Dairy Technol.* **2008**, *61*, 379–384.

30. Salvador, A.; Fiszman, S.M. Textural and sensory characteristics of whole and skimmed flavored set-type yogurt during long storage. *J. Dairy Sci.* **2004**, *87*, 4033–4041.
31. Roefs, S.P.F.M.; de Groot-Mostert, A.E.A.; van Vliet, T. Structure of acid casein gels 1. Formation and model of gel network. *Coll. Surf.* **1990**, *50*, 141–159.
32. Parnell-Clunies, E.M.; Kakuda, Y.; Deman, J.M. Influence of heat treatment of milk on the flow properties of yoghurt. *J. Food Sci.* **1986**, *51*, 1459–1462.
33. Mezger, T.G. *Das Rheologie Handbuch: Für Anwender von Rotations- und Oszillations-Rheometern*, 2nd ed.; Vincentz Network: Hannover, Germany, 2007 (in German).
34. Serra, M.; Trujillo, A.J.; Guamis, B.; Ferragut, V. Evaluation of physical properties during storage of set and stirred yogurts made from ultra-high pressure homogenization-treated milk. *Food Hydrocoll.* **2009**, *23*, 82–91.
35. Damin, M.R.; Minowa, E.; Alcântara, M.R.; Oliveira, M.N. Effect of cold storage on culture viability and some rheological properties of fermented milk prepared with yogurt and probiotic bacteria. *J. Texture Stud.* **2008**, *39*, 40–55.
36. Chandan, R.C. Dairy: Yoghurt. In *Food Processing: Principles and Applications*, 1st ed.; Smith, J.S., Hui, Y.H., Eds.; Blackwell Publishing: Oxford, UK, 2004; pp. 297–318.
37. Lucey, J.A.; Munro, P.A.; Singh, H. Rheological properties and microstructure of acid milk gels as affected by fat content and heat treatment. *J. Food Sci.* **1998**, *63*, 660–664.
38. Robinson, R.K.; Itsaranuwat, P. Properties of yoghurt and their appraisal. In *Fermented Milks*; Tamime, A.Y., Ed.; Blackwell Publishing: Oxford, UK, 2008; pp. 79–84.

13

Multi-Pulsed High Hydrostatic Pressure Treatment of Foods

Sencer Buzrul

Tütün ve Alkol Piyasası Düzenleme Kurumu (TAPDK), Ankara 06520, Turkey;
E-Mail: sencer.buzrul@tapdk.gov.tr

Academic Editor: Carl Joseph Schaschke

Abstract: Multi-pulsed high hydrostatic pressure (mpHHP) treatment of foods has been investigated for more than two decades. It was reported that the mpHHP treatment, with few exceptions, is more effective than the classical or single-pulsed HHP (spHHP) treatment for inactivation of microorganisms in fruit juice, dairy products, liquid whole egg, meat products, and sea foods. Moreover, the mpHHP treatment could be also used to inactivate enzymes in foods and to increase the shelf-life of foods. The effects of the mpHHP treatment of foods are summarized and the differences between the mpHHP and spHHP are also emphasized.

Keywords: high hydrostatic pressure; multi-pulsed high hydrostatic pressure; compression and decompression rates; microbial inactivation; pulse holding time

---

1. Introduction

High hydrostatic pressure (HHP) treatment is an effective technique to destroy microorganisms and inactivate enzymes in order to enhance safety and shelf-life of foods. Therefore, HHP has become a reality in the food industry and has spread world wide [1]. After 2000, the number of installed HPP machines for the food industry increased exponentially [2].

Classical HHP or single-pulsed HHP (spHHP) treatment can be applied as: compression to target pressure, holding for a certain period of time at the target pressure, and decompression to atmospheric pressure (Figure 1). On the other hand, it may be also possible to apply successive application of HHP in which more than one compression, holding, and decompression periods exist (Figure 2). This type of treatment is called multi-pulsed HHP (mpHHP).

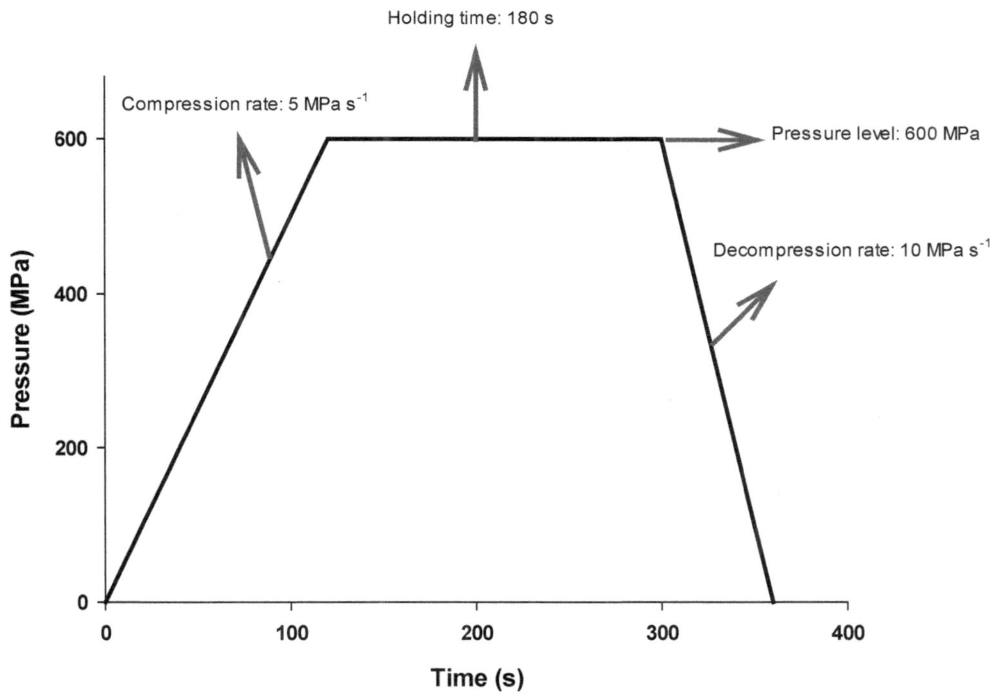

Figure 1. Classical or single-pulsed high hydrostatic pressure (HHP) treatment consisting of compression rate (5 MPa·s^{-1}), holding time (180 s) at a constant pressure level (600 MPa), and decompression rate (10 MPa·s^{-1}). Note that total duration of the treatment is 360 s (6 min).

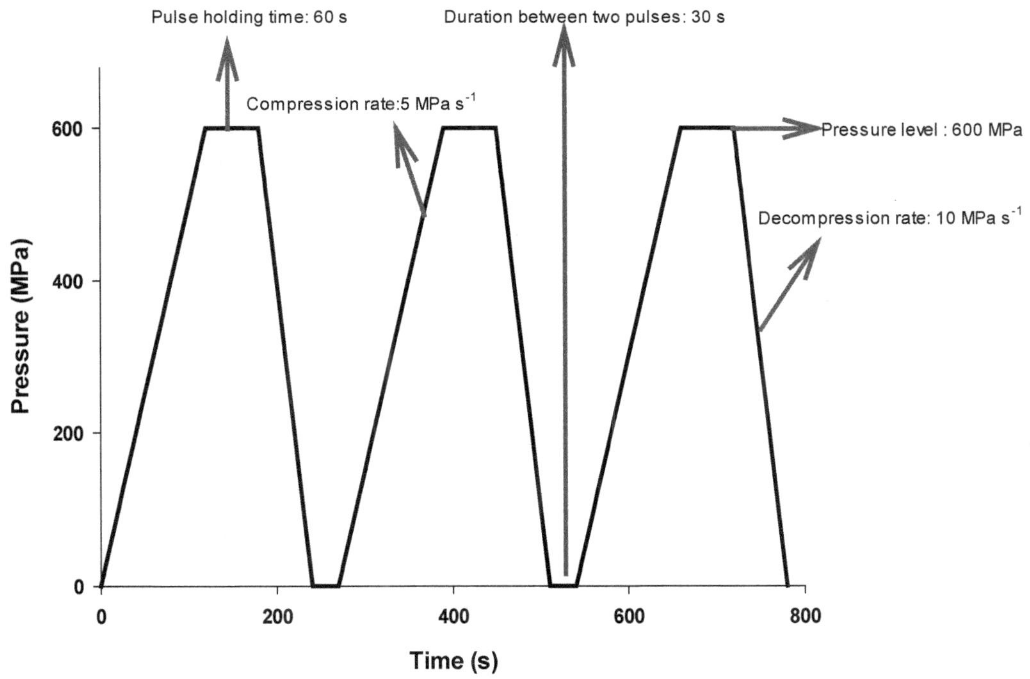

Figure 2. Multi-pulsed HHP treatment: 3 pulses × 60 s (30 s between each pulse) at 600 MPa. Compression and decompression rates are 5 and 10 MPa·s^{-1}, respectively. Note that total duration of the treatment is 780 s (13 min).

The mpHHP treatment, for the same holding time, is more effective than the spHHP treatment for enzyme [3–5], yeast cells [6] bacterial cells [7,8], and bacterial spores [9–11] inactivation. It was also reported that there was less recovery from injury for *Escherichia coli* for the mpHHP treatment compared to the spHHP treatment [12]. However, some researchers reported that the use of the mpHHP treatment did not considerably enhance pressure inactivation of virus [13], and bacteria [14,15] as compared to the sHHP treatment.

The mpHHP treatment inactivation of microorganisms in laboratory media, foods, blood plasma, vaccines, and drugs is well documented by a recent review [16]. This review, however, provides information not only about microorganisms but also enzymes, food quality and shelf-life.

2. Process Parameters of the mpHHP Treatment

It is known that pressure, temperature, and (holding) time are the most important process parameters of the spHHP treatment. However, more parameters should be taken into account before applying the mpHHP. Pressure and temperature are also the most important parameters for the mpHHP treatment. Besides, pulse duration, *i.e.*, pulse holding time, number of pulses, off-pressure time (duration between the pulses), compression and decompression rates or times, and pulse shape (ramp, square, sinusoidal) may also affect the outcomes of the mpHHP treatment (Figure 2). The effect of these process parameters on microbial inactivation was given by Buzrul [16] and will not be deeply investigated here.

3. Application of the mpHHP on Foods

3.1. Fruit Juices

Studies on fruit juices by the application of mpHHP began about two decades ago. The mpHHP treatment was reported to be more effective than the spHHP treatment for the same holding time for the inactivations of *Saccharomyces cerevisiae* in pineapple juice [17], *Byssochlamys nivea* ascospores in apple and cranberry juices [18]. On the other hand, Alemán *et al.* [17] observed no inactivation of *S. cerevisiae* in pineapple juice after 40–4000 fast sinusodial pulses (10 cycles/s) at 4–400 s total holding time in the range of 235–270 MPa (total processing time was 0.39–39 min) indicating that pulse shape is (step pressure pulse was effective, but sinusodial pulses had no effect on inactivation of yeasts in fruit juice) also an important parameter for the mpHHP treatment.

Donsì *et al.* [19] found that efficiency of the mpHHP treatment depends on the combination of pulse holding time and number of pulses for the inactivation of *S. cerevisiae* in pineapple and orange juices. They also observed higher reduction for slow compression rate (2.5 $MPa \cdot s^{-1}$) than that of faster compression rates (10.5 and 25 $MPa \cdot s^{-1}$) if several pulses (3 to 10 pulses) were applied. Buzrul *et al.* [20] found that increasing the pulse number did not effect the inactivation of *Escherichia coli* and *Listeria innocua* to great extends in kiwifruit juice (high inactivations were already obtained by application of the spHHP treatment in kiwifruit juice); however, in pineapple juice especially after 5 pulses inactivation increased significantly for both bacteria.

The mpHHP treatment up to 3 pulses with no holding time, *i.e.*, compression followed by decompression was also applied to inactivate pectin methyl esterase (PME) in single strength and concentrated orange juices [21]. The mpHHP has a significant contribution to inactivation of PME in both juices.

A comprehensive study by Donsì *et al.* [22] indicated that the effectiveness of the mpHHP treatment on apple and orange juices depends on the combination of pressure, temperature, and pulse number. Optimum conditions applied to apple (300 MPa, 50 °C, 6 pulses × 1 min) and orange (250 MPa, 45 °C, 6 pulses × 1 min) juices resulted in a minimum shelf-life of 21 days at 4 °C.

3.2. Dairy Products

Milk, cheese and yogurt are the dairy products treated with the mpHHP. The mpHHP was considerably more effective than the spHHP for the same total treatment time for inactivation of *E. coli* in skim milk [23], *E. coli* and *L. innocua* in whole milk [24,25].

The mpHHP treatment up to 4 pulses with no holding time was applied to inactivate *E. coli* O157:H7 and *L. monocytogenes* in raw milk cheese [26]. Significant microbial and enzyme inactivation could be possible by the application of the mpHHP (at room temperature, 3 pulses × 5 min) at higher pressures (600 and 800 MPa) in three different types of cheese which were at different ripening stages [27]. Storage of cheeses at 5 °C for 12 weeks revealed that microorganisms inactivated by the mpHHP were also absent during storage. López-Pedemonte *et al.* [28] obtained low inactivation (about 1.6 $\log_{10}$) for spores of *Bacillus cereus* in cheese by the mpHHP inactivation with 2 pulses (first pulse with low pressure (60 MPa) to germinate the spores and the second one is with high pressure (400 MPa) to inactivate the vegetative cells).

Applications of the spHHP (400 MPa for 15, 30, and 45 min) and mpHHP (400 MPa for 3 pulses × 5 min, 3 pulses × 10 min, and 3 pulses × 15 min) treatments in yogurt revealed that *Lactobacillus delbruecki* sp. *bulgaricus* was completely inactivated under all conditions whereas *Streptoccocus salivarius* sp. *thermophilus* was little reduced, maximum by one $\log_{10}$ [29].

3.3. Liquid Whole Egg

A few studies on microbial inactivation in liquid whole egg (LWE) by the mpHHP treatment revealed that the mpHHP treatment showed greater effectiveness than the spHHP treatment for inactivations of *Samonella* Enteritidis [30–32] and *E. coli* [33].

The effect of temperature during the mpHHP treatment is well documented in these studies. For example, Ponce *et al.* [33] applied the spHHP (350 MPa, 10 or 15 min) and the mpHHP (350 MPa, 2 or 3 pulses × 5 min) treatments at different temperatures (2, 20, or 50 °C) for the inactivation of *E. coli* in LWE. The highest reduction was achieved at 50 °C for both treatments; however, at lower temperatures, especially at 20 °C, the mpHHP treatment was clearly more effective than the spHHP treatments. Ponce *et al.* [30] observed the strongest effectiveness at 50 °C, followed by 20, 2, and −15 °C for the inactivation of *S.* Enteritidis in LWE after the application of mpHHP (2 or 3 pulses × 5 min at 350 and 450 MPa).

3.4. Meat Products

The mpHHP applied to mechanically recovered poultry meat showed that the mpHHP treatment was slightly better than the spHHP treatment for psychrotrophs, but the mpHHP treatment did not offer better results than the spHHP treatment for mesophiles [15,34]. On the other hand, the use of the mpHHP treatment instead of the spHHP treatment showed to be more advantageous for the inactivation of *E. coli* O157:H7 in ground beef [35] and *S.* Enteritidis in chicken breast fillets especially at higher pressures [36].

Morales *et al.* [35] and Del Olmo *et al.* [37] studied the effect of the spHHP and mpHHP treatments on color and texture of beef patties and chicken breast fillets, respectively. Changes in the color and texture of ground beef caused by spHHP and mpHHP treatments of the same lethality for *E. coli* O157:H7 (20 min for spHHP and 4 pulses × 1 min for mpHHP) were similar [35]. Color parameters (L *, a * and b *) were significantly higher for both treatments than for vacuum-packaged control fillets. Similarly, the texture of chicken breast fillets was also significantly affected by both treatments [37].

3.5. Sea Foods

The effect of the spHHP (400 MPa, 7 °C, 10 min) and the mpHHP (400 MPa, 7 °C, 2 pulses × 5 min) on microbial flora, total volatile bases, pH, and texture of purified and unpurified oysters was studied by López-Caballero *et al.* [14]. The mpHHP produced no apparent advantages over the spHHP based on any of the indices used.

The mpHHP treatment reduced the microbial load in octopus arm muscle more effectively than the spHHP treatment; however, the mpHHP treatment was not so effective in reducing autolytic activity [38,39]. Inactivations of *S.* Enteritidis and *Staphyloccoccus aureus* in sturgeon and trout caviar also studied [40]. Results indicated that the mpHHP treatment (350 MPa for *S.* Enteritidis and 450 MPa for *Staphyloccoccus aureus* at room temperature for 3 pulses × 5 min) were as effective as the spHHP treatment (400 MPa for *S.* Enteritidis and 500 MPa for *Staphyloccoccus aureus* at room temperature for 15 min).

3.6. Other Food Products

Similar results were also obtained for other food products: the mpHHP treatment was more effective than the spHHP treatment for the inactivation of *S. cerevisiae* in fresh cut pineapple [41], *S.* Enteritidis in raw almonds [42], *E. coli* in egg white [43]. Meyer [44] reported sterility in macaroni and cheese with spore load of *Clostridium sporogenes* and *B. cereus* by the mpHHP treatment (690 MPa, 90 °C, 2 pulses × 1 min; 1 min pause between the pulses).

A summary of the mpHHP inactivation of microorganisms in foods and a summary of studies on the effect of mpHHP treatment on quality, shelf-life, microbial and enzyme inactivation of foods are provided in Tables 1 and 2, respectively.

Table 1. Summary of multi-pulsed high hydrostatic pressure (mpHHP) inactivation of microorganisms in foods.

Microorganism	Product	CR or CT [a]	DR or DT [b]	Process Conditions [c]	Log Reduction	Reference
Saccharomyces cerevisiae	Pineapple juice	0.5 s	0.2 s	270 MPa, 23 °C, 10 pulses × 10 s	3.3	[17]
				270 MPa, 23 °C, 100 pulses × 1 s	3.5	
		0.34 s	0.18 s	270 MPa, 23 °C, 167 pulses × 0.6 s (0.2 s between the pulses)	3.9	
Byssochlamys nivea ascospores	Cranberry juice	2.4 MPa·s^{-1}	<10 s	689 MPa, 60 °C, 3 pulses × 1 s	>4.0 *	[18]
	Apple juice			689 MPa, 60 °C, 3 pulses × 1 s	>4.0 *	
S. cerevisiae	Pineapple juice	10.5 MPa·s^{-1}	ND [d]	250 MPa, 25 °C, 10 pulses × 1 min	4.0	[19]
	Orange juice			250 MPa, 25 °C, 6 pulses × 1 min	>4.5	
				250 MPa, 25 °C, 10 pulses × 1 min	>5.0	
				200 MPa, 45 °C, 6 pulses × 1 min	>5.0	
				200 MPa, 45 °C, 10 pulses × 1 min	≈5.5	
		2.5 MPa·s^{-1}	ND	200 MPa, 25 °C, 10 pulses × 1 min	≈2.7	
		25 MPa·s^{-1}	ND	200 MPa, 25 °C, 10 pulses × 1 min	≈2.2	
Escherichia coli	Pineapple juice	5 MPa·s^{-1}	5 MPa·s^{-1}	300 MPa, 20 °C, 10 pulses × 30 s	2.8	[20]
				350 MPa, 20 °C, 5 pulses × 60 s	2.6	
Listeria innocua				300 MPa, 20 °C, 10 pulses × 30 s	3.4	[20]
				350 MPa, 20 °C, 5 pulses × 60 s	3.6	
E. coli	Kiwifruit juice			300 MPa, 20 °C, 10 pulses × 30 s	4.7	[20]
				350 MPa, 20 °C, 5 pulses × 60 s	5.5	
L. innocua				300 MPa, 20 °C, 10 pulses × 30 s	4.8	[20]
				350 MPa, 20 °C, 5 pulses × 60 s	5.6	
E. coli	Skim milk	ND	ND	550 MPa, 20 °C, 3 pulses × 10 min	6.0	[23]
	Whole milk	5 MPa·s^{-1}	5 MPa·s^{-1}	400 MPa, 20–25 °C, 10 pulses × 1 min	4.0	[25]
				400 MPa, 20–25 °C, 10 pulses × 2 min	4.6	
L. innocua				400 MPa, 20–25 °C, 5 pulses × 4 min	3.9	[25]
				400 MPa, 20–25 °C, 10 pulses × 2 min	4.3	
E. coli K-12	Raw milk cheese	2.25 MPa·s^{-1}	< 3s	400 MPa, 25 °C, 4 pulses × 0 min	≈3.4	[26]
E. coli O157:H7				400 MPa, 25 °C, 4 pulses × 0 min	≈1.4	[26]
L. monocytogenes				400 MPa, 25 °C, 4 pulses × 0 min	≈3.8	[26]
Bacillus cereus spores	Cheese	ND	ND	60 MPa, 30 °C, 210 min + 400 MPa 30 °C, 15 min	1.6	[28]
S. Enteritidis	Liquid whole egg	180 s	90 s	350 MPa, 50 °C, 2 pulses × 5 min	7.8 *	[30]
		240 s	120 s	450 MPa, 20 °C, 2 pulses × 5 min	7.3 *	
		ND	ND	138 MPa, 20 °C, 2 pulses × 4 min	1.3	[31]
		45 s	6 s	350 MPa, 50 °C, 4 pulses × 2 min	>8.0 *	[32]
E. coli O157:H7	Ground beef	2.2 min	0.3 min	400 MPa, 12 °C, 3 pulses × 5 min	≈3.0	[35]
S. Enteritidis	Chicken breast fillets	96 s	16.2 s	300 MPa, 12 °C, 2 pulses × 5 min	2.5	[36]
		132 s	19.2 s	400 MPa, 12 °C, 3 pulses × 3 min	4.6	
Enterobacteriaceae	Octopus muscle	4 min	≈2 s	400 MPa, 7 °C, 3 pulses × 5 min	≈3.0	[39]
				400 MPa, 40 °C, 3 pulses × 5 min	≈3.0	

Table 1. *Cont.*

Microorganism	Product	CR or CT [a]	DR or DT [b]	Process Conditions [c]	Log Reduction	Reference
S. Enteritidis	Sturgeon caviar	ND	ND	450 MPa, 20 °C, 3 pulses × 5 min	>4.1	[40]
	Trout caviar			450 MPa, 20 °C, 3 pulses × 5 min	>2.7	
S. aureus	Sturgeon caviar	ND	ND	450 MPa, 20 °C, 3 pulses × 5 min	>3.5	[40]
	Trout caviar			450 MPa, 20 °C, 3 pulses × 5 min	>3.7	
S. Enteriditis	Raw almonds	≈3.6 min	1 min	414 MPa, 50 °C, 6 pulses × 20 s (30 s between the pulses)	1.3	[42]
E. coli	Egg white	ND	ND	300 MPa, 20 °C, 3 pulses × 2 min	>7.0 *	[43]
B. cereus spores	Macaroni and cheese	ND	ND	690 MPa, 90 °C, 2 pulses × 1 min (1 min between the pulses)	>6.0 *	[44]
Clostridium sporogenes spores				690 MPa, 90 °C, 2 pulses × 1 min (1 min between the pulses)	> 6.0 *	[44]

[a] CR: Compression rate; CT: Compression time; [b] DR: Decompression rate; DT: Decompression time; [c] The temperature given is either the initial or the process temperature of the treatment; [d] ND: Not determined; * Total inactivation.

Table 2. Summary of studies on the effect of mpHHP treatment on quality, shelf-life, microbial and enzyme inactivation of foods.

Product	CR or CT [a]	DR or DT [b]	Process conditions [c]	Achievement	Reference
Oyster	2.5 MPa·s^{-1}	15 s	400 MPa, 7 °C, 2 pulses × 5 min	No apparent advantages over spHHP treatment	[14]
Orange juice	2.8 min	≈10 s	400 MPa, 20 °C, 3 pulses × 0 s	92.4% inactivation of PME	[21]
	10.5 MPa·s^{-1}	4 s	250 MPa, 45 °C, 6 pulses × 60 s	21 days of shelf-life at 4 °C	[22]
Apple juice	10.5 MPa·s^{-1}	5 s	300 MPa, 50 °C, 6 pulses × 60 s	21 days of shelf-life at 4 °C	
Cheese	ND [d]	ND	800 MPa, ND, 3 pulses × 5 min	4–6 $\log_{10}$ inactivation of microorganisms Inactivation of proteases No growth of inactivated microorganisms at 5 °C for 12 weeks	[27]
Yogurt	ND [d]	ND	400 MPa, ND, 3 pulses × 5 min	Complete inactivation of *Lactobacillus bulgaricus* No acidity change at 1 and 20 °C for 3 weeks	[29]
Ground beef	2.2 min	0.3 min	400 MPa, 12 °C, 2 pulses × 60 s	Significant color and texture changes	[35]
Chicken breast fillets	2.2 min	17 s	400 MPa, 5 °C, 2 pulses × 60 s	Significant color and texture changes	[37]

[a] CR: Compression rate; CT: Compression time; [b] DR: Decompression rate; DT: Decompression time; [c] The temperature given is either the initial or the process temperature of the treatment; [d] ND: Not determined.

4. Commercial Application of the mpHHP

Although there is now enough evidence that the mpHHP treatment is an effective way of inactivating microorganisms and enzymes, there is no commercial application of the mpHHP treatment up to date. One and most important reason for this is that the mpHHP treatment is a longer application and thus more expensive than the spHHP treatment [8,13,16,25,41]—see Figures 1 and 2. Besides, most probably a more complicated HHP equipment which can withstand fast compression and decompression rates (to reduce the total duration of the treatment) is needed for commercial applications.

However, as the technology improves it may be possible to have faster compression and decompression rates hence it may be possible to reach compatible total treatment times for the commercial applications of the mpHHP treatment. To the best of the author's knowledge there is no study on the cost and optimization of process parameters of the mpHHP treatment. Optimization between pulse number, pulse holding time, pressure level, compression and decompression rates as well as initial and target temperature in a lab scale equipment will accelerate the application of commercial mpHHP treatment. Moreover, the differences in effectiveness of mpHHP and spHHP treatments must be weighed against the design capabilities of added wear on the HHP equipment, and possible additional time required for pulse treatment [45].

5. Conclusions

The mpHHP treatment could be used to inactivate microorganisms and enzymes in foods. It could also be used to contribute the quality and shelf-life of foods. However, it should be noted that optimization between the pressure, temperature, pulse number, pulse holding time, and compression and decompression rates can increase the effectiveness of the mpHHP treatment. However, more studies are needed especially on the cost of the mpHHP treatment.

Conflicts of Interest

The author declares no conflict of interest.

References

1. Bermúdez-Aguirre, D.; Barbosa-Cánovas, G.V. An update on high hydrostatic pressure, from laboratory to industrial applications. *Food Eng. Rev.* **2011**, *3*, 44–61.
2. Tonello, C. Case studies on high-pressure processing of foods. In *Nonthermal Processing Technologies for Food*; Zhang, H.Q., Barbosa-Cánovas, G.V., Balasubramaniam, V.M., Dunne, C.P., Farkas, D.F., Yuan, J.T.C., Eds.; Woodhead Publishing Limited: Cambridge, London, UK, 2011; pp. 36–50.
3. Curl, A.L.; Jansen, E.F. Effect of high pressures on trypsin and chymotrypsin. *J. Biol. Chem.* **1949**, *184*, 45–54.
4. Curl, A.L.; Jansen, E.F. The effect of high pressures on pepsin and chymotrypsinogen. *J. Biol. Chem.* **1950**, *185*, 713–723.
5. Ludikhuyze, L.R.; van den Broeck, I.; Weemaes, C.A.; Hendrickx; M.E. Kinetic parameters for pressuretemperature inactivation of *Bacillus subtilis* α-amylase under dynamic conditions. *Biotechnol. Prog.* **1997**, *13*, 617–623.
6. Palou, E.; López-Malo, A.; Barbosa-Cánovas, G.V.; Welti-Chanes, J.; Swanson, B.G. Oscillatory high hydrostatic pressure inactivation of *Zygosaccharomyces bailii*. *J. Food Prot.* **1998**, *61*, 1213–1215.

7. Masschalck, B.; García-Graells, C.; van Haver, E.; Michiels, C.W. Inactivation of high pressure resistant *Escherichia coli* by lysozyme and nişin under high pressure. *Innov. Food Sci. Emerg. Technol.* **2000**, *1*, 39–47.
8. Rivalain, N.; Roquain, J.; Boiron, J.M.; Maurel, J.P.; Largeteau, A.; Ivanovic, Z.; Demazeau, G. High hydrostatic pressure treatment for the inactivation of *Staphylococcus aureus* in human blood plasma. *New Biotechnol.* **2012**, *29*, 409–414.
9. Hayakawa, I.; Kanno, T.; Yoshiyama, K.; Fujio, Y. Oscillatory compared with continuous high pressure sterilization on *Bacillus stearothermophilus* spores. *J. Food Sci.* **1994**, *59*, 164–167.
10. Furukawa, S.; Nakahara, A.; Hayakawa, I. Effect of reciprocal pressurization on germination and killing of bacterial spores. *Int. J. Food Sci. Technol.* **2000**, *35*, 529–532.
11. Ahn, J.; Balasubramaniam, V.M. Effects of inoculum level and pressure pulse on the inactivation of *Clostridium sporogenes* spores by pressure-assisted thermal processing. *J. Microbiol. Biotechnol.* **2007**, *17*, 616–623.
12. Pilavtepe-Çelik, M.; Buzrul, S.; Alpas, H.; Largeteau, A.; Demazeau, G. Multi-pulsed high hydrostatic pressure treatment for inactivation and injury of *Escherichia Coli*. *J. Verbrauch. Lebensm.* **2011**, *6*, 343–348.
13. Kingsley, D.H.; Guan, D.; Hoover, D.G.; Chen, H. Inactivation of Hepatitis A virus by h igh-pressure processing: The role of temperature and pressure oscillation. *J. Food Prot.* **2006**, *69*, 2454–2459.
14. López-Caballero, M.E.; Pérez-Mateos, M.; Montero, P.; Borderías, A.J. Oyster preservation by high-pressure treatment. *J. Food Prot.* **2000**, *63*, 196–201.
15. Yuste, J.; Pla, R.; Capellas, M.; Sendra, E.; Beltran, E.; Mor-Mur, M. Oscillatory high pressure processing applied to mechanically recovered poultry meat for bacterial inactivation. *J. Food Sci.* **2001**, *66*, 482–484.
16. Buzrul, S. Multi-pulsed high hydrostatic pressure treatment of microorganisms: A review. *Innov. Food Sci. Emerg. Technol.* **2014**, *26*, 1–11.
17. Alemán, G.D.; Ting, E.Y.; Mordre, S.C.; Hawes, A.C.O.; Walker, M.; Farkas, D.F.; Torres, J.A. Pulsed ultra high pressure treatments for pasteurization of pineapple juice. *J. Food Sci.* **1996**, *61*, 388–390.
18. Palou, E.; López-Malo, A.; Barbosa-Cánovas, G.V.; Welti-Chanes, J.; Davidson, P.M.; Swanson, B.G. Effect of oscillatory high hydrostatic pressure treatments on *Byssochlamys nivea* ascospores suspended in fruit juice concentrates. *Lett. Appl. Microbiol.* **1998**, *27*, 375–378.
19. Donsì, G.; Ferrari, G.; Maresca, P. Pulsed high pressure treatment of *Saccharomyces cerevisiae*: The effect of process parameters. *J. Food Eng.* **2007**, *78*, 984–990.
20. Buzrul, S.; Largeteau, A.; Alpas, H.; Demazeau, G. Inactivation of *Escherichia coli* and *Listeria innocua* in kiwifruit and pineapple juices by high hydrostatic pressure. *Int. J. Food Microbiol.* **2008**, *124*, 275–278.
21. Basak, S.; Ramaswamy, H.S. Pulsed high pressure inactivation of pectin methyl esterase in single strength and concentrated orange juices. *Can. Biosyst. Eng.* **2001**, *43*, 3.25–3.29.
22. Donsì, G.; Ferrari, G.; Maresca, P. Pasteurization of fruit juices by means of a pulsed high pressure process. *J. Food Sci.* **2010**, *75*, E169–E177.

23. García-Graells, C.; Masschalck, N.; Moonjai, N.; Michiels, C. High pressure inactivation and survival of pressure-resistant *Escherichia coli* mutants in milk. In *Advances in High Pressure Bioscience and Biotechnology*; Ludwig, H., Ed.; Springer-Verlag: Berlin, Germany, 1999; pp. 133–136.
24. Buzrul, S.; Largeteau, A.; Alpas, H.; Demazeau, G. Pulsed pressure treatment for inactivation of *Escherichia coli* and *Listeria innocua* in whole milk. *J. Phys. Conf. Ser.* **2008**, *121*, Art. No: 142001, doi:10.1088/1742-6596/121/4/142001.
25. Buzrul, S.; Alpas, H.; Largeteau, A.; Demazeau, G. Efficiency of pulse pressure treatment for inactivation of *Escherichia coli* and *Listeria innocua* in whole milk. *Eur. Food Res. Technol.* **2009**, *229*, 127–131.
26. Shao, Y.; Ramaswamy, H.S.; Zhu, S. High-pressure destruction kinetics of spoilage and pathogenic bacteria in raw milk cheese. *J. Food Proc. Eng.* **2007**, *30*, 357–374.
27. Reps, A.; Kołakowski, P.; Dajnowiec, F. The effect of high pressure on microorganisms and enzymes of ripening cheeses. In *High Pressure Food Science, Bioscience & Chemistry*; Isaacs, N.S., Ed.; The Royal Society of Chemistry: Cambridge, London, UK, 1998; pp. 265–270.
28. López-Pedemonte, T.J.; Roig-Sagués, A.X.; Trujillo, A.J.; Capellas, M.; Guamis, B. Inactivation of spores of *Bacillus cereus* in cheese by high hydrostatic pressure with the addition of nisin or lysozyme. *J. Dairy Sci.* **2003**, *86*, 3075–3081.
29. Reps, A.; Warminska-Radyko, I.; Dajnowiec, F. Effect of high pressure on yogurt. In *Advances in High Pressure Bioscience and Biotechnology*; Ludwig, H., Ed.; Springer-Verlag: Berlin, Germany, 1999; pp. 453–456.
30. Ponce, E.; Pla, R.; Sendra, E.; Guamis, B.; Mor-Mur, M. Destruction of *Salmonella enteritidis* inoculated in liquid whole egg by high hydrostatic pressure: Comprative study in selective and non-selective media. *Food Microbiol.* **1999**, *16*, 357–365.
31. Huang, E.; Mittal, G.S.; Griffiths, M.W. Inactivation of Salmonella enteritidis in liquid whole egg using combination of pulsed electrical field, high pressure and ultrasound. *Biosyst. Eng.* **2006**, *94*, 403–413.
32. Bari, M.L.; Ukuku, D.O.; Mori, M.; Kawamoto, S.; Yamamoto, K. Effect of hydrostatic pressure pulsing on the inactivation of *Salmonella* Enteritidis in liquid whole egg. *Foodborne Path. Dis.* **2008**, *5*, 175–182.
33. Ponce, E.; Pla, R.; Capellas, M.; Guamis, B.; Mor-Mur, M. Inactivation of *Escherichia coli* inoculated in liquid whole egg by high hydrostatic pressure. *Food Microbiol.* **1998**, *15*, 265–272.
34. Yuste, J.; Mor-Mur, M.; Capellas, M.; Guamis, B.; Pla, R. Microbiological quality of mechanically recovered poultry meat treated with high hydrostatic pressure and nisin. *Food Microbiol.* **1998**, *15*, 407–414.
35. Morales, P.; Calzada, J.; Ávila, M.; Nuñez, M. Inactivation of *Escherichia coli* O157:H7 in ground beef by single-cycle and multiple-cycle high-pressure treatments. *J. Food Prot.* **2008**, *71*, 811–815.
36. Morales, P.; Calzada, J.; Rodríguez, B.; de Paz, M.; Nuñez, M. Inactivation of *Salmonella* Enteritidis in chicken breast fillets by single-cycle and multiple-cycle high pressure treatments. *Foodborne Path. Dis.* **2009**, *6*, 577–581.

37. Del Olmo, A.; Morales, P.; Ávila, M.; Calzada, J.; Nuñez, M. Effect of single-cycle and multiple-cycle high-pressure treatments on the colour and texture of chicken breast fillets. *Innov. Food Sci. Emerg. Technol.* **2010**, *11*, 441–444.
38. Hurtado, J.L.; Montero, P.; Borderías, A.J. Influence of high isostatic pressure on muscle of octopus (*Octopus vulgaris*). In *Advances in High Pressure Bioscience and Biotechnology*; Ludwig, H., Ed.; Springer-Verlag: Berlin, Germany, 1999; pp. 409–412.
39. Hurtado, J.L.; Montero, P.; Borderías, J.; Solas, M. High-pressure/temperature treatment effect on the characteristics of octopus (*Octopus vulgaris*) arm muscle. *Eur. Food Res. Technol.* **2001**, *213*, 22–29.
40. Fioretto, F.; Cruz, C.; Largeteau, A.; Sarli, T.A.; Demazeau, G.; El Moueffak, A. Inactivation of *Staphylococcus aureus* and *Salmonella enteritidis* in tryptic soy broth and caviar samples by high pressure processing. *Braz. J. Med. Biol. Res.* **2005**, *38*, 1259–1265.
41. Alemán, G.D.; Ting, E.Y.; Farkas, D.F.; Mordre, S.C.; Hawes, A.C.O.; Torres, J.A. Comparison of static and step-pulsed ultra-high pressure on the microbial stability of fresh cut pineapple. *J. Sci. Food Agric.* **1998**, *76*, 383–388.
42. Goodridge, L.D.; Willford, J.; Kalchayanand, N. Destruction of *Salmonella* Enteriditis inoculated onto raw almonds by high hydrostatic pressure. *Food Res. Int.* **2006**, *39*, 408–412.
43. Homma, K.; Haga, N. Effect of high pressure-treatment on sterilization and physical properties of egg white. In *High Pressure Science for Food*; Hayashi, R., Ed.; San-Ei Publishing: Kyoto, Japan, 1990; pp. 317–324.
44. Meyer, R. Ultra High Pressure, High Temperature Food Preservation Process. US 6,017,572A, 25 January 2000.
45. Farkas, D.R.; Hoover, D.G. High pressure processing. *J. Food Sci.* **2000**, *65*, 47–64.

14

Incidence, Antimicrobial Susceptibility, and Toxin Genes Possession Screening of *Staphylococcus aureus* in Retail Chicken Livers and Gizzards

Lubna S. Abdalrahman and Mohamed K. Fakhr *

Department of Biological Science, The University of Tulsa, Tulsa, OK 74104, USA;
E-Mail: Lubna-abdalrahman@utulsa.edu

* Author to whom correspondence should be addressed; E-Mail: Mohamed-fakhr@utulsa.edu

Academic Editor: Christopher J. Smith

Abstract: Few recent outbreaks in Europe and the US involving *Campylobacter* and *Salmonella* were linked to the consumption of chicken livers. Studies investigating *Staphylococcus aureus* in chicken livers and gizzards are very limited. The objectives of this study were to determine the prevalence, antimicrobial resistance, and virulence of *S. aureus* and MRSA (Methicillin-Resistant *Staphylococcus aureus*) in retail chicken livers and gizzards in Tulsa, Oklahoma. In this study, 156 chicken livers and 39 chicken gizzards samples of two brands were collected. While one of the brands showed very low prevalence of 1% (1/100) for *S. aureus* in chicken livers and gizzards, the second brand showed prevalence of 37% (31/95). No MRSA was detected since none harbored the *mecA* or *mecC* gene. Eighty seven *S. aureus* isolates from livers and 28 from gizzards were screened for antimicrobial resistance to 16 antimicrobials and the possession of 18 toxin genes. Resistance to most of the antimicrobials screened including cefoxitin and oxacillin was higher in the chicken gizzards isolates. While the prevalence of enterotoxin genes *seg* and *sei* was higher in the gizzards isolates, the prevalence of hemolysin genes *hla*, *hlb*, and *hld* was higher in the livers ones. The lucocidin genes *lukE-lukD* was equally prevalent in chicken livers and gizzards isolates. Using *spa* typing, a subset of the recovered isolates showed that they are not known to be livestock associated and, hence, may be of a human origin. In conclusion, this study stresses the importance of thorough cooking of chicken livers and gizzards since it might contain multidrug resistant enterotoxigenic *S. aureus*. To our knowledge this is the first study to specifically investigate the prevalence of *S. aureus* in chicken livers and gizzards in the US.

Keywords: *Staphylococcus aureus*; antibiotic resistance; toxins; toxin genes; prevalence; chicken livers; chicken gizzards; antimicrobials; foodborne pathogens; retail meat

---

1. Introduction

Recent studies highlighted the importance of chicken livers as a food safety hazard. A US study found that 77% of retail chicken livers and 33% of chicken gizzards were contaminated with *Campylobacter* [1]. Several recent outbreaks involving *Campylobacter* from poultry livers have occurred in Europe [2–5] and in the US [6,7]. A US outbreak of *Salmonella* Heidelberg was linked to Kosher Broiled Chicken Livers [8]. One study—as a part of a larger study on retail raw chicken meat throughout Japan—reported a high prevalence of *S. aureus* in chicken livers (63.8%) and chicken gizzards (58.1%) after enrichment [9]. Another study in Turkey reported 9/30 of chicken giblets (30%) to be positive for the bacterium [10].

Staphylococcus aureus is the fifth pathogen causing domestically acquired foodborne illnesses annually in the United States [11] and its food poisoning is an important cause of food-borne diseases worldwide [12]. Several staphylococcal enterotoxins and enterotoxin-like superantigens have been described [13–15]. Staphylococcal food poisoning is an intoxication resulting from the consumption of food that has been contaminated with the toxin itself and most of its symptoms are self-limiting within one to two days after consumption [16]. *Staphylococcus aureus* has the ability to destroy red blood cells by producing three types of hemolysins, known as alpha, beta and delta toxins [17]. The beta hemolysin gene is considered dangerous since it encodes the beta toxins that have the ability to inhibit the ciliary activity of human lungs and cornea [18–20].

Most studies in the literature specifically reporting prevalence of MRSA (Methicillin-Resistant *Staphylococcus aureus*) in food animals were conducted in pigs [21–24]. *S. aureus* and MRSA have been also isolated and was highly prevalent in poultry in the Netherlands [25]. It was also detected in retail chicken meat in Japan [9]. In the last few years, few studies were conducted to determine the prevalence of *S. aureus* and MRSA in retail meats in several states in the United States that includes North Dakota [26,27], Georgia [28], Minnesota, and New Jersey [29], Iowa [30], Detroit, Michigan [31], Maryland [32] and Louisiana, [33]. Only few of those studies included data from poultry such as ground turkey [32], chicken and/or turkey [26,27,30,31]. While most studies conducted in Europe identified MRSA isolated from retail meats as belonging to Livestock associated MRSA strains (LA-MRSA), the majority of MRSA isolates detected in the US retail meats were human associated strains except few that were LA-MRSA and were mostly reported in pork meat. So, opposite to the case in Europe, US retail meats when contaminated with MRSA the origin is most likely from human handling the meat products.

Studies investigating the prevalence of *S. aureus* and MRSA in chicken livers and gizzards are very limited [9] and to our knowledge none is available to date from the US. Chicken livers in particular are usually undercooked to preserve taste and texture which can be risky if contaminated with enterotoxin producing strains of *S. aureus* specially that most of these enterotoxins are heat stable. The objectives of this study were to determine the prevalence of *S. aureus* and MRSA in retail chicken livers and gizzards collected in Tulsa, Oklahoma and to characterize the recovered isolates for their antimicrobial susceptibility and possession of toxin genes. To our knowledge this study is the first to specifically investigate the prevalence of *S. aureus* in chicken livers and gizzards in the US.

2. Experimental Section

2.1. Isolation of Staphylococcus aureus from Retail Chicken Livers and Gizzards

Chicken livers and gizzards samples were collected from several different grocery stores in the Tulsa, Oklahoma area for a period of six months starting January of 2010. A total of 195 chilled retail chicken liver and chicken gizzard samples were used in this study (156 chicken livers and 39 chicken Gizzards) (Table 1). Meat samples were purchased from nine grocery stores that belong to six different franchises chains at variable locations in the city. The chicken livers and gizzards belonged to two major brands, which are designated brand A and brand B (Table 1). Samples were selected to be as variable as possible with different expiration and production dates. Chicken livers and gizzards samples were added to 10 mL of buffered peptone water (BPW) (BPW; EMD, Gibbstown, NJ, USA) in sterile plastic bags (VWR Scientific, Radnor, PA, USA) and massaged by hand for approximately 5 min. Ten milliliters was then transferred from the bag and added to 10 mL of enrichment broth of 2× Trypticase Soy Broth with 10% sodium chloride and 1% sodium pyruvate, then incubated at 37 °C for 24 h. A loopful was then streaked to Baird Parker (BP) selective media plates and incubated at 37 °C for 48 h [33]. Four suspected *S. aureus* colonies (those that have black colonies surrounded by 2 to 5 mm clear zones) were selected and streaked to Trypticase Soy Agar (TSA) plates and subcultured for confirmation on MSA (Mannitol Salt Agar) plates. Pure prospective *S. aureus* cultures were kept at −80 °C until PCR confirmation.

Table 1. PCR primers and their references for *Staphylococcus aureus* and MRSA identification.

Gene	Size (bp)	Primer sequences (5′–3′)	Bacterium	References
Sa4221-1 *Sa4221-2*	108	AAT CTT TGT CGG TAC ACG ATA TTC TTC ACG CGT AAT GAG ATT TCA GTA GAT AAT ACA ACA	*S. aureus*	[34]
mecA-F *mecA-R*	312	GTT GTA GTT GTC GGG TTT GG CTT CCA CAT ACC ATC TTC TTT AAC	*MRSA*	[21]
mecA1F *mecA2R*	533	AAA ATC GAT GGT AAA GGT TGG C AGT TCT GCA GTA CCG GAT TTG C	*MRSA*	[35]
MecHomFW *MecHomRV*	356	TCA CCA GGT TCA AC[Y] CAA AA CCT GAA TC[W] GCT AAT AAT ATT TC	*MRSA*	[36]
mecAM10/0061 F1 *mecAM10/0061 R1*	1800	CCA GAT ATA GTA GCA TTA TA AAA GAT GAC GAT ATT GAG	*MRSA*	[37]

2.2. DNA Extraction

DNA was extracted from the prospective *S. aureus* strains using the single cell lysing buffer (SCLB) method [38]. One day old colonies were picked and suspended in 40 μL of single cell lysing buffer (SCLB) solution (1.0 mL of TE buffer (10 mM Tris-HCL and 1 mM EDTA) and 10 µl of 5 mg/mL proteinase K) in a 0.2 mL microtube. In a thermocycler, bacterial cells were lysed by initial incubation at 80 °C for 10 min, followed by 55 °C for 10 min, and then 95 °C for 10 min [38]. DNA extracted by the above mentioned method was stored at −20 °C until used as a DNA template for PCR.

2.3. PCR Identification

A multiplex PCR reaction was used to identify the isolated suspected *S. aureus* by using specific primers for *S. aureus* and MRSA to amplify a 108 bp and a 312 bp fragments respectively (Table 4). Twenty microliters PCR reactions, which included 10 µL of Qiagen Master Mix (Qiagen Inc., Valencia, CA, USA), 4 µL of sterile water (Qiagen), 1 µL of each forward and reverse primer (IDT, Coralville, IA, USA), and 2 µL of DNA template, were performed. The PCR protocol was as follows: initial denaturing at 95 °C for 5 min (followed by 35 cycles of denaturing at 95 °C for 1 min, annealing at 55 °C for 1 min, and extension at 72 °C for 1 min) and ending with extension at 72 °C for 10 min. PCR amplicons were subjected to agarose gel electrophoresis and DNA bands were visualized and recorded using a gel documentation system. Isolates showing resistance to cefoxitin and/oxacillin were subjected to PCR confirmation using a second set of MRSA primers that amplify a 533 bp *mecA* fragment and two other variant MRSA *mecA* primer sets (also known as *mecC*) that amplify 356 bp and 1800 bp fragments to confirm the MRSA phenotype (Table 1).

2.4. Antimicrobial Susceptibility Testing

A total of 115 *S. aureus* recovered isolates (87 chicken liver isolates and 28 chicken gizzard isolates) were subjected to antimicrobial resistance profiling against sixteen different antimicrobials that belong to ten different antibiotic classes (Table 2). Isolates were grown on Mueller-Hinton (MH) agar (Difco) and incubated for 48 h at 37 °C. Cultures were then added to Mueller-Hinton broth (Difco), adjusted to turbidity equal to a 0.5 McFarland standard, and inoculated onto 6-inch MH agar plates supplemented with the appropriate antimicrobial at different concentrations (Table 2) including the breakpoint established for each antimicrobial according to the Clinical and Laboratory Standards Institute (CLSI) when available [39]. Plates were then incubated at 37 °C for 48 h and results were read for growth or no growth and denoted as resistant or susceptible, respectively according to the breakpoints for each of the tested antimicrobials (Table 2).

Table 2. A list of the 16 antimicrobials, their classes, concentrations used for susceptibility testing, and the breakpoints used for each antimicrobial.

Antimicrobial Class	Antimicrobials	Conc. 1 * (μg/mL)	Conc. 2 (μg/mL) (Break point)	Conc. 3 (μg/mL)	Conc. 4 (μg/mL)
β-Lactams	penicillin	0.125	0.25	0.5	1
	ampicillin	0.25	0.5	1	2
	oxacillin + 2% Nacl	2	4	8	16
	cefoxitin + 2% Nacl	4	8	16	32
Tetracyclines	tetracycline	8	16	32	64
	doxycycline	8	16	32	64
Macrolides	azithromycin	4	8	16	32
	erythromycin	4	8	16	32
Aminoglycosides	kanamycin	32	64	128	256
	gentamicin	8	16	32	64
Fluoroquinolones	ciprofloxacin	2	4	8	16
Lincosamides	clindamycin	2	4	8	16
Phenicols	chloramphenicol	16	32	64	128
Glycopeptides	vancomycin	16	32	64	128
Rifamycines	rifampin	2	4	8	16
Sulfonamides	trimethoprim/sulfamethoxazole	2/38	4/76	8/152	16/304

* Conc.: Concentration.

2.5. Detection of Toxin Genes

A total of 115 *Staphylococcus aureus* isolates (87 chicken liver isolates and 28 chicken gizzard isolates) were screened for eighteen different toxin genes that belong to six different toxin gene groups (Table 3). Multiplex PCR was used to detect 18 different toxin genes of *S. aureus* isolates that include enterotoxins, toxic shock syndrome toxin 1, exfoliative toxins, leucocidins, Panton-Valentine leucocidin (PVL), and hemolysins (Table 3). Three multiplex reactions (A, B, and C), each of which included six toxin genes, were performed (Table 3). The multiplex PCR targeting the toxin genes were performed in a 20 μL reaction solution that contained 10 μL of Green Master Mix (Promega), 2μL of sterile water, 2 μL of the DNA template and 0.5 μL of each of the toxin gene primers. The PCR protocol included an initial denaturation at 95 °C for 5 min, followed by 30 cycles of denaturation (94 °C for 1 min), annealing (57 °C for 1 min), and extension (72 °C for 1 min), ending with an extension at 72 °C for 7 min. PCR amplicons were subjected to agarose gel electrophoresis and DNA bands were visualized and recorded using a gel documentation system. The expected amplicon band sizes of *S. aureus* toxin genes are shown in Table 3. Several representative amplicons of each positive toxin were sequenced in house using the same amplifying primers to confirm PCR accuracy.

Table 3. Multiplex PCR primers, reaction sets, references, and toxin groups for the screened toxin genes.

Toxin Gene (Toxin group)	Size (bp)	Primer sequences (5′–3′)	Multiplex PCR reaction set	References
sea (Enterotoxins)	521	GCA GGG AAC AGC TTT AGG C GTT CTG TAG AAG TAT GAA ACA CG	A	[40]
seb-sec (Enterotoxins)	665	ATG TAA TTT TGA TAT TCG CAG TG TGC AGG CAT CAT ATC ATA CCA	A	[40]
sec (Enterotoxins)	284	CTT GTA TGT ATG GAG GAA TAA CAA TGC AGG CAT CAT ATC ATA CCA	A	[40]
sed (Enterotoxins)	385	GTG GTG AAA TAG ATA GGA CTG C ATA TGA AGG TGC TCT GTG G	A	[40]
see (Enterotoxins)	171	TAC CAA TTA ACT TGT GGA TAG AC CTC TTT GCA CCT TAC CGC	A	[40]
seg (Enterotoxins)	328	CGT CTC CAC CTG TTG AAG G CCA AGT GAT TGT CTA TTG TCG	A	[40]
seh (Enterotoxins)	359	CAA CTG CTG ATT TAG CTC AG GTC GAA TGA GTA ATC TCT AGG	B	[40]
sei (Enterotoxins)	466	CAA CTC GAA TTT TCA ACA GGT AC CAG GCA GTC CAT CTC CTG	B	[40]
sej (Enterotoxins)	142	CAT CAG AAC TGT TGT TCC GCT AG CTG AAT TTT ACC ATC AAA GGT AC	B	[40]
tst (Toxic Shock Syndrome Toxin 1)	560	GCT TGC GAC AAC TGC TAC AG TGG ATC CGT CAT TCA TTG TTA A	B	[40]
eta (Exfoliative toxins)	93	GCA GGT GTT GAT TTA GCA TT AGA TGT CCC TAT TTT TGC TG	B	[41]
etb (Exfoliative toxins)	226	ACA AGC AAA AGA ATA CAG CG GTT TTT GGC TGC TTC TCT TG	B	[41]
lukS-lukF (Panton-Valentine leucocidin (PVL))	433	ATC ATT AGG TAA AAT GTC TGG ACA TGA TCC A GCA TCA AST GTA TTG GAT AGC AAA AGC	C	[42]
lukE-lukD (Leucocidin)	269	TGA AAA AGG TTC AAA GTT GAT ACG AG TGT ATT CGA TAG CAA AAG CAG TGC A	C	[42]
lukM (Leucocidin)	780	TGG ATG TTA CCT ATG CAA CCT AC GTT CGT TTC CAT ATA ATG AAT CAC TAC	C	[42]
hla (Hemolysins)	209	CTG ATT ACT ATC CAA GAA ATT CGA TTG CTT TCC AGC CTA CTT TTT TAT CAG T	C	[42]
hlb (Hemolysins)	309	GTG CAC TTA CTG ACA ATA GTG C GTT GAT GAG TAG CTA CCT TCA GT	C	[42]
hld (Hemolysins)	111	AAG AAT TTT TAT CTT AAT TAA GGA AGG AGT G TTA GTG AAT TTG TTC ACT GTG TCG A	C	[42]

2.6. Molecular Typing Using spa Genotyping

A subset of the recovered *Staphylococcus aureus* isolates were subjected to molecular typing using *spa* typing. Molecular typing using *spa* was done according to published primers and protocols [43] and *spa* types were assigned using the BioNumerics Software (Applied Math, Austin, TX, USA) through the Ridom Spa Server.

3. Results and Discussion

3.1. Prevalence of Staphylococcus aureus and MRSA in Chicken Livers and Gizzards

A total of 195 chilled retail chicken liver and chicken gizzard samples were purchased from several Tulsa area grocery stores starting January 2010 for a period of 6 months. The number of chicken liver samples was 156 and the number of chicken gizzard samples was 39 (Table 1). The chicken livers and gizzards collected in this study belonged to two major brands, which were designated brand A and brand B (Table 4). As shown in Table 4, the overall prevalence of *S. aureus* in chicken livers and gizzards including the two brands together was 36/195 (18.5%). While 27/156 (17.3%) of chicken livers were contaminated with *S. aureus*, 9/39 (23.1%) of chicken gizzards were positive for the bacterium. The prevalence of *S. aureus* in brand A chicken livers was 26/71 (36.6%) while it was 9/24 (37.5%) in chicken gizzards of the same brand (Table 1). Only one out of 85 chicken liver samples (1.2%) of brand B was positive for *S. aureus* and none of the chicken gizzards of this brand was positive for *S. aureus*. No isolates of chicken livers and gizzards were positive for MRSA since none of them carried *mecA* or *mecC* genes.

Even though the overall prevalence of *S. aureus* in chicken livers and gizzards was 36/195 (18.5%) in our study, the 36.6% and the 37.5% prevalence in brand A chicken livers and gizzards respectively is alarming (Table 4). While there were no available studies in the literature that specifically determined the prevalence of *S. aureus* in chicken livers and gizzards, a study in Turkey reported contamination in 9/30 (30%) of chicken giblets as a part of a larger study on chicken meat [10]. A second study in Japan reported a higher prevalence of *S. aureus* in chicken livers (63.8%) and chicken gizzards (58.1%) after enrichment while it was 47.9% and 22.6% respectively before enrichment [9]. The higher prevalence in the Japanese study might be due to differences between the US and the Japanese retail poultry markets. It can also be due to the methods used for identification since the Japanese study used only biochemical tests for identification of the *S. aureus* strains while molecular identification was used in our study. *S. aureus* was isolated from 56% of ground turkey collected from Maryland, USA [32] and was found in 25% of retail chicken in Detroit, Michigan where 3.9% were MRSA [31]. In another study in Iowa, 17.8% of retail chicken was contaminated with *S. aureus* [30], while in North Dakota a higher prevalence of *S. aureus* (67.6%) in retail chicken was reported [26].

The big difference between the overall prevalence of *S. aureus* in chicken livers and gizzards in brand A in our study (36.8%) and only 1% in brand B (Table 4) might be due to the difference in food safety and microbiology quality control handling protocols at the two production companies. While not conclusive, this data might suggest that chicken livers and gizzards contamination with *S. aureus* most probably occurs during handling at the slaughter house or at the retail packaging step.

Table 4. Prevalence of *Staphylococcus aureus* in the 195 collected chicken liver and gizzards samples.

Prevalence of *Staphylococcus aureus*								
Chicken Livers			Chicken Gizzards			Chicken Livers and Gizzards		
Brand A	Brand B	Total	Brand A	Brand B	Total	Brand A	Brand B	Total
np/n * (%)	*np/n* (%)	*np/n* (%)	*np/n* (%)	*np/n* (%)	*np/n* (%)	*np/n* (%)	*np/n* (%)	*np/n* (%)
26/71 (36.6)	1/85 (1.2)	27/156 (17.3)	9/24 (37.5)	0/15 (0)	9/39 (23.1)	35/95 (36.8)	1/100 (1)	36/195 (18.5)

* *np*: number of positive samples, *n*: number of samples collected.

3.2. Antimicrobial Susceptibility of the Recovered Isolates

A total of 115 *Staphylococcus aureus* isolates (87 chicken liver isolates and 28 chicken gizzard isolates) were subjected to antimicrobial resistance profiling against sixteen different antimicrobials that belong to ten different antibiotic classes (Tables 2 and 5). As shown in Table 5, the percentage of resistance of the 115 *S. aureus* isolates from chicken livers and gizzards to the sixteen tested antimicrobials were as follow: ampicillin (88.9%), tetracycline (71.3%), doxycycline (63.5%), penicillin (60.9%), erythromycin (45.2%), azithromycin (40,9%), vancomycin (39.1%), oxacillin with 2% NaCl (32.2%), ciprofloxacin (29.6%), trimethoprim/sulfamethazole (24.3%), rifampin (23.5%), cefoxitin with 2% NaCl (19.1%), clindamycin (12.2%), kanamycin (12.2%), chloramphenicol (10.4%), and gentamicin (10.4%). As shown in Table 5, the percentage of resistance to the sixteen tested antimicrobials varied between chicken livers and chicken gizzards isolates. The percentage of resistance found in the chicken gizzards was higher than chicken livers isolates for the following 10 antimicrobials: azithromycin, ciprofloxacin, oxacillin, cefoxitin, tetracycline, vancomycin, doxycycline, penicillin, kanamycin, and erythromycin (Table 5). On the other hand, for gentamycin, ampicillin, trimethoprim/sulfamrthoxazole, clindamycin, rifampin, chloramphenicol the chicken livers isolates showed a higher resistance (Table 5). This variability in antimicrobial resistance between isolates from chicken livers and gizzards might be attributed to the concentration of different antimicrobials in the liver and/or the fact that chicken gizzards often have more fats that would make some highly lipid soluble antimicrobials like azithromycin get to higher concentrations in the gizzards. Overall 35/115 (30%) of the screened isolates from chicken livers and gizzards were multidrug resistant to more than seven antimicrobials (data not shown) which is worrisome.

Resistance to vancomycin was relatively high in isolates from chicken livers and gizzards in our study (Table 5).Vancomycin Resistant *Enterococcus faecium* (VRE) was previously reported in swine in Michigan, USA and was thought to be widespread despite the historical absence of the use of agricultural glycopeptides like avoparcin. Screening our phenotypically vancomycin strains for the presence of the *vanA* gene is currently underway as a part of a larger study focusing on vancomycin resistant *S. aureus* strains isolated from various US retail meats. Chicken livers and gizzards isolates in our study were highly resistant to ampicillin, tetracycline, doxycycline, penicillin, and erythromycin (Table 5). The literature is lacking data about antimicrobial resistance of *S. aureus* strains isolated from chicken livers and gizzards. One study in Turkey [10] reported that *S. aureus* isolates from chicken giblets were resistant to penicillin G (22.2%) and erythromycin (33.3%). *S. aureus* recovered strains in our study that showed resistance to cefoxitin and/or oxacillin (highly prevalent in chicken gizzards as shown in Table 5) were subjected to additional PCR protocols to check for the presence of a *mecA* homologue (Table 1). None of these isolates showed the presence of *mecA* gene or it homologues (*mecC*). Phenotypic MRSA isolates that do not harbor the *mecA* gene were reported [32,44]. This might be due to over production of Beta-lactamase enzymes or the presence of a variant *mecA* gene that does not amplify with the known PCR primers. The recent advancement in whole genome sequencing through next generation sequencing can help identifying such homologues.

The high number of multidrug resistant *S. aureus* detected in our study is alarming. It raises concerns about inappropriate practices including the use of antimicrobials as growth promotors in food animal production and the frequent use of antimicrobials in poultry husbandry. Genes coding for antimicrobial resistance can move through horizontal gene transfer to clinical pathogenic strains and contribute to the creation of superbugs. Death in hospitals are often attributed to sepsis resulting from infections caused by multidrug resistant pathogens like MRSA, *Pseudomonas aeruginosa* or *Candida albicans* rather than the original cause of the hospitalization.

Table 5. Antimicrobial resistance of the 115 *Staphylococcus aureus* chicken liver and gizzard isolates against 16 different antimicrobials.

	Antimicrobial Resistance		
Antibiotic	**Chicken Livers** ***np/n*** **(%) ***	**Chicken Gizzards** ***np/n*** **(%)**	**Chicken Livers and Gizzard** ***np/n*** **(%)**
azithromycin	25/87 (28.7)	22/28 (78.6)	47/115 (40.9)
ciprofloxacin	13/87 (14.9)	21/28 (75.0)	34/115 (29.6)
gentamicin	10/87 (11.5)	2/28 (7.1)	12/115 (10.4)
oxacillin	19/87 (21.8)	18/28 (64.3)	37/115 (32.2)
cefoxitin	10/87 (11.5)	12/28 (42.9)	22/115 (19.1)
tetracycline	59/87 (67.8)	23/28 (82.1)	82/115 (71.3)
vancomycin	30/87 (34.5)	15/28 (53.6)	45/115 (39.1)
doxycycline	49/87 (56.3)	24/28 (85.7)	73/115 (63.5)
trimethoprim/sulfamethoxazole	22/87 (25.3)	6/28 (21.4)	28/115 (24.3)
clindamycin	13/87 (14.9)	1/28 (3.6)	14/115 (12.2)
penicillin	47/87 (54.0)	23/28 (82.1)	70/115 (60.9)
ampicillin	83/87 (95.4)	19/28 (67.9)	102/115 (88.9)
kanamycin	17/87 (19.5)	7/28 (25.0)	14/115 (12.2)
erythromycin	31/87 (35.6)	21/28 (75.0)	52/115 (45.2)
rifampin	25/87 (28.7)	2/28 (7.1)	27/115 (23.5)
chloramphenicol	11/87 (12.6)	1/28 (3.6)	12/115 (10.4)

* *np*: number of positive isolates, *n*: number of isolates collected.

3.3. Toxin Genes Possession Screening of the Recovered Isolates

A total of 115 *Staphylococcus aureus* isolates (87 chicken liver isolates and 28 chicken gizzard isolates) were screened for eighteen different toxin genes that belong to six different toxin gene groups (Tables 3 and 6). As shown in Table 6, the prevalence of toxin genes in the 115 *S. aureus* isolates from chicken livers and gizzards to the eighteen tested toxin genes were as follow: *hla* (94.5%), *hld*

(94.5%), *hlb* (48.7%), *sei* (42.6%), *lukE-lukD* (36.5%), *seg* (29.6%), *seh* (4.3%), *sed* (0.9%), *sea* (0%), *seb-sec* (0%), *sec* (0%), *see* (0%), *sej* (0%), *tst* (0%), *eta* (0%), *etb* (0%), *lukM* (0%), and *lukS-lukF* (0%). *S. aureus* hemolysin genes were found at a higher percentage in the chicken livers and gizzards than other groups of toxin genes screened. Also no isolates harbored enterotoxin genes *sed, sea, seb-sec, sec, see*, or *sej*, the toxic shock syndrome toxin 1 gene *tst*, the exfoliative toxin genes *eta, etb*, or the Leucocidin gene *lukM* (Table 6). The prevalence of enterotoxin genes *seg* (71.43%) and *sei* (92.9%) in chicken gizzards was higher than in chicken livers, where *seg* was 16.1% and *sei* was 26.4%. One isolate from chicken livers was positive for the entoretoxin gene *sed* (1.2%) and 23/87 of chicken livers isolates were positive for the entoretoxin gene *sei* (26.4%). The prevalence of hemolysin genes *hla* (97.5%), *hld* (97.5%) and *hlb* (64.4%) in chicken livers was higher than in chicken gizzards when it was 85.7%, 85.7% and 0% respectively. The Hemolysin gene *hlb* was present only in the chicken livers. Both chicken livers and gizzards isolates had similar prevalence for *lukE-lukD*.

Table 6. The prevalence of 18 different toxin genes in the 115 *Staphylococcus aureus* chicken liver and gizzard isolates.

	Prevalence of Toxin Genes		
Toxin Gene	**Chicken Livers** ***np/n*** **(%) ***	**Chicken Gizzards** ***np/n*** **(%)**	**Chicken Livers and Gizzards** ***np/n*** **(%)**
sea	0/87 (0)	0/28 (0)	0/115 (0)
seb-sec	0/87 (0)	0/28 (0)	0/115 (0)
sec	0/87 (0)	0/28 (0)	0/115 (0)
sed	1/87 (1.2)	0/28 (0)	1/115 (0.9)
see	0/87 (0)	0/28 (0)	0/115 (0)
seg	14/87 (16.1)	20/28 (71.43)	34/115 (29.6)
seh	5/87 (5.8)	0/28 (0)	5/115 (4.3)
sei	23/87 (26.4)	26/28 (92.9)	49/115 (42.6)
sej	0/87 (0)	0/28 (0)	0/115 (0)
tst	0/87 (0)	0/28 (0)	0/115 (0)
eta	0/87 (0)	0/28 (0)	0/115 (0)
etb	0/87 (0)	0/28 (0)	0/115 (0)
lukE-lukD	32/87 (36.8)	10/28 (35.7)	42/115 (36.5)
lukM	0/87 (0)	0/28 (0)	0/115 (0)
hla	85/87 (97.7)	24/28 (85.7)	109/115 (94.5)
hlb	56/87 (64.4)	0/28 (0)	56/115 (48.7)
hld	85/87 (97.7)	24/28 (85.7)	109/115 (94.5)
lukS-lukF	0/87 (0)	0/28 (0)	0/115 (0%)

* *np*: number of positive isolates, *n*: number of isolates collected.

The literature is lacking data related to the prevalence of toxin genes in *S. aureus* isolated from chicken livers and gizzards. Even recent studies discussing *S. aureus* in US retail poultry in general lacks such toxin genes prevalence data. A study in Japan reported that 25.3% of their chicken liver isolates were enterotoxigenic while 36.4% of their chicken gizzards produced enterotoxins [9]. Even though they have not used PCR to detect enterotoxin genes in the Japanese study, their chicken gizzards strains showed a higher prevalence of enterotoxins that their chicken livers ones which is in agreement with our findings. The higher prevalence of hemolysin genes in chicken livers isolates than in the chicken gizzards ones might be due to the availability of blood in the liver which might select for *S. aureus* strains with blood lysing abilities. Chicken livers and gizzards should be cooked thoroughly since enterotoxins of *S. aureus* are known for their heat tolerance. So even if the cooking temperature was high enough to kill the pathogen, enterotoxins produced on the chicken livers or gizzards could tolerate such temperature increasing the risk of food poisoning.

3.4. Genotyping Using spa Typing

A subset of *Staphylococcus aureus* recovered strains (6 from chicken livers, and 5 from chicken gizzards) were subjected to molecular typing by *spa* typing (Figure 1). As it is shown in Figure 1, *spa* types were grouped into two major clusters with the majority of isolates in each cluster belonging to one source. As it is also shown in figure 1, the tested isolates showed higher diversity in regards to their *spa* types since 7 different *spa* types were detected among a subset of 11 isolates. The detected *spa* types (t1081, t064, t002, and t091) were not known to be livestock associated and hence, maybe of a human origin [45]. This is in agreement with what we discussed earlier in the introduction section about that the origin of *S. aureus* strains detected in US retail meats is mostly of a human origin rather than livestock associated as it is the case in Europe.

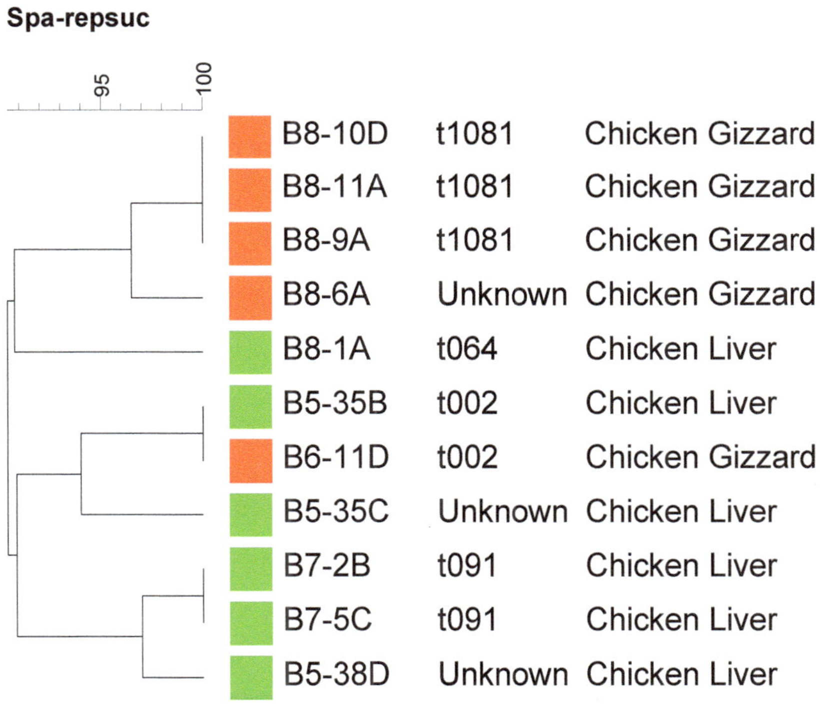

Figure 1. A dendrogram showing *spa* typing for a subset of the recovered *Staphylococcus aureus* strains representing chicken livers and chicken gizzards sources. Strains isolated from the same meat source are labeled by the same color square.

4. Conclusions

The prevalence of *S. aureus* in retail chicken livers and gizzards tested in this study varied between the two brands tested. While one of the brands showed very low prevalence of *S. aureus*, the second brand showed prevalence close to 37%. The percentage of resistance to most of the antimicrobials screened was generally higher in isolates recovered from chicken gizzards. While no isolate harbored the *mecA* or *mecC* gene, a higher percentage of the chicken gizzards isolates were resistant to cefoxitin and/or oxacillin making them phenotypically similar to MRSA. A high percentage of *S. aureus* recovered strains particularly from chicken gizzards harbored enterotoxins *seg* and *sei*. The lucocidin genes *lukE-lukD* was equally prevalent in chicken livers and gizzards isolates. The hemolysin *hlb* gene was only prevalent in the chicken livers strains while *hla* and *hld* were prevalent in chicken livers and gizzards strains. Using *spa* typing, a subset of the recovered isolates showed that they are not known to be livestock associated and hence, maybe of a human origin. Data obtained from this study stress the importance of thorough cooking of chicken livers and gizzards since it might contain multidrug resistant enterotoxigenic *S. aureus*.

Acknowledgments

L. A. is grateful to the Libyan Government for financial support during her M.S. study in the US. The authors would like to acknowledge financial support from The University of Tulsa Office of Research and Sponsored Programs (The Student Research Grant Program) in subsidizing the purchase of the retail chicken livers and gizzards used in this study. The authors are also grateful to Peggy Hill for the English language editing of the manuscript.

Author Contributions

M. K. F. suggested and designed the experiments. L. A. performed the experiments. L. A. and M. K. F. wrote and revised the manuscript.

Conflicts of Interest

The authors declare no conflict of interest.

References

1. Noormohamed, A.; Fakhr, M.K. Incidence and antimicrobial resistance profiling of *Campylobacter* in retail chicken livers and gizzards. *Foodborne Pathog. Dis.* **2012**, *9*, 617–624.
2. Little, C.L.; GormLy, F.J.; Rawal, N.; Richardson, J.F. A recipe for disaster: Outbreaks of campylobacteriosis associated with poultry liver pâté in England and Wales. *Epidemiol. Infect.* **2010**, *138*, 1691–1694.
3. Abid, M.; Wimalarathna, H.; Mills, J.; Saldana, L.; Pang, W.; Richardson, J.F.; Martin, C.J.M.; McCarthy, N.D. Duck liver-associated outbreak of campylobacteriosis among humans, United Kingdom, 2011. *Emerg. Infect. Dis.* **2013**, *19*, 1310.
4. O'Leary, M.C.; Harding, O.; Fisher, L.; Cowden, J. A continuous common-source outbreak of campylobacteriosis associated with changes to the preparation of chicken liver pâté. *Epidemiol. Infect.* **2009**, *137*, 383–388.

5. Forbes, K.J.; GormLey, F.J.; Dallas, J.F.; Labovitiadi, O.; MacRae, M.; Owen, R.J.; Richardson, J.; McGuigan, C.C. *Campylobacter* immunity and coinfection following a large outbreak in a farming community. *J. Clin. Microbiol.* **2009**, *47*, 111–116.
6. Centers for Disease Control and Prevention (CDC). Multistate outbreak of *Campylobacter jejuni* infections associated with undercooked chicken livers—Northeastern United States, 2012. *Morb. Mortal. Wkly. Rep.* **2013**, *62*, 874.
7. Terry, L. Outbreak Tied to Chicken Livers: 6 Sick, High-End Restaurants "Mortified". Available online: http://www.oregonlive.com/health/index.ssf/2014/02/post_85.html (accessed on 12 August 2014).
8. Centers for Disease Control (CDC). Multistate Outbreak of Human *Salmonella* Heidelberg Infections Linked to "Kosher Broiled Chicken livers" from Schreiber Processing Corporation. Available online: http://www.cdc.gov/salmonella/heidelberg-chickenlivers/011112/index.htmL (accessed on 12 August 2014).
9. Kitai, S.; Shimizu, A.; Kawano, J.; Sato, E.; Nakano, C.; Uji, T.; Kitagawa, H. Characterization of methicillin-resistant *Staphylococcus aureus* isolated from retail raw chicken meat in Japan. *J. Vet. Med. Sci.* **2005**, *67*, 107–110.
10. Gundogan, N.; Citak, S.; Yucel, N; Devren, Y.A. A note on the incidence and antibiotic resistance of *Staphylococcus aureus* isolated from meat and chicken *samples. Meat Sci.* **2005**, 69, 807–810.
11. Centers for Disease Control (CDC). CDC Estimates of Foodborne Illness in the United States, 2011. Available online: http://www.cdc.gov/foodborneburden/2011-foodborne-estimates.htmL (accessed on 16 May 2013).
12. Genigeorgis, C.A. Present state of knowledge on staphylococcal intoxication. *Int. J. Food Microbiol.* **1989**, *9*, 327–360.
13. Zschöck, M.; Kloppert, B.; Wolter, W.; Hamann, H.P.; LämmLer, C. Pattern of enterotoxin genes *seg*, *seh*, *sei* and *sej* positive *Staphylococcus aureus* isolated from bovine mastitis. *Vet. Microbiol.* **2005**, *108*, 243–249.
14. Ono, H.K.; Omoe, K.; Imanishi, K.; Iwakabe, Y.; Hu, D.L.; Kato, H.; Saito, N.; Nakane, A.; Uchiyama, T.; Shinagawa, K. Identification and characterization of two novel staphylococcal enterotoxin, types S and T. *Infect. Immunol*. **2008**, *76*, 4999–5005.
15. Wang, S.C.; Wu, C.M.; Xia, S.C.; Qi, Y.H.; Xia, L N.; Shen, J.Z. Distribution of superantigenic toxin genes in *Staphylococcus aureus* isolates from milk samples of bovine subclinical mastitis cases in two major diary regions of China. *Vet. Microbiol.* **2009**, *137*, 276–281.
16. Argudín, M.A.; Mendoza, M.C.; Rodicio, M.R. Food poisoning and *Staphylococcus aureus* enterotoxins. *Toxins* **2010**, *2*, 1751–1773.
17. Burnside, K.; Lembo, A.; de los Reyes, M.; Iliuk, A.; Binhtran, N.-T.; Connelly, J.E.; Lin, W.-J.; Schmidt, B.Z.; Richardson, A.R.; Fang, F.C.; *et al.* Regulation of hemolysin expression and virulence of *Staphylococcus aureus* by a serine/threonine kinase and phosphatase. *PLoS ONE* **2010**, *5*, e11071.
18. O'Callaghan, R.J.; Callegan, M.C.; Moreau, J.M.; Green, L.C.; Foster, T.J.; Hartford, O.M.; Engel, L.S.; Hill, J.M. Specific roles of alpha-toxin and beta-toxin during *Staphylococcus aureus* corneal infection. *Infect. Immun.* **1997**, *65*, 1571–1578.

19. Kim, C.S.; Jeon, S.Y.; Min, Y.G.; Rhyoo, C.; Kim, J.W.; Yun, J.B.; Park, S.W.; Kwon, T.Y. Effects of betatoxin of *Staphylococcus aureus* on ciliary activity of nasal epithelial cells. *Laryngoscope* **2000**, *110*, 2085–2088.
20. Hayashida, A.; Bartlett, A.H.; Foster, T.J.; Park, P.W. *Staphylococcus aureus* betatoxin induces *lung injury through* syndecan-1. *Am. J. Pathol.* **2009**, *174*, 509–518.
21. De Neeling, A.J.; van Leeuwen, W.J.; Schouls, L.M.; Schot, C.S.; van Veen-Rutgers, A.; Beunders, A.J.; Buiting, A.G.; Hol, C.; Ligtvoet, E.E.; Petit, P.L.; *et al.* Resistance of staphylococci in The Netherlands: Surveillance by an electronic network during 1989–1995. *J. Antimicrob. Chemoth.* **1998**, *41*, 93–101.
22. Khanna, T.; Friendship, R.; Dewey, C.; Weese, J.S. Methicillin resistant *Staphylococcus aureus* colonization in pigs and pig farmers. *Vet. Microbiol.* **2008**, *128*, 298–303.
23. Voss, A.; Loeffen, F.; Bakker, J.; Klaassen, C.; Wulf, M. Methicillin resistant *Staphylococcus aureus* in pig farming. *Emerg. Infect. Dis.* **2005**, *11*, 1965–1966.
24. Van den Eede, A.; Hermans, K.; Lipinska, U.; Struelens, M.; Deplano, A.; Denis, O.; Gasthuys, F.; Haesebrouck, F.; Martens, A. Nasal carriage of methicillin-resistance *Staphylococcus aureus* in the equine population: Prevalence, typing and antimicrobial resistance, p. 20. In Proceedings of the 2nd Symposium on Antimicrobial Resistance in Animals and the Environment, ARAE-2007, Tours, France, 17–19 December 2007.
25. De Boer, E.; Zwartkruis-Nahuis, J.T.M.; Wit, B.; Huijsdens, X.W.; de Neeling, A.J.; Bosch, T.; van Oosterom, R.A.A.; Vila A.; Heuvelink, A.E. Prevalence of methicillin-resistant *Staphylococcus aureus* in meat. *Int. J. Food Microbiol.* **2009**, *134*, 52–56.
26. Buyukcangaz, E.; Velasco, V.; Sherwood, J.S.; Stepan, R.M.; Koslofsky, R.J.; Logue, C.M. Molecular typing of *Staphylococcus aureus* and methicillin-resistant *S. aureus* (MRSA) isolated from animals and retail meat in North Dakota, United States. *Foodborne Pathog. Dis.* **2013**, *10*, 608–617.
27. Velasco, V.; Sherwood, J.S.; Rojas-García, P.P.; Logue, C.M. Multiplex real-time PCR for detection of *Staphylococcus aureus*, *mecA* and Panton-Valentine Leukocidin (PVL) genes from selective enrichments from animals and retail meat. *PloS ONE* **2014**, *9*, e97617.
28. Jackson, C.R.; Davis, J.A.; Barrett, J.B. Prevalence and characterization of methicillin-resistant *Staphylococcus aureus* isolates from retail meat and humans in Georgia. *J. Clin. Microbiol.* **2013**, *51*, 1199–1207.
29. O'Brien, A.M.; Hanson, B.M.; Farina, S.A.; Wu, J.Y.; Simmering, J.E.; Wardyn, S.E.; Smith, T.C. MRSA in conventional and alternative retail pork products. *PLoS ONE* **2012**, *7*, e30092.
30. Hanson, B.M.; Dressler, A.E.; Harper, A.L.; Scheibel, R.P.; Wardyn, S.E.; Roberts, L.K.; Smith, T.C. Prevalence of *Staphylococcus aureus* and methicillin-resistant *Staphylococcus aureus* (MRSA) on retail meat in Iowa. *J. Infect. Public Health* **2011**, *4*, 169–174.
31. Bhargava, K.; Wang, X.; Donabedian, S.; Zervos, M.; da Rocha, L.; Zhang, Y. Methicillin-resistant *Staphylococcus aureus* in retail meat, Detroit, Michigan, USA. *Emerg. Infect. Dis.* **2011**, *17*, 1135.
32. Kelman, A.; Soong, Y.A.; Dupuy, N.; Shafer, D.; Richbourg, W.; Johnson, K.; Meng, J. Antimicrobial susceptibility of *Staphylococcus aureus* from retail ground meats. *J. Food Prot.* **2011**, *74*, 1625–1629.

33. Pu, S.; Wang, F.; Ge, B. Isolation and characterization of methicillin-resistant *Staphylococcus aureus* strains from Louisiana retail meats. *Appl. Environ. Microb.* **2009**, *75*, 265–267.
34. Martineau, F.; Picard, F.J.; Roy, P.H.; Ouellette, M.; Bergeron, M.G. Species-specific and ubiquitous-DNA-based assays for rapid identification of *Staphylococcus aureus. J. Clin. Microbiol.* **1998**, *36*, 618–623.
35. Murakami, K.; Minamide, W.; Wada, K.; Nakamura, E.; Teraoka, H.; Watanabe, S. Identification of methicillin-resistant strains of staphylococci by polymerase chain reaction. *J. Clin. Microbiol.* **1991**, *29*, 2240–2244.
36. Alvarez, L.G.; Holden, M.T.G.; Lindsay, H.; Webb, C.R.; Brown, D.F.J.; Curran, M.D.; Walpole, E.; Brooks, K.; Pickard, D.J.; Teale, C.; *et al.* Methicillin-resistant *Staphylococcus aureus* with a novel *mecA* homologue in human and bovine populations in the UK and Denmark: A descriptive study. *Lancet* **2011**, *11*, 595–603.
37. Shore, A.C.; Deasy, E.C.; Slickers, P.; Brennan, G.; Connell, B.O.; Monecke, S.; Ehricht, R.; Coleman, D.C. Detection of staphylococcal cassette chromosome *mec* Type XI carrying highly divergent *mecA*, *mecI*, *mecR1*, *blaZ*, and *ccr* genes in human clinical isolates of clonal complex 130 Methicillin-Resistant *Staphylococcus aureus*. *Antimicrob. Agents Chemother.* **2011**, *55*, 3765–3773.
38. Marmur, J. A procedure for the isolation of deoxyribonucleic acid from microorganisms. *J. Mol. Bio.* **1961**, *3*, 298–218.
39. Clinical and Laboratory Standards Institute. Performance Standards for Antimicrobial Susceptibility Testing; Twenty-First Informational Supplement. *CLSI* **2011**, *31*, 1–172.
40. Monday, S.R.; Bohach, G.A. Use of multiplex PCR to detect classical and newly described pyrogenic toxin genes in staphylococcal isolates. *J. Clin. Microbiol.* **1999**, *37*, 3411–3414.
41. Mehrotra, M.; Wang, G.; Johnson, W.M. Multiplex PCR for detection of genes for *Staphylococcu aureus* enterotoxins, exfoliative toxins, toxic shock syndrome toxin 1, and methicillin resistance. *J. Clin. Microbiol.* **2000**, *38*, 1032–1035.
42. Jarraud, S.; Mougel, C.; Thioulouse, J.; Lina, G.; Meugnier, H.; Forey, F.; Nesme, X.; Etienne, J.; Vandenesch, F. Relationships between *Staphylococcus aureus* genetic background, virulence factors, *agr* groups (alleles), and human disease. *Infect. Immun.* **2002**, *70*, 631–641.
43. Shopsin, B.; Gomez, M.; Montgomery, S.O.; Smith, D.H.; Waddington, M.; Dodge, D.E.; Bost, D.E.; Riehman, M.; Naidich, S.; Kreiswirth, B.N. Evaluation of protein A gene polymorphic region DNA sequencing for typing of *Staphylococcus aureus* strains. *J. Clin. Microbiol.* **1999**, *37*, 3556–3563.
44. Pereira, V.; Lopes, C.; Castro, A.; Silva, J.; Gibbs, P.; Teixeira, P. Characterization for enterotoxin production, virulence factors, and antibiotic susceptibility of *Staphylococcus aureus* isolates from various foods in Portugal. *Food Microbiol.* **2009**, *26*, 278–282.
45. Hetem, D.J.; Bootsma, M.C.; Troelstra, A.; Bonten, M.J. Transmissibility of livestock-associated methicillin-resistant *Staphylococcus aureus*. *Emerg. Infect. Dis.* **2013**, 19, 1797.

15

Combined Effect of Pressure-Assisted Thermal Processing and Antioxidants on the Retention of Conjugated Linoleic Acid in Milk

Sergio I. Martinez-Monteagudo and Marleny D.A. Saldaña *

Department of Agricultural, Food and Nutritional Science, University of Alberta, Edmonton, AB T6G 2P5, Canada; E-Mail: marleny.saldana@ualberta.ca

* Author to whom correspondence should be addressed; E-Mail: marleny.saldana@ualberta.ca

Academic Editor: Carl Joseph Schaschke

Abstract: The effect of pressure-assisted thermal processing (PATP) in combination with seven synthetic antioxidants was evaluated on the retention of conjugated linoleic acid (CLA) in enriched milk. Milk rich in CLA was first saturated with oxygen, followed by the addition of either catechin, cysteine, ascorbic acid, tannic acid, gallic acid, caffeic acid or *p*-coumaric acid (500 mg kg^{-1} untreated milk). Samples were treated at 600 MPa and 120 °C up to 15 min of holding time. During PATP, CLA not only oxidized at a slower rate, but also less oxygen was consumed compared to the control (0.1 MPa and 120 °C). In addition, phenolic antioxidants were able to quench dissolved oxygen in samples treated with PATP. For those samples added with gallic acid and catechin, 85% and 75% of the CLA was retained after 15 min of holding time at 600 MPa and 120 °C, respectively. The retention of CLA was enhanced by the application of PATP in combination with gallic acid.

Keywords: antioxidants; conjugated linoleic acid; dissolved oxygen; pressure-assisted thermal processing; phenolics

1. Introduction

Pressure-assisted thermal processing (PATP) has become a valuable alternative to traditional thermal treatments. Although sterilized products treated with PATP have not been commercialized yet, this technology has the potential to deliver a variety of novel products, and its further implementation at the industrial level is expected in the short to medium term [1].

There are three important characteristics of PATP. First, the rise in the sample temperature due to the heat of compression is used to reach the target or sterilization temperature, reducing the thermal damage that takes place in traditional sterilization processes [2]. Second, pressure reduces interatomic distance, affecting interactions, the bond energy which is distance dependent. Such interactions are van der Waals forces, electrostatic forces, hydrogen bonding and hydrophobic interactions of proteins. In contrast, covalent bonds are unlikely to be affected by pressure, because the bonding distance can be hardly further compressed [3]. This has been the main hypothesis in preserving the biological activity of functional compounds, such as ascorbic acid [4,5], folates [6,7], anthocyanins [8], lycopene [9] and conjugated linoleic acid [10]. Finally, the rate of a chemical reaction can be increased or decreased by pressure, according to whether the molar volume of the intermediate state (activated complex) is less or more voluminous. For a chemical reaction, the effect of pressure favours those reactions with a negative reaction volume and those reaction pathways with a negative activation volume [3].

Some authors have suggested that PATP can be used to produce superior quality products in cases where the traditional thermal treatments have failed to deliver high-quality products, such as egg-based and milk-based products, baby foods, desserts, gravies, soups and sauces [3,11].

In an effort to develop high-value milk products, various investigations have highlighted the health benefits of conjugated linoleic acid (CLA), a bioactive component naturally found in milk fat [12]. Unfortunately, CLA in milk suffers significant losses during heat treatment, including ultra-high temperature (UHT), as discussed elsewhere [13–16]. Interestingly, CLA was retained up to 90% when catechin, a potent antioxidant, was added at 1 g per kg of milk, regardless of the processing conditions used (60–120 °C and 100–600 MPa) [17]. These findings suggest that the use of antioxidants can enhance the retention of CLA in milk. However, their effectiveness and the way in which they capture free radicals or react with the dissolved oxygen to retard the free radical reactions have not been studied under PATP conditions. Understanding the antioxidant mechanisms under PATP conditions is important for product development. Thus, the objective of this study was to evaluate the combined effect of PATP and seven commonly-used antioxidants on the retention of CLA in milk.

2. Experimental Section

2.1 Obtaining CLA-Enriched Milk and Sample Preparation

Milk rich in CLA was obtained from the Dairy Research and Technology Centre at the University of Alberta (Edmonton, AB, Canada), following the protocol provided elsewhere [18]. Briefly, 12 lactating Holstein cows were first fed with a control diet for 3 days. Then, the cows were fed for 16 days with a diet supplemented with sunflower ground seed at 6% of dry matter. The diet consists of 60% forage and 40% concentrate. Cows were housed in tie stalls, and water was available at all times. The CLA-enriched milk was stored at −20 °C until needed. Prior to the PATP experiments, the CLA-enriched milk was thawed with running water at room temperature. Then, the milk rich in CLA was saturated with oxygen by bubbling oxygen for 45 min at room temperature. After that, five hundred milligrams of either catechin (CAT), cysteine (Cyst), ascorbic acid (AA), tannic acid (TAN), gallic acid (GA), caffeic acid (CAF) or *p*-coumaric acid (CUM) were added per one kg of untreated milk. All antioxidants used were reagent grade and were purchased from Sigma-Aldrich (Saint Louis, MO, USA). The antioxidants were dissolved using a bench homogenizer Diax 900 (Rose Scientific Ltd., Edmonton, AB, Canada).

2.2. Pressure-Assisted Thermal Processing

PATP experiments were carried out in 3-mL polypropylene tubes (Cryogenic vial, Fisher Scientific, Edmonton, AB, Canada). The tubes were filled with either raw CLA-enriched milk or CLA-enriched milk added with an antioxidant, capped and shaken before immersion in an oil bath at 102 °C. The preheating temperature was fixed, considering that the temperature of milk rises 3 °C per 100 MPa [19]. After 2 min, the pre-heated samples were transferred to the high pressure multivessel system (Apparatus U111 Unipress, Warszawa, Poland), where the vessels were already heated at 120 °C. The vessels were heated with a thermostat (Lauda Proline RP 855 Low Temperature, Lauda-Königshofen, Germany) using propylene glycol, which was also the pressure transmission fluid. The unit has four high pressure vessels with an internal volume of 8 mL. The temperature of the transmission fluid was recorded using a Type K thermocouple located at the bottom of each vessel. Then, the samples were pressurized at 600 MPa using a rate of 10 MPa s^{-1}. Once 600 MPa was reached, the samples were held for 1, 5, 10 and 15 min. At the end of the holding time, the vessels were decompressed, and the samples were removed immediately from the high pressure vessels and cooled down with ice to avoid any further CLA degradation. For the control treatments, samples of raw CLA-milk in closed polypropylene tubes were pre-heated at 102 °C and subsequently transferred to an oil bath at 120 °C to imitate the PATP experimental run. For simplicity, the control treatments are referred to as samples treated at 0.1 MPa. PATP-treated and control samples were kept at −18 °C until further analysis. An additional set of experiments was conducted using different concentrations of CAT and GA (0, 100, 200, 300 and 500 mg kg^{-1}). The samples were treated at 120 °C for 15 min at either 0.1 or 600 MPa. The CLA was expressed in terms of normalized (CLA_t/CLA_o) retained fraction from 0 to 1. The experimental protocol has been validated for kinetics studies elsewhere [10]. All experimental data were obtained in triplicate.

2.3. Analytical Determinations

2.3.1. CLA Determination

The CLA content was determined by gas chromatography using a 100-m SP-2560 fused-silica capillary column, adopting the methodology previously described elsewhere [20]. The CLA content measured by the GC represented the total CLA.

2.3.2. Dissolved Oxygen

The dissolved oxygen (DO_2) in milk was measured using an oxygen meter (OM-4, Microelectrodes, Inc., Bedford, NH, USA). The oxygen meter was calibrated using a two-point calibration (0% and 100% of saturation). The electrode was immersed in 0% solution obtained by boiled and cooled water under nitrogen flushing. The 100% of saturation solution was obtained by bubbling oxygen for 45 min at room temperature in 500 mL of enriched milk. At each experimental point, the electrode was allowed to stabilize to the sample temperature before the DO_2 measurement was recorded. The measurements of DO_2 were performed in duplicate.

2.4. Data Analysis

2.4.1. Retention of CLA

The retained CLA in the presence of oxygen during processing was described according to reaction as follows:

$$\mathrm{CLA} + \mathrm{O}_2 \overset{k_1}{\rightarrow} \text{oxidation products}$$

where k_1 is the oxidation rate constant (min^{-1}) for each antioxidant, the k_1 value was obtained by solving the following differential equation:

$$\frac{d\mathrm{CLA}}{dt} = \frac{d\mathrm{O}_2}{dt} = -k_1 \cdot \mathrm{CLA} \cdot \mathrm{O}_2 \quad (1)$$

2.4.2. Quenching Ability

An additional set of experiments was performed to determine the quenching ability of phenolic antioxidants in milk treated at 120 °C and 600 MPa. Quenching oxygen by phenolic antioxidants was described by a bimolecular mechanism that follows:

$$\mathrm{O}_2 + \mathrm{Q} \overset{k_2}{\rightarrow} \text{products}$$

where Q is the quenching agent or phenolic antioxidant; and k_2 is the apparent quenching rate constant (min^{-1}). Total phenolic content was considered as the quenching agent, and it was determined by the Folin-Ciocalteu method reported elsewhere [21]. Then, Equations (2)–(4) were solved simultaneously to obtain k_1 and k_2 values.

$$\frac{d\mathrm{CLA}}{dt} = -k_1 \cdot \mathrm{CLA} \cdot \mathrm{O}_2 \quad (2)$$

$$\frac{d\mathrm{O}_2}{dt} = -k_1 \cdot \mathrm{CLA} \cdot \mathrm{O}_2 - k_2 \cdot \mathrm{O}_2 \cdot \mathrm{Q} \quad (3)$$

$$\frac{d\mathrm{Q}}{dt} = -k_2 \cdot \mathrm{O}_2 \cdot \mathrm{Q} \quad (4)$$

The differential equations were solved using DYNAFIT (BioKin Ltd., Madison, WI, USA, www.biokin.com/dynafit).

3. Results and Discussion

3.1. Pressure, Temperature and Time History

Figure 1 shows the pressure and temperature history during PATP of CLA-enriched milk treated at 600 MPa and 120 °C for 15 min of holding time. The time needed to insert the preheated sample (102 °C), adjust the transmission fluid volume and close the high pressure vessel was considered as the loading time (t_l). Likewise, the time required to reach isothermal and isobaric conditions was labelled as the compression time (t_c). The temperature of the transmission fluid dropped 13 °C during compression, because the temperature of the pressurizing fluid that comes from the intensifier was lower (~75 °C) compared to the fluid temperature inside the vessel (120 °C). The temperature of the sample and the medium increased due to adiabatic heating, reaching a final temperature of 120 °C. The point where both isothermal and isobaric conditions have been reached was considered the start of the holding time (t_h). It can be assumed that the temperature of the transmission fluid, sample and vessel are the same (120 °C) during the holding time. Knowledge of the temperature and pressure history during PATP is needed for a correct interpretation of the experimental findings. The reduction of CLA can be attributed to a combination of pressure, temperature and time rather than temperature itself. The processing conditions of 600 MPa and 120 °C for at least 5 min were enough to inactivate seven-log of *Bacillus amyloliquefaciens* in CLA-enriched milk and to inactivate 99.9% of alkaline phosphatase [22].

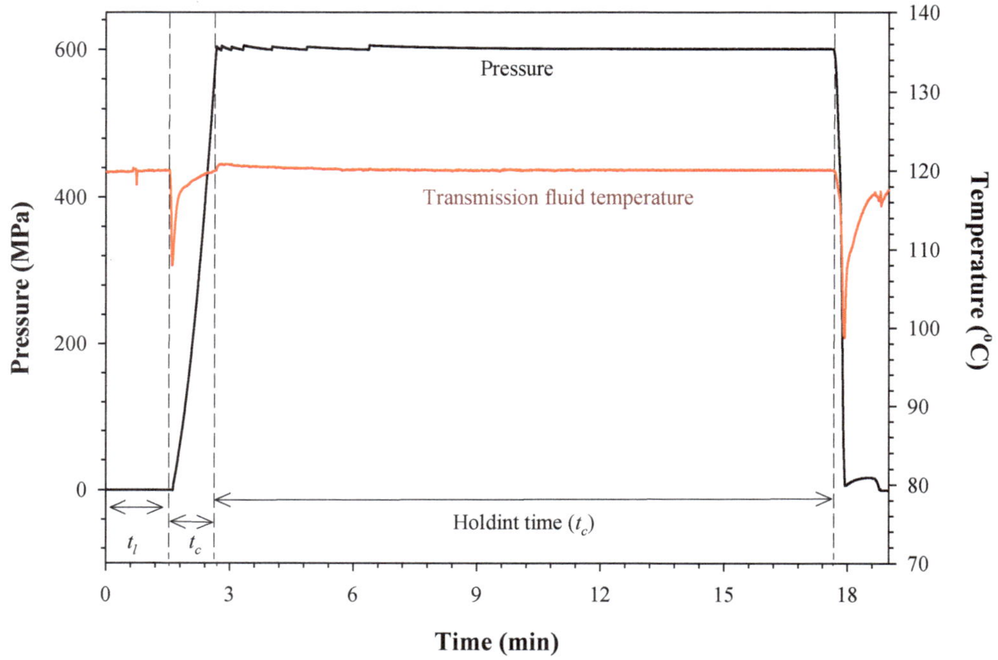

Figure 1. Typical pressure, temperature and time history during pressure-assisted thermal processing of conjugated linoleic acid (CLA)-enriched milk treated at 600 MPa/120 °C (t_l, loading time; t_c, compression time; t_h, holding time).

3.2. CLA-Enriched Milk

Milk saturated with oxygen and treated at 120 °C for up to 15 min was used to induce oxidation and therefore to evaluate the role of different antioxidants. Figure 2a shows the CLA changes in CLA-enriched milk treated at 0.1 MPa/120 °C or 600 MPa/120 °C up to 15 min. At 0.1 MPa, the retained fractions of CLA were 0.44 (44%) and 0.35 (35%) after 10 and 15 min, respectively. Similarly, Martinez-Monteagudo and Saldaña [15] reported that 15%–21% of CLA was retained in CLA-enriched milk heated at 120 °C /0.1 MPa. The difference in the retained fractions can be attributed to the different batch milk samples used. In milk fortified with CLA (2% of CLA in the total fat), the retained CLA ranged from 79% to 95%, depending on the processing conditions (125–145 °C/2–20 s) [13,14]. The high temperature used (120 °C) induces oxidation of CLA via the formation of free radicals through thermolysis. In addition, the dissolved oxygen reacts with oxygen species to form oxygen radicals, such as hydroxyl (HO•), peroxyl (ROO•) and hydroperoxyl (HOO•). These radicals can easily act as initiators for the oxidation of CLA. This is in agreement with the changes in DO_2 observed in Figure 2b. At 0.1 MPa, the remaining fractions of DO_2 after 10 and 15 min were 0.61 and 0.37, respectively. At 600 MPa, CLA was remarkably stable compared to the control treatment at 0.1 MPa, retaining 0.76 and 0.74 after 10 and 15 min, respectively. Based on these results, pressure may influence the reaction mechanism by which CLA is oxidized. Experimental data from the literature showed that thermal decomposition through homolysis of single bond and two-bond scission is delayed by the application of pressure up to 395 MPa at 100 °C [23,24]. A deceleration in the decomposition or homolysis of unsaturated fatty acids might result in a protective effect of the CLA.

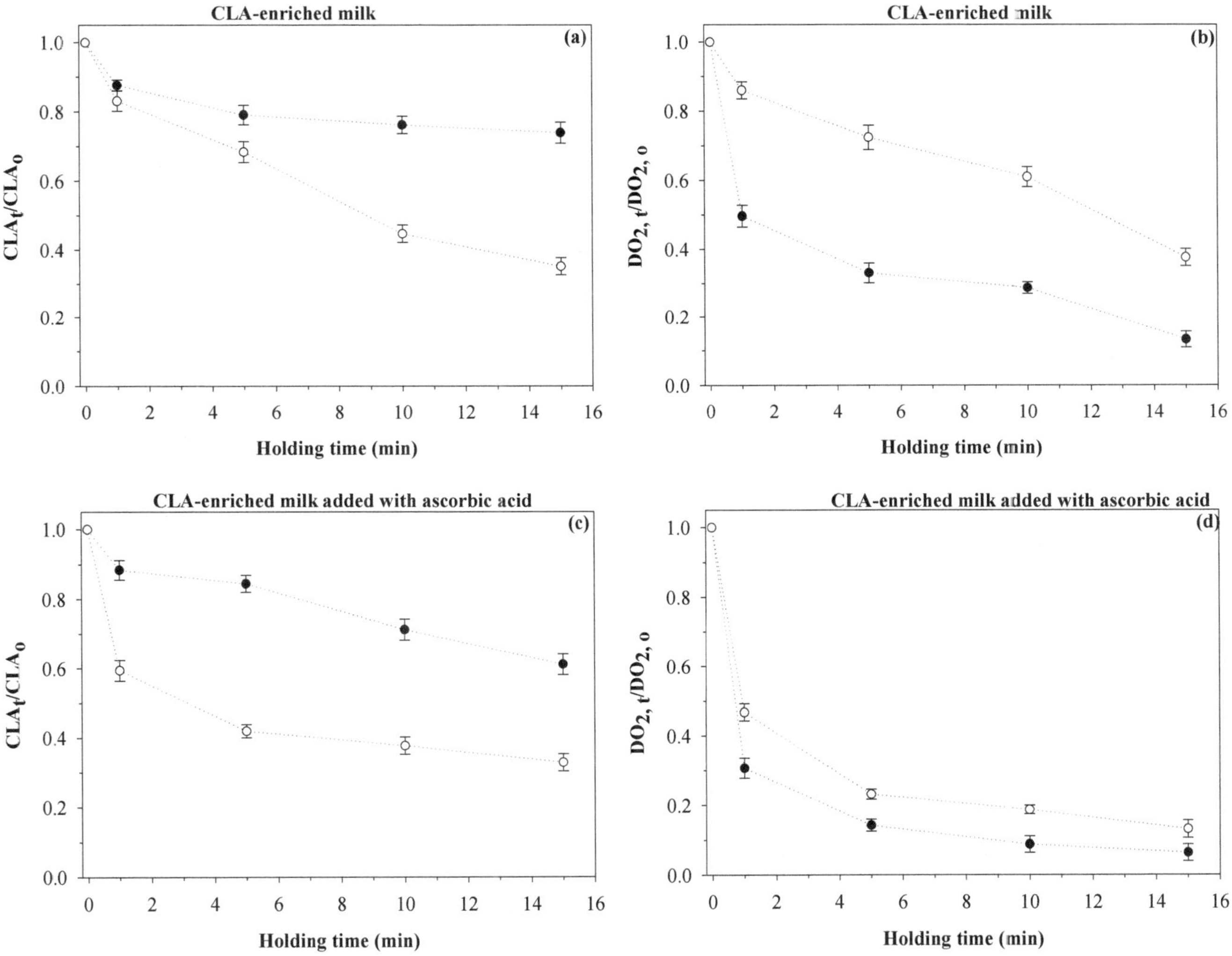

Figure 2. Changes in conjugated linoleic acid (CLA) and dissolved oxygen (DO_2) in CLA-enriched milk treated at 120 °C and either 600 MPa (●) or 0.1 MPa (○).

Figure 2b shows the normalized DO_2 in CLA-enriched milk treated at PATP. The remaining fractions of DO_2 at 600 MPa/120 °C were 0.28 and 0.13 after 10 and 15 min, respectively. Table 1 shows the k_I values calculated from Equations (1) and (2). The k_I value accounts for two contributions, CLA reduction and DO_2 consumption, and its determination can be used to infer possible reaction mechanisms. The

k_1 value for milk was lower at 600 MPa/120 °C compared with that at 0.1 MPa/120 °C (0.059 and 0.091 min^{-1}, respectively), indicating that pressure had a protective effect. Similarly, Martinez-Monteagudo and Saldaña [10] showed that the combination of 600 MPa/120 °C induces isomerisation rather than oxidation of CLA in milk rich in CLA. Isomerization only changes the isomer distribution of CLA without changing the total amount of CLA. In this study, total CLA was quantified, but milk fat contains over 20 different CLA isomers for which *cis*-9/*trans*-11 is the predominant isomer (≥85%). There are two possible mechanisms for the isomerization of CLA, and the predominant mechanism depends on the oxygen content [25]. The first mechanism consists of the migration of the double bond, known as sigmatropic molecular rearrangements, which is induced by temperature in the absence of oxygen. The second mechanism occurs in the presence of oxygen through free radical chain reaction, similar to the initiation step of the autoxidation of linoleic acid (C18:2) [26]. Interestingly, the initiation and propagation steps of free radical polymerization are indeed accelerated by pressure [27]. Therefore, it is reasonable to hypothesize that isomerization of CLA through the free radical mechanism is the predominant reaction at 600 MPa/120 °C. Contrary to this, oxidation of CLA rather than isomerization is the predominant reaction at 0.1 MPa/120 °C.

Table 1. Oxidation rate constant (k_1, min^{-1}) of CLA-enriched milk treated at 120 °C and either 600 or 0.1 MPa.

Antioxidant	**600 MPa**		**0.1 MPa**	
	k_1, min^{-1}	**95% CI**	k_1, min^{-1}	**95% CI**
Milk	0.059	0.006	0.091	0.010
Ascorbic acid	0.081	0.009	0.366	0.050
Cysteine	0.030	0.005	0.194	0.051
Caffeic acid	0.041	0.015	0.078	0.020
Catechin	0.051	0.009	0.139	0.015
Gallic acid	0.092	0.018	0.150	0.036
p-Coumaric acid	0.042	0.011	0.118	0.011
Tannic acid	0.069	0.019	0.084	0.017

k_1 is obtained by solving Equations (1) and (2); 95% CI, 95% confidence interval.

3.3. Ascorbic Acid

Changes in CLA content in enriched milk added with AA and treated at sterilization conditions are shown in Figure 2c. The retained CLA contents after 10 and 15 min were 0.37 and 0.32 at 0.1 MPa/120 °C, respectively. After 15 min, the retained CLA was similar to that in CLA-enriched milk without antioxidant addition (~0.35; Figure 2a). The processing conditions of 120 °C and up to 15 min are rather severe and could oxidize the ascorbic acid, causing it to lose its antioxidant activity. In UHT milk, ascorbic acid is lost by up to 50%, depending on the severity of the treatment time (135–150 °C/4 s) [28]. On the other hand, the combination of AA and a pressure of 600 MPa at 120 °C enhanced the retention of CLA (0.61) after 15 min of treatment compared with the treatment at 0.1 MPa/120 °C (0.32) and milk without added antioxidant (0.35). One possible explanation is that pressure protects the AA, which further reacts with DO_2 and free radicals. AA was used to inhibit the formation of aldehydes in milk treated at 655 MPa and 75 °C for 5 min [29]. An investigation on the retention of AA in pressurized phosphate buffer systems showed that AA remained unchanged at a moderate temperature of 50 °C and 850 MPa, while at a higher temperature of 80 °C and 850 MPa, the AA degradation became notorious. More importantly, AA can be used to enhance the stability of folates in pressurized orange and tomato juices [4]. Another investigation showed that the combination of 700 MPa and 110 °C accelerated the degradation rate of ascorbic acid up to 20 times in raspberries [5]. These variations can be attributed to differences in the food matrix, processing conditions and dissolved oxygen.

The changes in DO_2 in milk with added AA are shown in Figure 2d. For those samples treated at 0.1 MPa, the DO_2 rapidly decreased (0.23 remaining fraction) within 5 min of holding time. As the experiment proceeded, the remaining fractions of DO_2 were 0.18 and 0.13 after 10 and 15 min, respectively. In the case of PATP, the retained fractions of DO_2 were 0.08 and 0.06 after 10 and 15 min, respectively. A close examination of the k_l values (Table 1) indicated that more oxygen was consumed by those samples with added AA and treated at 0.1 MPa/120 °C (0.366 min^{-1}) compared with those samples with added AA and treated at 600 MPa/120 °C (0.081 min^{-1}). In the presence of oxygen, the AA is converted to dehydroascorbic acid (DHA) by abstracting a proton, making it unstable [30]. At the beginning of the holding time, the AA might be acting as a scavenging agent due to its relatively high initial concentration (0.5 g kg^{-1}) [31]. Moreover, the initial molar concentration of the AA ($[AA]_o$ = 2.8 mM) is four-times higher than the initial molar concentration of oxygen ($[DO_2]_o$ = 0.68 mM). A molar ratio of at least 2 (AA/O_2) was reported to protect other compounds from oxidation [4]. When most of the DO_2 was consumed, the oxidation of CLA was evident. One possible reason is that the AA can be ionized when pressure is applied to form ascorbate, which is more reactive than the AA. Both compounds oxidize to produce hydrogen peroxide, which might trigger the late oxidation of CLA [32].

3.4. Cysteine

The remaining CLA in samples with added cysteine (Cyst) is shown in Table 2. At 0.1 MPa and 120 °C, the retained fractions of CLA were 0.59 and 0.56 after 10 and 15 min of holding time, respectively. An investigation showed that Cyst can be used to inhibit the formation of off-flavor compounds in UHT milk [29]. At 600 MPa, the retained fractions of CLA were 0.73 and 0.70 after 10 and 15 min, respectively. The combination of Cyst and pressure yielded a higher remaining fraction of CLA compared with the control treatment (0.1 MPa). Likewise, Cyst inhibited the formation of aldehydes and hydrogen sulfide in milk pressurized at 655 MPa and heated at 75 °C for 10 min [29].

On the other hand, the DO_2 rapidly decreased with increasing holding time at 0.1 MPa, resulting in a value of 0.05 after 15 min of holding time. At 600 MPa, the DO_2 fraction was considerable higher than that at 0.1 MPa with a value of 0.42 after 15 min of holding time. In the presence of oxygen, Cyst reacts with the hydroxyl radical to form sulfonic acid [30]. This could be the reason for the substantial reduction in the DO_2 fraction observed at 0.1 MPa (Table 2). The k_1 values reported in Table 1 at 600 and 0.1 MPa were 0.030 and 0.194 min^{-1}, respectively, meaning that a small amount of oxygen was needed to oxidize CLA. Cyst has the ability to quench oxygen. An investigation on the reactivity of the O_2 with amino acids showed that oxidation of amino acids was the predominant reaction rather than O_2 quenching. Cyst possesses a thiol group, which is believed to act as an antioxidant by donating hydrogen to neutralize free radicals [33]. The ability of Cyst to quench free radicals depends on the state of protonation of the functional group [34]. An investigation on the oxidation of milk fat with different amino acids added showed that a non-protonated (NH_2) amino group, such as Cyst, inhibited the oxidation [35]. In PATP experiments, the reaction system suffers a temporary and reversible modification in the pH, which might change the reaction mechanism [36].

Table 2. Changes in conjugated linoleic acid (CLA) and dissolved oxygen (DO_2) in CLA-enriched milk added with either cysteine, caffeic acid or catechin and treated at 120 °C and 0.1 or 600 MPa.

Holding time (min)	Normalized retention of conjugated linoleic acid (CLA)					
	Cysteine		Caffeic acid		Catechin	
	0.1 MPa	600 MPa	0.1 MPa	600 MPa	0.1 MPa	600 MPa
0	1.00 ± 0.00a	1.00 ± 0.00a	1.00 ± 0.00a	1.00 ± 0.00a	1.00 ± 0.00a	1.00 ± 0.00a
1	0.73 ± 0.03b	0.81 ± 0.03b	0.91 ± 0.03a	0.74 ± 0.03b	0.95 ± 0.03b	0.87 ± 0.03b
5	0.63 ± 0.02c	0.72 ± 0.02c	0.79 ± 0.02b	0.71 ± 0.02c	0.77 ± 0.02c	0.81 ± 0.02c
10	0.59 ± 0.03d	0.73 ± 0.03cd	0.70 ± 0.02c	0.71 ± 0.03c	0.69 ± 0.02d	0.79 ± 0.03c
15	0.56 ± 0.02e	0.70 ± 0.03d	0.69 ± 0.02c	0.66 ± 0.03d	0.39 ± 0.02e	0.75 ± 0.03d
Holding time (min)	**Normalized dissolved oxygen (DO_2)**					
	Cysteine		**Caffeic acid**		**Catechin**	
	0.1 MPa	**600 MPa**	**0.1 MPa**	**600 MPa**	**0.1 MPa**	**600 MPa**
0	1.00 ± 0.00a	1.00 ± 0.00a	1.00 ± 0.00a	1.00 ± 0.00a	1.00 ± 0.00a	1.00 ± 0.00a
1	0.13 ± 0.02b	0.79 ± 0.03b	0.88 ± 0.03b	0.78 ± 0.03b	0.78 ± 0.02b	0.71 ± 0.03b
5	0.11 ± 0.03c	0.64 ± 0.01c	0.54 ± 0.01c	0.65 ± 0.01c	0.65 ± 0.03c	0.30 ± 0.04c
10	0.11 ± 0.01c	0.61 ± 0.03d	0.29 ± 0.03d	0.56 ± 0.02d	0.56 ± 0.01d	0.14 ± 0.03d
15	0.05 ± 0.02d	0.42 ± 0.04e	0.25 ± 0.02e	0.18 ± 0.03e	0.18 ± 0.02e	0.09 ± 0.03e

Means ± standard deviation ($n = 3$); columns with different letters (a–f) are significantly different ($p < 0.05$).

3.5. Phenolic Antioxidants

Tables 2 and 3 show the retention of CLA in milk added with different phenolic antioxidants. The highest retention fraction of CLA in milk was obtained after adding CAF (0.69), followed by TAN (0.64), CUM (0.43), CAT (0.39) and GA (0.36) after 15 min of holding time at 120 °C/0.1 MPa. On the other hand, after 15 min and 120 °C/600 MPa, the highest retention fraction was obtained with GA (0.85), followed by CAT (0.75), CAF (0.66) CUM (0.63) and TAN (0.62). In general, phenolic antioxidants possess the ability to inhibit the autoxidation of lipids by trapping intermediate radicals and quenching dissolved oxygen due to their low ionization potential [37]. Tables 2 and 3 also show the changes in DO_2 in milk with added phenolic antioxidants. After 15 min of holding time at 120 °C/0.1 MPa, the consumption of DO_2 was larger in those samples with added TAN and GA (0.11), followed by CAT (0.18), CUM (0.21) and CAF (0.25). After 15 min of holding time at 120 °C/600 MPa, the consumption of DO_2 was greater in samples with added CUM (0.10), CAT (0.11) and GA (0.11), followed by TAN (0.13) and CAF (0.18). Upon thermal treatment, TAN is hydrolyzed and formed gallic acid and polygalloyls, which enhanced the antioxidant capacity. An investigation on the oxidation of soybean oil showed that the addition of hydrolyzed TAN extended the oxidation induction period by up to 84% compared with the control treatment [35]. These authors thermally treated TAN (121 °C up to 60 min) and found that 15 min of thermal treatment was enough to hydrolyze TAN. The ability of TAN to react with oxygen was demonstrated during the oxidation of ascorbate in the presence of copper, where TAN not only reacted with oxygen, but also chelated metal ions, delaying the oxidation process [38]. In PATP-treated samples, the k_1 value was 0.069 min^{-1}, suggesting a change in the reaction mechanism induced by pressure in the presence of TAN (Table 1).

The value of k_1 (Table 1) in milk with CAF addition obtained at PATP was 0.041 min^{-1}, indicating that the addition of CAF enhanced the retention of CLA. In the case of samples added with CAT, the k_1 values were 0.051 and 0.139 min^{-1} at 600 and 0.1 MPa, respectively (Table 1). The k_1 values indicated that a significant amount of oxygen was needed to oxidize CLA. For those samples added with GA, the k_1 values were 0.092 and 0.150 min^{-1} at 600 and 0.1 MPa, respectively. Finally, those samples added with CUM yielded k_1 values of 0.042 and 0.118 min^{-1} at 600 and 0.1 MPa, respectively. During pressure treatment, free phenols are probably ionized, and therefore, their ability to quench oxygen is significantly enhanced. Phenols are susceptible to ionization by pressure due to charge delocalization between the oxygen and aromatic ring, yielding molar activation volumes between −8 and −20 cm^3 mol^{-1} [37]. Phenols form hydrogen bonds with surrounding molecules. Free phenols, on the other hand, are likely to be ionized, and the donated proton rapidly neutralizes the free radicals. In PATP, the use of phenolic antioxidants not only enhances the retention of CLA, but also quenches oxygen, which avoids isomerization, therefore protecting the biological activity of CLA.

Table 3. Changes in conjugated linoleic acid (CLA) and dissolved oxygen (DO_2) in CLA-enriched milk added with either gallic acid, *p*-coumaric acid or tannic acid and treated at 120 °C and 0.1 or 600 MPa.

Holding time (min)	**Normalized retention of conjugated linoleic acid (CLA)**					
	Gallic acid		***p*-Coumaric acid**		**Tannic acid**	
	0.1 MPa	**600 MPa**	**0.1 MPa**	**600 MPa**	**0.1 MPa**	**600 MPa**
0	1.00 ± 0.00a	1.00 ± 0.00a	1.00 ± 0.00a	1.00 ± 0.00a	1.00 ± 0.00a	1.00 ± 0.00a
1	0.69 ± 0.03b	0.98 ± 0.02b	0.91 ± 0.03b	0.87 ± 0.03b	0.91 ± 0.03b	0.78 ± 0.03b
5	0.61 ± 0.02c	0.94 ± 0.02c	0.76 ± 0.04c	0.75 ± 0.02c	0.78 ± 0.02c	0.76 ± 0.02c
10	0.51 ± 0.02d	0.89 ± 0.03d	0.53 ± 0.02d	0.72 ± 0.03d	0.75 ± 0.02d	0.72 ± 0.03d
15	0.36 ± 0.02e	0.85 ± 0.03e	0.43 ± 0.03e	0.63 ± 0.03e	0.64 ± 0.02e	0.62 ± 0.03e
Holding time (min)	**Normalized dissolved oxygen (DO_2)**					
	Gallic acid		***p*-Coumaric acid**		**Tannic acid**	
	0.1 MPa	**600 MPa**	**0.1 MPa**	**600 MPa**	**0.1 MPa**	**600 MPa**
0	1.00 ± 0.00a	1.00 ± 0.00a	1.00 ± 0.00a	1.00 ± 0.00a	1.00 ± 0.00a	1.00 ± 0.00a
1	0.18 ± 0.03b	0.68 ± 0.02b	0.84 ± 0.02b	0.47 ± 0.03b	0.63 ± 0.03b	0.27 ± 0.03b
5	0.17 ± 0.02b	0.41 ± 0.04c	0.27 ± 0.04c	0.18 ± 0.01c	0.33 ± 0.02c	0.21 ± 0.01c
10	0.13 ± 0.02c	0.12 ± 0.03d	0.23 ± 0.03d	0.11 ± 0.02d	0.30 ± 0.03d	0.19 ± 0.02d
15	0.11 ± 0.02c	0.11 ± 0.03d	0.21 ± 0.02e	0.10 ± 0.02d	0.11 ± 0.02e	0.13 ± 0.02e

Means ± standard deviation ($n = 3$); columns with different letter (a–f) are significantly different ($p < 0.05$).

3.6. Different Concentrations of Catechin and Gallic Acid

An additional set of experiments were conducted using different concentrations of CAT and GA (0, 100, 200, 300 and 500 mg kg^{-1}, Figure 3). The samples were treated at 120 °C for 15 min at either 0.1 or 600 MPa. Using catechin and gallic acid, better CLA retention was obtained at 600 MPa than at

0.1 MPa. In the case of milk with added CAT and treated at 0.1 MPa (Figure 3a), the retention of CLA was enhanced from 0.27 to 0.44 as the concentration increased up to 500 mg kg^{-1}. At 600 MPa, the retention of CLA was remarkably enhanced, reaching values of 0.82 when 500 mg kg^{-1} of CAT was added. Similarly, the combined effect of 600 MPa and GA resulted in a retention of 0.85, which is considerably higher than that obtained at 0.1 MPa (0.41). Catechin has been used for delaying non-enzymatic browning in UHT milk [39]. It was demonstrated that a concentration of 0.1 mmol L^{-1} was enough to inhibit the browning of UHT non-enriched milk at 145 °C for 15 s during prolonged storage up to 36 days. More importantly, consumer sensory analysis showed no difference with respect to the control samples (samples without added catechin) [39]. In our study, a similar concentration of CAT was used (500 mg kg^{-1}, 0.17 mmol L^{-1}); therefore, changes in sensory perception due to the addition of CAT are not expected, but further research is needed to confirm this assumption.

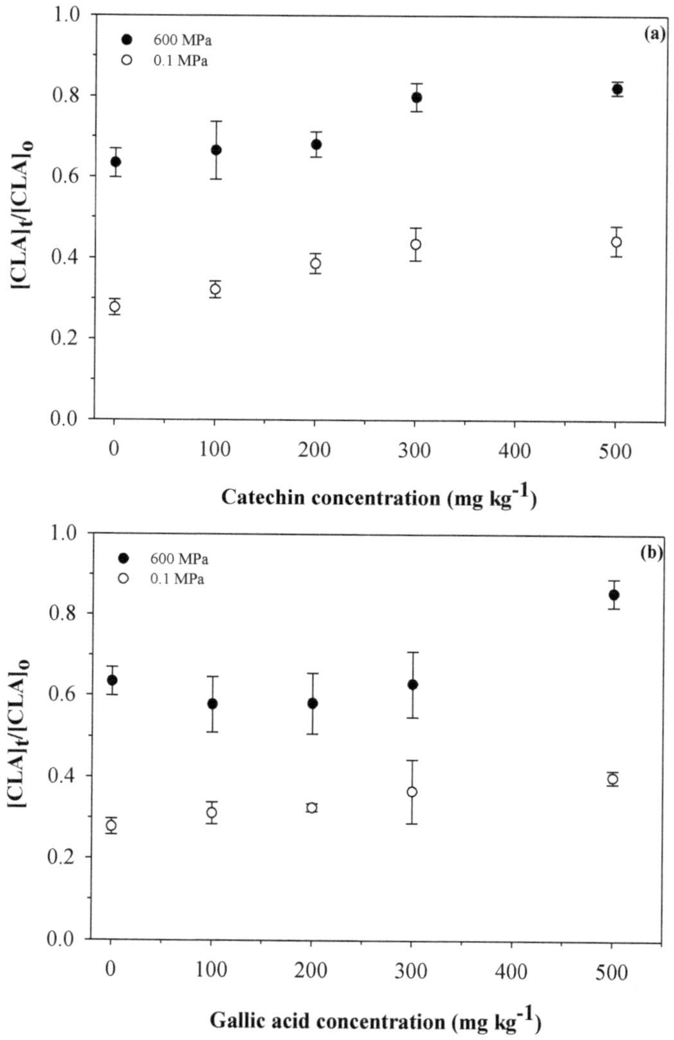

Figure 3. Effect of the concentration of (**a**) catechin and (**b**) gallic acid on the CLA retention in CLA-enriched milk treated at 120 °C for 15 min at either 0.1 or 600 MPa.

3.7. Possible Reaction Mechanisms of Phenolic Antioxidants

The use of phenolic antioxidants (500 mg kg^{-1}) in combination with high pressure yielded a high retention of CLA (~85% in the case of GA), even at a prolonged holding time of 15 min. In general, the effectiveness of a given antioxidant is usually associated with the structure-function relation, in which the number of –OH groups are related to its effectiveness. In the case of the retention of CLA upon pressure, the number of –OH groups did not reflect the antioxidant effectiveness. While TAN possess 27 –OH groups in its structure, the retained normalized CLA was only 0.633. In contrast, GA and CAT with four and five –OH groups, respectively, yielded higher retention values (>0.75) than that obtained using TAN. The stability of antioxidants played an important role in enhancing the retention of CLA. The GA is considered a stable molecule, as proven by its thermogravimetric analysis, which showed that decarboxylation and further degradation started only at 260 °C [40]. A closer observation of Tables 2 and 3 revealed that the oxygen was gradually consumed while the CLA remained unchanged. Phenolic antioxidants are known for their ability to quench oxygen, preventing the initiation stage of autoxidation [34]. Table 4 shows the changes in total phenolic content for milk samples with added 500 mg kg^{-1} of GA, CAT and CAF and treated at 120 °C/600 MPa.

Table 4. Normalized phenolic content in CLA-enriched milk added with different phenolic antioxidants and treated at 120 °C and 600 MPa.

Holding time (min)	Antioxidant added (500 mg kg^{-1}) Gallic acid	Catechin	Caffeic acid
0	1.00 ± 0.00	1.00 ± 0.00	1.00 ± 0.00
1	0.94 ± 0.02	0.72 ± 0.03	0.94 ± 0.02
5	0.81 ± 0.01	0.53 ± 0.04	0.81 ± 0.01
10	0.61 ± 0.02	0.44 ± 0.05	0.61 ± 0.04
15	0.48 ± 0.07	0.40 ± 0.02	0.48 ± 0.05
k_1 (min^{-1})	0.029 ± 0.006	0.078 ± 0.005	0.014 ± 0.005
k_2 (min^{-1})	0.081 ± 0.006	0.051 ± 0.019	0.061 ± 0.025

The variability of k_1 and k_2 correspond to the 95% confidence interval.

In general, the phenolic content followed a similar pattern to that observed for the DO_2, reaching normalized retention values in the range of 0.40–0.48 after 15 min of holding time. Table 4 also shows the apparent quenching rate constant. Interestingly, GA yielded the highest k_2 value (quenching constant rate) and the lowest k_1 value (oxidation constant rate of CLA) followed by CAT and CAF at 120 °C/600 MPa up to 15 min. The use of GA resulted in the highest CLA retention followed by CAT. It seems that GA has the ability to quench oxygen and, therefore, to delay the oxidation of CLA. Another important factor to consider is the ratio of DO_2 to the antioxidant. Ratios for GA and CAT were calculated, as these antioxidants yielded the highest retention of CLA. After 10 min of holding time, ratios of 0.22 and 1.28 were obtained for GA and CAT, respectively. A ratio of DO_2/antioxidant higher than one means that the antioxidant is consumed faster than the available oxygen, indicating that there is an excess of oxygen that can further react. On the other hand, a DO_2/antioxidant ratio of less than one could be viewed as the amount of oxygen being limited with respect to the amount of antioxidant (GA). These calculated ratios further demonstrated that the use of GA significantly enhanced the retention of CLA during PATP of milk. The processing conditions of temperature, pressure and holding time not only influenced the retention of CLA, but also the effectiveness of the tested phenolic antioxidants. Further studies should determine the content of each antioxidant used after PATP treatments to evaluate its stability in the enriched milk.

4. Conclusions

In the presence of oxygen, CLA was remarkably stable in samples treated with PATP compared to the control. The analysis of remaining CLA and consumed oxygen suggests that pressure-induced isomerization of CLA, which might occur through a free radical mechanism. The application of PATP in milk samples containing phenolic antioxidants enhanced the retention of CLA. Gallic acid yielded the highest retention of CLA followed by catechin. During PATP, the rate of oxygen quenching was higher than the oxidation rate of CLA, which might avoid isomerization and, therefore, preserve the biological activity of CLA. Thus, the combination of PATP and the use of phenolic antioxidants can be applied to produce milk-based beverages rich in CLA, addressing the growing demand for functional drinks.

Acknowledgments

The authors thank the Natural Sciences and Engineering Research Council of Canada (NSERC) and the Food and Health Innovation Initiative of the Faculty of Agricultural, Life and Environmental Sciences at University of Alberta for funding this project. Martinez-Monteagudo expresses his gratitude to Consejo Nacional de Ciencia y Tecnologia (CONACYT, Mexico) and Instituto de Inovacion y Transferencia Tecnologica (I^2T^2, Mexico) for the financial support (No. 187497).

Author Contributions

This investigation was performed by SI Martinez-Monteagudo as part of his PhD thesis under the supervision of MDA Saldaña.

Conflicts of Interest

The authors declare no conflict of interest.

References

1. Balasubramaniam, V.M.; Martínez-Monteagudo, S.I.; Gupta, R. Principles and application of high pressure-based technologies in the food industry. *Ann. Rev. Food Sci. Tech.* **2015**, *6*, 1–28.
2. Gupta, R.; Kopec, R.E.; Schwartz, S.J.; Balasubramaniam, V.M. Combined pressure-temperature on carotenoid retention and bioaccessibility in tomato juice. *J. Agric. Food Chem.* **2011**, *59*, 7808–7817.
3. Martinez-Monteagudo, S.I.; Saldaña, M.D.A. Chemical reactions in food systems at high hydrostatic pressure. *Food Eng. Rev.* **2014**, *6*, 105–127.
4. Oey, I.; Verlinde, P.; Hendrickx, M.E.; van Loey, A. Temperature and pressure stability of L-ascorbic acid and/or 6s 5-methyltetrahydrofolic acid: A kinetic study. *Eur. Food Res. Tech.* **2006**, *223*, 71–77.
5. Verbeyst, L.; Bogaerts, R.; van der Plancken, I.; Hendrickx, M.E.; van Loey, A. Modelling of vitamin C degradation during thermal and high-pressure treatments of red fruits. *Food Bioprocess. Tech.* **2013**, *6*, 1015–1023.

6. Butz, P.; Serfert, Y.; Garcia, A.F.; Dieterich, S.; Lindauer, R.; Bognar, A.; Tauscher, B. Influence of high-pressure treatment at 25 degrees C and 80 degrees C on folates in orange juice and model media. *J. Food Sci.* **2004**, *69*, S117–S121.
7. Matser, A.A.; Krebbers, B.; van den Berg, R.W.; Bartels, P.V. Advantages of high pressure sterilisation on quality of food products. *Trends Food Sci. Technol.* **2004**, *15*, 79–85.
8. Verbeyst, L.; Oey, I.; van der Plancken, I.; Hendrickx, M.; van Loey, A. Kinetic study on the thermal and pressure degradation of anthocyanins in strawberries. *Food Chem.* **2010**, *123*, 269–274.
9. Gupta, R.; Balasubramaniam, V.M.; Schwartz, S.J.; Francis, D.M. Storage stability of lycopene in tomato juice subjected to combined pressure-heat treatments. *J. Agric. Food Chem.* **2010**, *58*, 8305–8313.
10. Martinez-Monteagudo, S.I.; Saldaña, M.D.A. Modeling the retention kinetic of conjugated linoleic acid during high-pressure sterilization of milk. *Food Res. Int.* **2014**, *62*, 169–176.
11. Sevenich, R.; Kleinstueck, E.; Crews, C.; Anderson, W.; Pye, C.; Riddellova, K.; Hradecky, J.; Moravcova, E.; Reineke, K.; Knorr, D. High-pressure thermal sterilization: Food safety and food quality of baby food puree. *J. Food Sci.* **2014**, *79*, M230–M237.
12. Cook, M.E.; Pariza, M. The role of conjugated linoleic acid (CLA) in health. *Int. Dairy J.* **1998**, *8*, 459–462.
13. Campbell, W.; Drake, M.A.; Larick, D.K. The impact of fortification with conjugated linoleic acid (CLA) on the quality of fluid milk. *J. Dairy Sci.* **2003**, *86*, 43–51.
14. Herzallah, S.M.; Humeid, M.A.; Al-Ismai, K.M. Effect of heating and processing methods of milk and dairy products on conjugated linoleic acid and trans fatty acid isomer content. *J. Dairy Sci.* **2005**, *88*, 1301–1310.
15. Martinez-Monteagudo, S.I.; Saldaña, M.D.A. Retention of bioactive lipids in heated milk: Experimental and modelling. *Food Bioprod. Process.* **2015**, *94*, 290–296.
16. Martínez-Monteagudo, S.I.; Leal-Dávila, M.; Curtis, J.M.; Saldaña, M.D.A. Oxidative stability of ultra high temperature milk enriched in conjugated linoleic acid and trans-vaccenic acid. *Int. Dairy J.* **2015**, *43*, 70–77.
17. Martinez-Monteagudo, S.I.; Saldaña, M.D.A.; Torres, J.A.; Kennelly, J.J. Effect of pressure-assisted thermal sterilization on conjugated linoleic acid (CLA) content in CLA-enriched milk. *Innov. Food Sci. Emerg. Tech.* **2012**, *16*, 291–297.
18. Bell, J.A.; Griinari, J.M.; Kennelly, J.J. Effect of safflower oil, flaxseed oil, monensin, and vitamin E on concentration of conjugated linoleic acid in bovine milk fat. *J. Dairy Sci.* **2006**, *89*, 733–748.
19. Rasanavagam, V.; Balasubramaniam, V.M.; Ting, E.; Sizer, C.E.; Bush, C.; Anderson, C. Compression heating of selected fatty food materials during high pressure processing. *J. Food Sci.* **2003**, *68*, 2544–2559.
20. Martinez-Monteagudo, S.I.; Saldaña, M.D.A.; Kennelly, J.J. Kinetics of non-isothermal oxidation of anhydrous milk fat rich in conjugated linoleic acid using differential scanning calorimetry. *J. Therm. Anal. Calorim.* **2012**, *107*, 973–981.
21. Singh, P.P.; Saldaña, M.D.A. Subcritical water extraction of phenolic compounds from potato peel. *Food Res. Int.* **2011**, *44*, 2452–2458.

22. Martínez-Monteagudo, S.I.; Gänzle, M.G.; Saldaña, M.D.A. High-pressure and temperature effects on the inactivation of *Bacillus amyloliquefaciens*, alkaline phosphatase and storage stability of conjugated linoleic acid in milk. *Innov. Food Sci. Emerg. Tech.* **2014**, *16*, 291–297.
23. Neuman, R.C. Pressure effects as mechanistic probes of organic radical reactions. *Accounts Chem. Res.* **1972**, *5*, 381–387.
24. Neuman, R.C.; Amrich, M.J. High pressure studies. X. Activation volumes for homolysis of single bonds. *J. Am. Chem. Soc.* **1972**, *98*, 2730–2733.
25. Destaillats, F.; Angers, P. Thermally induced formation of conjugated isomers of linoleic acid. *Eur. J. Lipid Sci. Tech.* **2005**, *107*, 167–172.
26. Destaillats, F.; Angers, P. Evidence for [1,5] Sigmatropic rearrangements of CLA in heated oils. *Lipids* **2002**, *37*, 435–438.
27. Isaacs, N.S. *Liquid Phase High Pressure Chemistry*; John Wiley: Toronto, Canada, 1981.
28. Lindmark-Mansson, H.; Akesson, B. Antioxidative factors in milk. *Br. J. Nutr.* **2000**, *84*, S103–S110.
29. Vazquez-Landaverde, P.A.; Torres, J.A.; Qian, M.C. Effect of high-pressure-moderate-temperature processing on the volatile profile of milk. *J. Agr. Food Chem.* **2006**, *54*, 9184–9192.
30. Choe, E.; Min, D.B. Chemistry and reactions of reactive oxygen species in foods. *Crit. Rev. Food Sci. Nutr.* **2006**, *46*, 1–22.
31. Yen, G.C.; Duh, P.D.; Tsai, H.L. Antioxidant and pro-oxidant properties of ascorbic acid and gallic acid. *Food Chem.* **2002**, *79*, 307–313.
32. Chimi, H.; Cillard, J.; Cillard, P.; Rahmani, M. Peroxyl and hydroxyl radical scavenging activity of some natural phenolic antioxidants. *J. Am. Oil Chem. Soc.* **1991**, *68*, 307–312.
33. Khan, N.S.; Ahmad, A.; Hadi, S.M. Anti-oxidant, pro-oxidant properties of tannic acid and its binding to DNA. *Chem-Biol. Interact.* **2000**, *125*, 177–189.
34. Sánchez-Moreno, C.; Larrauri, J.A.; Saura-Calixto, F. Free radical scavenging capacity and inhibition of lipid oxidation of wines, grape juices and related polyphenolic constituents. *Food Res. Int.* **1999**, *32*, 407–412.
35. Kim, T.J.; Silva, J.L.; Kim, M.K.; Jung, Y.S. Enhanced antioxidant capacity and antimicrobial activity of tannic acid by thermal processing. *Food Chem.* **2010**, *118*, 740–746.
36. Huppertz, T.; Smiddy, M.A.; Vpadhyay, V.K.; Kelly, A.L. High-pressure-induced changes in bovine milk: A review. *Int. J. Dairy Technol.* **2006**, *59*, 58–66.
37. Hamann, S.D.; Linton, M. Influence of pressure on the ionization of substituted phenols. *J. Chem. Soc. Faraday Trans. 1* **1974**, *70*, 2239–2249.
38. Ohara, K.; Ichimura, Y.; Nagaoka, S. Kinetic study of singlet-oxygen quenching by caffeic acid and related phenols. *Bull. Chem. Soc. Jpn.* **2009**, *82*, 689–691.
39. Schamberger, G.P.; Labuza, T.P. Effect of green tea flavonoids on Maillard browning in UHT milk. *LWT Food Sci. Technol.* **2007**, *40*, 1410–1417.
40. Garro-Galvez, J.M.; Fechtal, M.; Riedl, B. Gallic acid as a model of tannins in condensation with formaldehyde. *Thermochim. Acta* **1996**, *274*, 149–163.

16

Quality Characteristics and Shelf-Life of Ultra-High Pressure Homogenized (UHPH) Almond Beverage

Victoria Ferragut *, Dora C. Valencia-Flores, Marianita Pérez-González, Joan Gallardo and Manuela Hernández-Herrero

Centre Especial de Recerca Planta de Tecnologia dels Aliments (CERPTA), XaRTA, TECNIO, Departament de Ciència Animal y dels Aliments, Facultat de Veterinària, Universitat Autònoma de Barcelona, Bellaterra 08193, Barcelona, Spain; E-Mails: docevaf@yahoo.com.mx (D.C.V.-F.); mar.perez.glez@gmail.com (M.P.-G.); joanjgch@gmail.com (J.G.); Manuela.Hernandez@uab.cat (M.H.-H.)

* Author to whom correspondence should be addressed; E-Mail: victoria.ferragut@uab.cat

Academic Editor: Carl Joseph Schaschke

Abstract: The effects of ultra-high-pressure homogenization (UHPH) at 200 MPa, in combination with different inlet temperatures (55 or 75 °C) during storage at 4 °C were studied and compared with pasteurized (90 °C, 90 s) almond beverage. Microbiological analysis of the physical (particle sedimentation and color) and volatile profile of the most relevant compound in almond beverages was performed at days 1, 7, 14, and 21 of cold storage. UHPH treatment 200 at 75 °C led to higher microbiological reduction after treatment and higher stability during cold storage in almond beverages than pasteurization or UHPH 200 at 55 °C. Physical characteristics of UHPH-treated samples exhibited a high stability during storage with a stable color. Volatile compounds extracted by solid-phase microextraction were identified by gas chromatography coupled with mass spectrometry. The effect of UHPH treatment significantly ($p < 0.05$) affected the volatile profile compared with pasteurized beverages, although UHPH conditions applied produced similar effects in almond beverages. Benzaldehyde was the most abundant compound detected in all treatments. Hexanal was more abundant in UHPH-treated samples, indicating a higher lipid oxidation compared to pasteurized almond beverages.

Keywords: ultra-high pressure homogenization (UHPH); almond beverage; shelf life; color; physical stability; volatile profile

1. Introduction

Almond beverage, like other vegetable beverages, is widely present in the European market as an alternative to cow's milk, although they are not really comparable foods. The increase in consumption of these products in the last decades is related to consumers' tendency to look for healthy foods, especially those from vegetal origin. Vegetable beverages have been promoted as healthy foods, mainly as protectors against cardiovascular diseases [1] due to the equilibrium of unsaturated and saturated fatty acids present in their composition. Moreover, they contain other beneficial bioactive compounds: antioxidants such as flavonoids, vitamin E and polyamines, fiber, and phytosterols, which reduce cholesterol absorption. In addition, a consumer sector with particular interest for vegetable beverages are those with lactose intolerance or an allergy to cow's milk proteins.

The most conventional treatment of commercial almond beverage is UHT (ultra-high temperature, a sterilization treatment above 100 °C for several seconds) and pasteurization (heat treatments below 100 °C for several minutes). Heat treatments may cause undesirable chemical changes, which include destruction of amino acids and vitamins, browning reactions, and development of cooked flavors. On the other hand, almond beverage is a colloidal system formed by dispersed particles such as oil droplets, solid particles from raw materials, proteins, and starch granules. This complexity makes it difficult to obtain a stable product to be stored, even for a not very long time. Commonly, in conventional heat-treated vegetable milks, such as soy milk or almond beverage, particles experience sedimentation and separate from the continuous phase, causing loss of quality.

In recent years, several emergent technologies have been developed, such as ultra-high pressure homogenization (UHPH) processing. The aim of this technology is to obtain liquid foods with better quality than conventional heat treatments, which is based on the coupled effects of UHPH on hygienic preservation and colloidal stabilization.

Pressure applied in UHPH processing is in the range of 100 to 400 MPa, producing physical phenomena such as cavitation, turbulence, impact, and shear forces [2] acting on treated food. The consequences of those mechanical forces, in addition to temperature increase experienced as a consequence of heat dissipation in the high pressure valve, are the production of fine and stable colloidal dispersions [2] and microbial destruction [3–5]. UHPH has great potential in the food industry. In dairy products, the UHPH-treated milk was shown not only to be an alternative for direct consumption of milk [6,7], but also for derived products such as yogurt [8] and fresh cheese [9].

In a previous study [10], a prospective investigation of UHPH conditions (pressure and inlet temperature combination) on almond beverage characteristics was made. Results showed that 200 MPa combined with different inlet temperature of the product were able to produce an overall good quality almond beverage, better than that processed by conventional pasteurization. In terms of heat treated foods, pasteurization is considered a "fresh-like" product compared to other more severe heat treatments like UHT or bath sterilization. Thus, two different UHPH treatments at 200 MPa were selected with the objective of assessing their overall quality characteristics and evolution during cold storage of almond beverages compared to pasteurized.

2. Experimental Section

2.1. Preparation and Processing of Almond Beverage

Ground almond seed used in this study was kindly provided by Nectina, S.A. (Ruidoms, Tarragona, Spain). To obtain the almond beverage, 4% (w/w) almond and 0.03% (w/w) lecithin (ADM Soy Lecithin, Decatur, Sabadell, Spain) in water at 60 °C was mixed in a tank for 8 min, circulated in a colloidal mill (E. Bachiller, B.S.A. Barcelona, Spain) for 12 min, and subsequently filtered to separate the liquid phase using a 0.100 mm steel sieve. The almond beverage obtained was the base product (BP) used for subsequent treatments and as a control for comparing the effects of pasteurization or UHPH treatments. Gross composition of BP was 3.40 ± 0.11 dry matter; 2.04 ± 0.10 fat; 1.18 ± 0.10 protein; 0.12 ± 0.02 ash; and 0.08 ± 0.22 carbohydrate. Mean pH value was 7.31 ± 0.013. BP was divided into portions; one of them was pasteurized in a multi beam tubular heat exchanger (Finamat heat exchanger, Model 6500/010, Gea Finnah Gmbh, Anhaus, Germany) at 90 °C, 90 s, with previous homogenization in a double effect homogenizer (Niro Soavi, Model X68P, Parma, Italy) at 18 MPa and 4 MPa. The two other portions were UHPH treated at two different inlet temperature-pressure combinations: 200 MPa, 55 °C inlet temperature (200, 55); and 200 MPa, 75 °C inlet temperature (200, 75).

UHPH treatments were performed with a high pressure homogenizer (model FPG11300, Stansted Fluid Power Ltd., Essex, UK). This device comprises a high-pressure ceramic valve able to support 400 MPa. Inlet and outlet temperatures of almond beverage were controlled by two heat exchangers (Garvía, Barcelona, Spain) located before the machine entrance and after the second homogenization valve, respectively. Inlet temperature, temperature after the high pressure valve, and outlet temperature were monitored in all productions. Almond beverage samples were collected in sterile bottles under a laminar flow cabin adapted to the UHPH system and stored at 4 °C to analyze quality parameters during cold storage at days 1, 7, 14, and 21. A portion of samples at those different storage periods were frozen at −80 °C for further analysis of volatile compounds.

2.2. Microbial Analysis

Microbiological quality of almond beverage was assessed by enumerating the following microorganisms: mesophilic aerobic bacteria were counted on PCA medium (Oxoid Ltd., 96 Basingstoke, Hampshire, UK) incubated for 48 h at 30 °C. Mesophilic aerobic spore counts were assessed by heat shock at 80 °C for 10 min, quickly cooled in ice, plated on PCA medium (Oxoid), and incubated for 48 h at 30 °C. *Bacillus cereus* was enumerated in brilliance *Bacillus cereus* medium (Oxoid), supplemented with *Bacillus cereus* selective supplement (Oxoid), and incubated at 30 °C for 48 h. Enterobacteria counts were determined in violet red bile glucose agar (VRBG, Oxoid) incubated at 37 °C for 24 h.

2.3. Physical Stability

Almond beverage (30 g) was poured into flexible plastic tubes (32 mm diameter, 115 mm length) and centrifuged (1000× *g*, 45 min, 20 °C). After removing the liquid phase, tubes were weighted and results were expressed as % (w/w) of solid deposition at the bottom of the tubes. Samples were preserved by adding 0.04% (w/v) sodium azide (NaN_3).

2.4. Color

Color determination was performed with a Hunter Lab colorimeter (MiniScan XE™, Hunter Associates Laboratory Inc., Reston, Virginia, USA), using the D65 illuminant with an angle of observation of 10°. Almond beverage samples (50 mL) were tempered to 20 °C before analysis and poured into an optically transparent glass cup that was covered with a black protector to avoid the ambient light interaction into the sample. ΔE (color difference; Equation (1)), L* (Lightness), a* (red-green component), b* (yellow-blue component), h* (hue; Equation (2)), and C* (chrome; Equation (3)) were obtained.

$$\Delta E=\sqrt{L^{*2}+a^{*2}+b^{*2}} \quad (1)$$

$$h^{*}=arc\tan\frac{b^{*}}{a^{*}} \quad (2)$$

$$C^{*}=\sqrt{a^{*2}+b^{*2}} \quad (3)$$

2.5. Volatile Profiles

To determine the volatile profile of almond beverage, we utilized the method described by [11]. Volatile compounds were extracted by solid-phase microextraction and were identified by gas chromatography coupled with mass spectrometry. Separation and identification of the volatile compounds were performed using an Agilent 6890 gas chromatograph couple to a 5975 MSD mass spectrometer (Agilent Technologies, Palo Alto, CA, USA). A J&W HP-INNOWAX capillary column (60 m × 0.25 mm × 0.25 μm) with a bonded polyethylene glycol (PEG) high polarity stationary phase was used. Tentative identification of volatile compounds was achieved by comparing their mass spectra with those of mass spectra libraries [12].

2.6. Statistical Analysis

All results are means of three independent almond beverage productions at the pilot plant. Each analysis was made in triplicate on days 1, 7, 14, and 21. Analysis of variance (ANOVA) by the Student-Newman-Keuls method was applied for multiple comparisons of the means. Data analysis was performed with SAS9.2.3 package [13]. Differences were considered to be significant at $p < 0.05$.

3. Results and Discussion

3.1. Effect of UHPH on Microorganism and Shelf-Life Estimation

Table 1 show the growth of total aerobic meshophilic bacteria and spores, and *Bacillus cereus* during 21 days of storage at 4 °C of almond beverage samples after applying different treatments of UHPH compared with base product (BP) and pasteurized samples.

Table 1. Growth of total mesophilic aerobic bacteria, mesophilic aerobic spores, and *B. cereus* counts (log CFU/mL) in base product (BP), pasteurized (PA), and UHPH (200, 55 °C and 200, 75 °C) almond beverages at 4 °C over 21 days.

		Day 1	Day 7	Day 14	Day 21
Total mesophilic aerobic bacteria	BP	5.25 ± 0.65 ^{a,B}	5.00 ± 0.14 ^{a,B}	6.72 ± 0.23 ^{a,A}	NE
	PA	3.16 ± 0.10 ^{b,D}	4.07 ± 0.17 ^{b,C}	4.47 ± 0.08 ^{b,B}	4.80 ± 0.14 ^{b,A}
	200, 55	2.92 ± 0.06 ^{c,C}	3.06 ± 0.16 ^{c,C}	4.90 ± 0.18 ^{b,B}	5.80 ± 0.10 ^{a,A}
	200, 75	2.50 ± 0.20 ^{d,C}	2.22 ± 0.08 ^{d,C}	2.90 ± 0.50 ^{c,C}	3.40 ± 0.20 ^{c,A}
Mesophilic aerobic spores	BP	3.31 ± 0.10 ^{a,B}	3.23 ± 0.14 ^{a,B}	4.58 ± 0.12 ^{a,A}	4.80 ± 0.14 ^{a,A}
	PA	2.65 ± 0.07 ^{b,B}	2.47 ± 0.24 ^{b,B}	2.45 ± 0.47 ^{b,A}	3.23 ± 0.38 ^{b,A}
	200, 55	2.28 ± 0.17 ^{c,B}	3.10 ± 0.27 ^{a,A}	2.80 ± 0.04 ^{b,A}	2.93 ± 0.21 ^{b,A}
	200, 75	1.60 ± 0.26 ^{d,A}	2.17 ± 0.16 ^{b,A}	2.41 ± 0.23 ^{a,b,A}	2.01 ± 0.56 ^{b,A}
B. cereus	BP	2.92 ± 0.64 ^{a,C}	4.89 ± 0.51 ^{a,B}	6.55 ± 0.52 ^{a,A}	6.92 ± 0.27 ^{a,A}
	PA	1.81 ± 0.04 ^{b,A}	1.91 ± 0.34 ^{b,A}	2.91 ± 0.09 ^{b,A}	2.16 ± 0.04 ^{b,A}
	200, 55	0.99 ± 0.08 ^{c,B}	1.56 ± 0.70 ^{b,A}	2.20 ± 0.70 ^{b,A}	2.10 ± 0.05 ^{b,A}
	200, 75	ND ^{d}	ND ^{c}	ND ^{c}	ND ^{c}

NE: Not Examined; ND: Not Detected (<0.5 CFU/mL); $^{a–d}$: Different superscripts in the same column are significantly different ($p < 0.05$). $^{A–C}$: Different superscripts in the same row are significantly different ($p < 0.05$).

All microbial groups showed a significant decrease at day 1 for almond beverage treated at 200 MPa, 55 °C and 200, 75 °C and pasteurized compared with base product (200, 75 °C > 200, 55 °C > PA > BP). Enterobacteria counts were not detected in any sample (detection limit < 0.5 CFU/mL) (data not shown), similar to what was found by other authors [5] in soy milk treated at the same conditions and in cow milk [14]. In this study coliforms were completely inactivated by both pasteurization and UHPH treatment (200 and 300 MPa). However, in milk, coliforms were not reduced after treatment of raw cow milk at 200 MPa and 55 °C [5].

The mean value of mesophilic bacteria counts observed in BP was 5.25 log CFU/mL. Among them, the predominant groups were probably spores and *B. cereus* (3.31 log and 2.92 log CFU/mL, respectively), which is in accordance with previous surveys reported in literature [5,10] in soy milk and almond beverage. Microbial counts of samples UHPH-treated 200, 55 °C were significantly different from those pasteurized samples although UHPH treatment 200, 75 °C was the most effective against all the bacterial groups. A total reduction of *B. cereus* was obtained and significant differences ($p < 0.05$) were observed among these samples and UHPH 200, 55 °C and pasteurized samples.

After cold storage, BP samples showed an increase in mesophilic bacterial total counts (from 5.25 to 6.72 log CFU/mL at day 14) as well as spores (from 3.31 to 4.8 log CFU/mL) and *B. cereus* (from 2.92 to 6.92 log CFU/mL). Total bacterial count, spores and *B. cereus*, increased progressively in pasteurized and UHPH 200, 55 °C samples during storage at 4 °C. In general, no significant differences ($p \geq 0.05$) were observed between counts in UHPH 200, 55 °C samples and those observed in the PA. An exception was found on day 21, when total mesophilic bacteria counts were significantly higher in UHPH 200, 55 °C samples than in pasteurized ones. The lack of difference between UHPH 200, 55 °C and PA was probably due to the recovery of healthy and sub-lethally injured vegetative cells during storage or by spore germination that remained after treatment. This recovery was not observed in UHPH 200, 75 °C. A complete microbial inactivation in soy milk at 200 MPa and inlet temperature 75 °C had been found previously [5]. Nevertheless, this treatment did not cause total inactivation in this study. This may be due to the different composition of soy milk and almond beverage and the different native microbiota.

We can conclude that increasing the inlet temperature to 75 °C is much more effective in preventing the proliferation of the spoilage microbiota of almond beverage and consequently in extending the shelf life of the product.

3.2. Physical Stability

This study was assessed by measuring the percentage (w/w) of solid sedimentation after centrifugation of samples. This parameter is related to the potential destabilization of dispersed particles and is the maximum value of sedimentation that could be produced in forced conditions. In parallel, the observation of spontaneous destabilization of almond beverages during storage was made.

Almond beverage is a water extract of almonds in the form of an oil-in-water emulsion. The most relevant functional property of proteins in this system is to cover oil droplets of the lipid fraction for maintaining a good dispersability of those in the continuous phase during storage. In spite of this, creaming of oil droplets and sedimentation of solid particles are the primary mode of destabilization of vegetable beverages. Both phenomena are dependent to a great extent on particle size distribution [15–17].

Particles in vegetable beverages include not only fat globules but also small particles, such as protein bodies and aggregates of protein and protein-fat globules. BP and PA almond beverage presented the highest values of particle size parameters, indicating that conventional homogenization at low pressure (18 MPa) applied before heat treatment did not produce an additional decrease compared to the colloidal mill in the grinding step of almond beverage elaboration. On the other hand, PA with single effect homogenization used in this study was not enough to disperse aggregates formed into small particles.

As expected, UHPH-treated soy milk showed higher colloidal stability (Table 2) compared to pasteurized almond beverage. Particle sedimentation measured by centrifugation is indirectly related to the stability of the system. Under the same conditions, centrifugation was applied to all samples, forcing particles and aggregates to separate from the bulk, either to the top, in the case of fat globules, or to the bottom, in the case of solid particles. However, in this product with low fat concentration, measurement of the fat layer in the top was very difficult to measure in BP and PA and no creaming layer was observed at all in UHPH-treated samples. This is why the percentage of solid sedimentation was considered as indicative of the colloidal stability. This analysis can be considered as indicative of the sedimentation potential of almond beverage during long storage periods and especially as a comparative measurement among treatments applied. PA almond beverage presented a higher amount of solids settled by centrifugation than the UHPH almond beverage throughout storage (Table 2). As storage time increased, a slight increase in the percentage of sediments was observed until day 14 (maximum shelf life of PA treatment). Solid sedimentation values in UHPH samples were low and an increase was only observed at day 21 ($p < 0.05$). UHPH treatments showed lower values of this parameter; 200, 75 °C was the treatment that exhibited the best stability within the whole storage period.

Table 2. Solid sedimentation [1] of base product (BP), pasteurized (PA), and UHPH (200, 55 °C and 200, 75 °C) almond beverages at 4 °C over 21 days.

Treatment	Day 1	Day 7	Day 14	Day 21
BP	1.83 ± 0.07 ^{a,A}	2.1 ± 0.05 ^{a,A}		
PA	1.69 ± 0.08 ^{b,B}	2.13 ± 0.62 ^{a,A,B}	2.27 ± 0.14 ^{a,A}	
200, 55	1.09 ± 0.11 ^{c,B}	1.26 ± 0.11 ^{b,B}	2.02 ± 0.28 ^{a,B}	3.22 ± 0.13 ^{a,A}
200, 75	0.94 ± 0.04 ^{d,C}	1.01 ± 0.04 ^{b,B,C}	1.15 ± 0.14 ^{b,B}	3.0 ± 0.2 ^{a,A}

$^{a–d}$: Different superscripts in the same column are significantly different ($p < 0.05$); $^{A–C}$: Different superscripts in the same row are significantly different ($p < 0.05$); [1]: Mean values ± SD (g/100g w/w) of solids.

3.3. Color Evaluation

It was observed [18] that the interaction between soy globulin fractions played an important role in the colloidal stability, mainly at high temperatures. They reported that the combination of denatured β-conglycinin and native glycinin caused lower stability of soy milk dispersion. Probably a number of different attractive particle interactions during storage are responsible for increasing solid aggregation, which was forced to sedimentation when centrifugation was applied. However, it has to be noted that those values obtained represent the maximum of destabilization. In fact, spontaneous sedimentation was negligible in UHPH samples. Solid sedimentation results indicated that the state of particle dispersion of the UHPH samples provided enough stability during 21 days of storage, revealed by solids accumulation in the bottom of the tube.

The quality of food is conditioned by many factors; some of them determine the sensory characteristics of a product. Color and appearance of foods are the first appreciations of the consumer, conditioning their preferences and influencing their choice. This characteristic is also related to chemical composition and physical changes produced during processing. Thus, color changes sometimes may determine the validity of a technology for a specific food product. Emulsion droplet characteristics have a marked influence on the appearance of this system. It was determined [19] that the lightness of an emulsion is correlated with the efficiency of the dispersion of droplets, which in turn is related to its size and its concentration of particles.

Color was measured in BP, PA, and UHPH samples of almond beverage. The effect of treatments on the CIEL*a*b* parameters, psychometric coordinates chrome (C*), and hue (h*) are shown in Table 3. Lightness (L*) is associated with the luminous intensity, which describes the light-reflecting or transmitting capacity of an object [20]. UHPH treatment caused a significant increase in the L* parameter in UHPH-treated samples compared to BP and PA almond beverages. This increase in lightness is related with the particle size reduction caused by the intensive homogenization [16], causing higher total light scatter, which has an effect on L* values' increase. During storage, UHPH-treated almond beverage always had a higher L* value compared with PA and BP. Over time, L* values of independent almond beverage treatments remained quite stable between 1 and 21 days, especially in UHPH-treated beverages. Similar results were observed by sensory analysis, with the UHPH almond beverage being perceived as lighter than PA samples by the panel (data not shown).

Table 3. Effect of treatments on CIEL*a*b* parameters1 in base product (BP), pasteurized (PA), and UHPH (200, 55 °C and 200, 75 °C) almond beverages at 4 °C over 21 days.

Treatment	Day	L*	a*	b*	h*	C*	ΔE_{PA}
BP	1	$81.4 \pm 0.2^{c,B}$	$-0.91 \pm 0.02^{c,A}$	$7.63 \pm 0.01^{a,B}$	$83 \pm 2^{a,A}$	$7.68 \pm 0.01^{a,B}$	$6.89 \pm 0.08^{a,C}$
PA	1	$87.58 \pm 0.09^{b,B}$	$-0.53 \pm 0.02^{a,A}$	$5.08 \pm 0.07^{b,A}$	$84 \pm 2^{a,A}$	$5.10 \pm 0.07^{b,A}$	
200,55	1	$89.3 \pm 0.02^{a,B}$	$-0.84 \pm 0.02^{b,A}$	$3.80 \pm 0.01^{c,A}$	$77 \pm 3^{b,B}$	$3.89 \pm 0.01^{c,A}$	$2.17 \pm 0.05^{b,B}$
200,75	1	$89.09 \pm 0.03^{a,A}$	$-0.96 \pm 0.01^{d,A}$	$3.47 \pm 0.02^{d,A}$	$74 \pm 4^{c,B}$	$3.60 \pm 0.02^{d,A}$	$2.25 \pm 0.10^{b,C}$
BP	7	$76.35 \pm 0.01^{c,D}$	$-1.12 \pm 0.01^{c,B}$	$6.25 \pm 0.02^{a,C}$	$79 \pm 3^{b,A}$	$6.35 \pm 0.01^{a,D}$	$8.35 \pm 0.08^{a,A}$
PA	7	$84.36 \pm 0.03^{b,C}$	$-0.72 \pm 0.02^{a,C}$	$3.93 \pm 0.02^{b,D}$	$79 \pm 2^{b,A}$	$4.00 \pm 0.02^{b,D}$	
200,55	7	$85.8 \pm 0.22^{a,C}$	$-0.96 \pm 0.07^{b,A,B}$	$2.66 \pm 0.01^{c,D}$	$67 \pm 4^{d,B}$	$2.83 \pm 0.02^{c,D}$	$1.94 \pm 0.03^{c,C}$
200,75	7	$85.71 \pm 0.05^{a,C}$	$-1.28 \pm 0.01^{d,C}$	$2.26 \pm 0.02^{d,D}$	$60 \pm 4^{d,B}$	$2.60 \pm 0.02^{d,D}$	$2.22 \pm 0.02^{b,C}$
BP	14	$79.85 \pm 0.37^{c,C}$	$-0.97 \pm 0.17^{a,b,A,B}$	$7.02 \pm 0.69^{a,B,C}$	$82 \pm 2^{a,A}$	$7.08 \pm 0.70^{a,C}$	$8.39 \pm 0.34^{a,A}$
PA	14	$87.85 \pm 0.02^{b,A}$	$-0.78 \pm 0.01^{a,D}$	$4.52 \pm 0.02^{b,C}$	$80 \pm 3^{a,A}$	$4.58 \pm 0.01^{b,C}$	
200,55	14	$89.34 \pm 0.02^{a,B}$	$-1.12 \pm 0.02^{b,c,B,C}$	$3.01 \pm 0.02^{c,C}$	$69 \pm 0^{d,B}$	$3.21 \pm 0.02^{c,C}$	$2.15 \pm 0.06^{b,B}$
200,75	14	$88.88 \pm 0.02^{a,B}$	$-1.22 \pm 0.02^{a,b,A,B}$	$2.56 \pm 0.02^{a,B,C}$	$66 \pm 0^{d,B}$	$2.83 \pm 0.02^{d,C}$	$2.26 \pm 0.10^{b,B}$
BP	21	$81.96 \pm 0.16^{d,A}$	$-0.88 \pm 0.02^{c,A}$	$8.63 \pm 0.06^{a,A}$	$84 \pm 2^{a,A}$	$8.68 \pm 0.06^{a,A}$	$7.11 \pm 0.02^{a,B}$
PA	21	$87.92 \pm 0.02^{c,A}$	$-0.65 \pm 0.01^{a,B}$	$4.77 \pm 0.04^{b,B}$	$82 \pm 2^{a,A}$	$4.82 \pm 0.04^{b,B}$	
200,55	21	$89.77 \pm 0.01^{a,A}$	$-0.83 \pm 0.01^{b,A}$	$3.24 \pm 0.02^{c,B}$	$76 \pm 3^{c,B}$	$3.35 \pm 0.02^{c,B}$	$2.40 \pm 0.10^{b,A}$
200,75	21	$89.25 \pm 0.04^{b,A}$	$-0.92 \pm 0.01^{d,A}$	$2.72 \pm 0.03^{d,B}$	$70 \pm 5^{d,B}$	$2.87 \pm 0.02^{c,B}$	$2.46 \pm 0.03^{b,A}$

[a–c]: Different letters in the same column are significantly different ($p < 0.05$); [A–D]: Different letters in the column for each storage period are significantly different ($p < 0.05$); [1]: Mean values ± SD of color parameters; ΔE was calculated taking into account PA as reference sample.

The particle size distribution and concentration of UHPH and PA almond beverages were different In UHPH, a great particle size reduction (oil droplets and dispersed solid particles) was observed, with the consequent increase of interface surface produced by the increase in the number of total particles caused by treatment conditions, which in turn affected the optical characteristics of almond beverages. a* and b* color parameters in all almond beverages had negative and positive values, respectively, indicating that green and yellow were the primary contribution to the color of that beverage with predominance of the yellow component. a* values were significantly ($p < 0.05$) smaller in PA samples than UHPH samples, with (200, 75 °C) treatment being significantly higher. Over the storage period, in general, this parameter experienced an increase in their values. The contribution of b* was smaller in UHPH samples than in BP and PA samples and values of this coordinate varied during storage while maintaining the initial tendency.

Coordinate values of a* and b* were transformed in psychometric coordinates C* and h* to have a more intuitive meaning of color assessment. Those values in Table 3 indicate that BP and PA had a more marked hue of yellow and exhibited more pure color than UHPH-treated beverages. Comparing UHPH treatments, the more intense conditions caused a reduction in h* and C* values. The evolution of this color attribute during storage varied without a specific trend while maintaining the initial tendency according to treatments applied.

Color difference takes into account differences between L*, a*, and b* of the sample and a standard, in this case the PA treatment to which UHPH treatment were compared. As shown in Table 3, ΔE values were higher for BP treatments than for UHPH almond beverage. These differences may be attributed primarily to the L* parameter contribution due to treatment applied. Different results of L* and ΔE were obtained [4,5] for soy milk treated by UHPH at 200 MPa and 55 °C, 75 °C, and 40 °C of inlet temperature, respectively [21]; a significant increase in the ΔE values was reported after 21 days of storage at 4 °C of different blends of soymilk treated at 142 °C for 4 s (UHT).

3.4. Volatile Profile Analysis

The volatile profile is an important quality of a food product. However, treatments applied in food processing may modify volatile components to some extent, no matter whether it is beneficial or detrimental to the overall quality of food, *i.e.*, fatty acid oxidation.

Around 100 compounds were identified in almond beverages [10]. For this study, 15 of them (Table 4) were chosen, taking into consideration their abundance in the chromatographic profile and their technical interest when analyzing the influence of treatments applied (PA or UHPH) to almond beverages and the evolution of volatile profile during storage. Among them, the most representative compounds in almond beverage head space were aldehydes (benzaldehyde, hexanal, and nonanal), alcohols (1-pentanol, 1-hexanol, 1-heptanol, 1-nonanol, 1-octen-3-ol, 1-nonanol, 2-etil-1-hexanol, benzeneethanol, and benzyl alcohol), and some from different families (toluene, styrene, and benzoic acid).

Table 4. Abundance[1] of selected volatile compounds in base product (BP), pasteurized (PA), and UHPH (200, 55 °C and 200, 75 °C) almond beverages at 4 °C over 21 days.

Compound	Day 1	Day 7	Day 14	Day 21
Benzaldehyde				
PA	283.45 ^{A,a}	194.22 ^{A,b}	181.60 ^{B,b}	
200, 55	304.28 ^{A,a}	289.02 ^{A,a}	206.20 ^{A,a}	177.85 ^{A,a}
200, 75	283.51 ^{A,a}	197.09 ^{A,a}	202.18 ^{A,a}	175.45 ^{A,a}
Hexanal				
PA	53.08 ^{A,a}	20.71 ^{B,b}	18.60 ^{B,b}	
200, 55	100.33 ^{A,a}	67.27 ^{A,a}	68.92 ^{A,a}	41.68 ^{A,a}
200, 75	114.80 ^{A,a}	94.52 ^{A,a}	96.69 ^{A,a}	52.32 ^{A,a}
Nonanal				
PA	3.37 ^{A,a}	2.85 ^{A,b}	2.23 ^{B,b}	
200, 55	4.12 ^{A,a}	3.51 ^{A,a}	4.74 ^{A,a}	4.10 ^{A,a}
200, 75	5.58 ^{A,a}	4.99 ^{A,a}	6.13 ^{A,a}	8.81 ^{A,a}
1-pentanol				
PA	23.03 ^{B,a}	22.78 ^{A,a}	19.77 ^{B,a}	
200, 55	29.26 ^{A,a}	24.16 ^{A,a}	27.98 ^{A,a}	20.16 ^{A,a}
200, 75	31.38 ^{A,a}	31.45 ^{A,a}	28.89 ^{A,a}	24.30 ^{A,a}
1-hexanol				
PA	60.66 ^{A,b}	99.28	96.18 ^{A,a}	
200, 55	42.89 ^{B,b}	39.50 ^{B,b}	114.4 ^{A,a}	93.45 ^{A,a}
200, 75	36.77 ^{B,a}	69.45 ^{A,B,a}	67.60 ^{A,a}	65.26 ^{A,a}
1-heptanol				
PA	11.74 ^{A,b}	11.19 ^{A,a,b}	8.71 ^{A,b}	
200, 55	18.88 ^{A,a}	15.43 ^{A,a}	20.09 ^{A,a}	16.63 ^{A,a}
200, 75	26.73 ^{A,a}	20.75 ^{A,a}	18.64 ^{A,a}	26.73 ^{A,a}
1-nonanol				
PA	3.27 ^{A,a}	3.03 ^{A,a}	2.16 ^{A,b}	
200, 55	1.66 ^{B,a}	1.70 ^{B,a}	2.62 ^{A,a}	1.92 ^{A,a}
200, 75	1.77 ^{B,a}	1.71 ^{B,a}	1.35 ^{B,a}	1.68 ^{A,a}
1-octen-3-ol				
PA	19.51 ^{B,ab}	19.60 ^{B,a}	14.39 ^{A,b}	
200, 55	22.72 ^{A,a}	21.14 ^{B,a}	26.99 ^{A,a}	21.38 ^{A,a}
200, 75	38.95 ^{A,B,a}	34.53 ^{A,a}	40.47 ^{B,a}	38.97 ^{A,a}
2-etil-1-hexanol				
PA	5.15 ^{A,b}	9.24 ^{A,a}	5.56 ^{A,b}	
200, 55	5.55 ^{A,a}	4.12 ^{B,a}	5.37 ^{A,a}	3.99 ^{A,a}
200, 75	3.89 ^{B,a}	5.29 ^{A,B,a}	5.08 ^{A,a}	4.46 ^{A,a}
Benzeneetanol				
PA	40.31 ^{A,b}	30.54 ^{A,c}	65.20 ^{A,a}	
200, 55	35.11 ^{A,B,a}	33.48 ^{A,a}	62.63 ^{A,a}	60.29 ^{A,a}
200, 75	30.33 ^{B,a}	29.85 ^{A,a}	26.94 ^{B,a}	28.61 ^{A,a}

$^{a–b}$: Different letters in the same row are significantly different ($p < 0.05$); $^{A–C}$: Different letters in the same column for a same compound are significantly different ($p < 0.05$); [1]: Integrated area counts. Mean value × 10^5.

Table 4. *Cont.*

Isoamyl alcohol				
PA	3.29 ^{A,a,b}	3.88 ^{A,a}	2.54 ^{A,b}	
200, 55	2.18 ^{B,b}	1.88 ^{B,b}	3.22 ^{A,a}	2.75 ^{A,a}
200, 75	1.70 ^{B,a}	1.96 ^{B,a}	1.79 ^{A,a}	1.71 ^{A,a}
Benzyl alcohol				
PA	4.50 ^{A,b}	28.55 ^{A,a}	28.88 ^{A,a}	
200, 55	8.62 ^{A,b}	12.67 ^{Ab}	66.44 ^{A,a}	55.62 ^{A,a}
200, 75	8.01 ^{A,a}	49.66 ^{A,a}	49.54 ^{A,a}	49.70 ^{A,a}
Tolueno				
PA	6.85 ^{A,a,b}	8.91 ^{A,b}	5.97 ^{A,a}	
200, 55	3.85 ^{B,a}	4.06 ^{A,a}	4.48 ^{A,a}	3.75 ^{A,a}
200, 75	4.04 ^{A,a}	6.34 ^{A,a}	3.94 ^{A,a}	3.74 ^{A,a}
Estyrene				
PA	0.67 ^{A,a}	0.70 ^{A,a}	0.59 ^{A,b}	
200, 55	1.60 ^{B,a}	1.41 ^{A,a}	1.44 ^{A,a}	1.43 ^{A,a}
200, 75	1.50 ^{A,B,a}	1.21 ^{A,a}	1.13 ^{A,a}	1.45 ± 1.06 ^{A,a}
Benzoic acid				
PA	3.57 ^{A,b}	1.84 ^{B,b}	1.69 ^{A,a}	
200, 55	3.37 ^{A,B,a}	3.15 ^{A,a}	1.65 ^{A,a}	1.62 ^{A,a}
200, 75	2.60 ^{B,a}	1.85 ^{A,B,a}	1.79 ^{A,a}	1.59 ^{A,a}

$^{a–b}$: Different letters in the same row are significantly different ($p < 0.05$); $^{A–C}$: Different letters in the same column for a same compound are significantly different ($p < 0.05$); [1]: Integrated area counts. Mean value × 10^5.

In general, UHPH conditions applied to almond beverages did not show significant differences in the volatile profile; it could be considered that both UHPH treatments (200, 55 °C and 200, 75 °C) were equivalent from this point of view. However, PA-treated samples exhibited differences in most compounds, making them different ($p < 0.05$) to UHPH-treated almond beverages. The effect of UHPH and PA treatments on volatile compounds evolution during storage is shown in Table 4. Benzaldehyde was the most abundant in the chromatographic profile of all samples, as expected since it is a "character-impact compound" in almonds, providing a pleasant almond-like flavor. Benzaldehyde can be produced naturally from the action of β-glucosidases, and can be thermally generated from phenylalanine [22]. Upon oxidation, benzaldehyde is transformed into benzoic acid and, via hydrogenation, into benzyl alcohol. This is coherent with the evolution of those compounds during storage, especially with benzyl alcohol evolution from day 1 on.

The generation of certain volatiles, such as hexanal and nonanal, may be used to monitor the oxidation of a food product. Hexanal is a good indicator of lipid oxidation for foods rich in omega-6 fatty acids [23]. The fatty acid composition of almonds is rich in linoleic acid, an omega-6 fatty acid. Analysis of hexanal pick area integration on day 1, as a first approximation, indicates that UHPH treatment cause higher oxidation than pasteurization, although statistical analysis show no differences ($p > 0.05$) among treatments, which can be attributed to the high variability found in the analysis of this compound, probably due to the effect of the different almond beverage productions. Taking into account the general production process of almond beverages, the most probable factors affecting lipid oxidation are high temperatures and the lipid exposition to catalyst agents. In UHPH treatments, and especially in the conditions applied in this study, the maximum temperature reached was 115 °C during a very short time (less than one second); thus the most probable negative effect on lipid oxidation was the high degree of exposure of the lipid fraction due to the fat droplet reduction and the resulting increase in surface area.

Over the storage period, a significant decrease ($p < 0.05$) was observed in hexanal content in all studied samples. Hexanal may be degraded to carboxylic acid in presence of oxygen [24] and may also be transformed in 1-hexanol by reduction [25], which is in line with the increase of this compound over storage.

Bezene etanol, 2-etil-1-hexanol, and benzyl alcohol also experienced a significant increase over the storage period ($p < 0.05$). These compounds probably originated as secondary products from lipid oxidation, as mentioned by the authors of [26], who found them in roasted almonds. The fat fraction in almonds is about 40% and the fatty acid profile is rich in mono- and polyunsaturated fatty acids. To produce almond beverage, the fat fraction has to be liberated by heat wet milling to obtain the coarse emulsion prior to homogenization and heat treatment for preservation of the final product. Thus, lipid degradation compounds contribute to the characteristic volatile profile of processed almonds.

4. Conclusions

UHPH is a potential alternative to conventional pasteurization, producing high stable vegetable beverages from a microbiological and physical point of view. Chemically, UHPH treatment negatively affected lipid oxidation, which has to be more extensively studied to evaluate the possibility of adding antioxidant agents and their impact on this quality parameter. In terms of microbiological quality, the mildest conditions of UHPH applied (200 MPa with 55 °C inlet temperature) were similar to pasteurization and 200 MPa, 75 °C considerably improved the hygienic characteristics of almond beverages, producing the best microbial stability during cold storage. At least one week of extended shelf life was obtained for almond beverage compared to that of pasteurized (90 °C, 90 s), with good overall quality.

Acknowledgments

The authors acknowledge the Ministerio de Ciencia e Innovación (AGL2008-05430-01) for financial support of the investigation.

Author Contributions

This investigation was performed by D. C. Valencia-Flores as part of her PhD thesis with the contribution of M. Pérez-González and J. Gallardo under the supervision of V. Ferragut and M. Hernández-Herrero.

Conflicts of Interest

The authors declare no conflict of interest.

References

1. Jenkins, D.J.; Kendall, C.W.C.; Josse, A.R.; Salvatore, S.; Brighenti, F.; Augustin, L.S.; Ellis, P.R.; Vidgen, E.; Rao, V. Almonds decrease postprandial glycemia, insulinemia, and oxidative damage in healthy individuals. *J. Nutr.* **2006**, *136*, 2987–2992.
2. Floury, J.; Bellettre, J.; Legrand, J.; Desrumaux, A. Analysis of a new type of high pressure homogeniser. A study of the flow pattern. *Chem. Eng. Sci.* **2004**, *59*, 843–853.
3. Sharma, V.; Singh, R.K.; Toledo, R.T. Microbial inactivation kinetics in soymilk during continuous flow high-pressure throttling. *J. Food Sci.* **2009**, *74*, 268–275.
4. Cruz, N.; Capellas, M.; Hernández, M.; Trujillo, A.J.; Guamis, B.; Ferragut, V. Ultra high pressure homogenization of soymilk: Microbiological, physicochemical and microstructural characteristics. *Food Res. Int.* **2007**, *40*, 725–732.
5. Poliseli-Scopel, F.H.; Hernández-Herrero, M.; Guamis, B.; Ferragut, V. Characteristics of soymilk pasteurized by ultra high pressure homogenization (UHPH). *Innov. Food Sci. Emerg. Technol.* **2013**, *20*, 73–80.
6. Hayes, M.G.; Fox, P.F.; Kelly, A.L. Potential applications of high pressure homogenization in processing of liquid milk. *J. Dairy Res.* **2005**, *72*, 25–33.
7. Pereda, J.; Ferragut, V.; Guamis, B.; Trujillo, A.J. Effect of ultra high-pressure homogenisation on natural-occurring micro-organisms in bovine milk. *Milchwissenschaft* **2006**, *61*, 245–248.
8. Serra, M.; Trujillo, A.J.; Guamis, B.; Ferragut, V. Evaluation of physical properties during storage of set and stirred yogurts made from ultra-high pressure homogenization-treated milk. *Food Hydrocoll.* **2009**, *23*, 82–91.
9. Zamora, A.; Ferragut, V.; Quevedo, J.M.; Guamis, B.; Trujillo, A.J. Ultra-high pressure homogenisation of milk: technological aspects of cheese-making and microbial shelf life of a starter-free fresh cheese. *J. Dairy Res.* **2012**, *79*, 168–175.
10. Valencia-Flores, D.C.; Hernández-Herrero, M.; Guamis, B.; Ferragut, V. Comparing the effects of ultra-high-pressure homogenization and conventional thermal treatments on the microbiological, physical, and chemical quality of almond beverages. *J. Food Sci.* **2013**, *78*, 199–205.
11. Pérez-Gonzalez, M.; Gallardo-Chacón, J.J.; Valencia-Flores, D.C.; Ferragut, V. Optimization of a head space SPME GC-MS methodology for the analysis of processed almond beverages. *Food Anal. Meth.* **2015**, *8*, 612–623.
12. MacLafferty, F.W. *Wiley Registry of Mass Spectral Data*; John Wiley & Sons: Hoboken, NJ, USA, 2000.
13. SAS Institute Inc. SAS/STAT® 9.1. 2004. Available online: http://www.sas.com/en_us/software/analytics/stat.html (accessed on 18 May 2015).
14. Pereda, J.; Ferragut, V.; Quevedo, J.M.; Guamis, B.; Trujillo, A.J. Effects of ultra-high pressure homogenization on microbial and physicochemical shelf life of milk. *J. Dairy Sci.* **2007**, *90*, 1081–1093.
15. Smiddy, M.A.; Martin, J.E.; Huppertz, T.; Kelly, A.L. Microbial shelf-life of high-pressure-homogenised milk. *Int. Dairy J.* **2007**, *17*, 29–32.

16. Poliseli-Scopel, F.H.; Hernández-Herrero, M; Guamis, B.; Ferragut, V. Comparison of ultra high pressure homogenization and conventional thermal treatments on the microbiological, physical and chemical quality of soymilk. *LWT Food Sci. Technol.* **2012**, *46*, 42–48.
17. Durand, A.; Franks, G.V.; Hosken, R.W. Particle sizes and stability of UHT bovine, cereal and grain milks. *Food Hydrocoll.* **2003**, *17*, 671–678.
18. Shimoyamada, M.; Tsushima, N.; Tsuzuki, K.; Asao, H.; Yamauchi, R. Effect of heat treatment on dispersion stability of soymilk and heat denaturation of soymilk protein. *Food Sci. Technol. Res.* **2008**, *14*, 32–38.
19. Chantrapornchai, W.; Clydesdale, F.; Mcclements, D.J. Influence of droplet size and concentration on the color of oil-in-water emulsions. *J. Agric. Food Chem.* **1998**, *46*, 2914–2920.
20. Kwok, K.C.; MacDougall, D.B.; Niranjan, K. Reaction kinetics of heat-induced colour changes in soymilk. *J. Food Eng.* **1999**, *40*, 15–20.
21. Achouri, A.; Boye, J.I.; Zamani, Y. Changes in soymilk suality as a function of composition and storage. *J. Food Qual.* **2007**, *30*, 731–744.
22. Chu, L.F.; Yaylayan, V.A. Model studies on the oxygen-induced formation of benzaldehyde from phenylacetaldehyde using pyrolysis GC-MS and FTIR. *J. Agric. Food Chem.* **2008**, *56*, 10697–10704.
23. Shahidi, F. Indicators for Evaluation of Lipid Oxidation and Off-flavor Development in Food. In *Food Flavors: Formation, Analysis and Packaging Influences*; Contis, E.T., Ho, C.T., Mussinan, C.J., Parliment, T.H., Shahidi, F., Spanier, A.M., Eds.; Elsevier Science: Amsterdam, The Netherlands, 1998.
24. Lin, M. Gas Chromatographic and Mass Spectral Analyses. *J. Agric. Food Chem.* **1970**, *13*, 333–336.
25. Kakumyan, P.; Kato, M.; Hajika, M.; Matsui, K. Development of a screening system for the evaluation of soybean volatiles. *Biosci. Biotechnol. Biochem.* **2009**, *73*, 1844–1848.
26. Aguila, A.; Barringuer, S. Effect of roasting conditions on color and volatile profile including HMF level in sweet almonds (*Prunus dulcis*). *J. Food. Sci.* **2012**, *77*, 461–468.

Permissions

All chapters in this book were first published in Foods, by MDPI; hereby published with permission under the Creative Commons Attribution License or equivalent. Every chapter published in this book has been scrutinized by our experts. Their significance has been extensively debated. The topics covered herein carry significant findings which will fuel the growth of the discipline. They may even be implemented as practical applications or may be referred to as a beginning point for another development.

The contributors of this book come from diverse backgrounds, making this book a truly international effort. This book will bring forth new frontiers with its revolutionizing research information and detailed analysis of the nascent developments around the world.

We would like to thank all the contributing authors for lending their expertise to make the book truly unique. They have played a crucial role in the development of this book. Without their invaluable contributions this book wouldn't have been possible. They have made vital efforts to compile up to date information on the varied aspects of this subject to make this book a valuable addition to the collection of many professionals and students.

This book was conceptualized with the vision of imparting up-to-date information and advanced data in this field. To ensure the same, a matchless editorial board was set up. Every individual on the board went through rigorous rounds of assessment to prove their worth. After which they invested a large part of their time researching and compiling the most relevant data for our readers.

The editorial board has been involved in producing this book since its inception. They have spent rigorous hours researching and exploring the diverse topics which have resulted in the successful publishing of this book. They have passed on their knowledge of decades through this book. To expedite this challenging task, the publisher supported the team at every step. A small team of assistant editors was also appointed to further simplify the editing procedure and attain best results for the readers.

Apart from the editorial board, the designing team has also invested a significant amount of their time in understanding the subject and creating the most relevant covers. They scrutinized every image to scout for the most suitable representation of the subject and create an appropriate cover for the book.

The publishing team has been an ardent support to the editorial, designing and production team. Their endless efforts to recruit the best for this project, has resulted in the accomplishment of this book. They are a veteran in the field of academics and their pool of knowledge is as vast as their experience in printing. Their expertise and guidance has proved useful at every step. Their uncompromising quality standards have made this book an exceptional effort. Their encouragement from time to time has been an inspiration for everyone.

The publisher and the editorial board hope that this book will prove to be a valuable piece of knowledge for researchers, students, practitioners and scholars across the globe.

List of Contributors

Pier Giorgio Righetti
Department of Chemistry, Materials and Chemical Engineering "Giulio Natta", Politecnico di Milano, Via Mancinelli 7, Milano 20131, Italy

Elisa Fasoli
Department of Chemistry, Materials and Chemical Engineering "Giulio Natta", Politecnico di Milano, Via Mancinelli 7, Milano 20131, Italy

Alfonsina D'Amato
Department of Chemistry, Materials and Chemical Engineering "Giulio Natta", Politecnico di Milano, Via Mancinelli 7, Milano 20131, Italy

Egisto Boschetti
EB Consulting, Paris, France

Sharmila Jayatilake
United Graduate School of Agricultural Sciences, Iwate University, Ueda 3-18-8, Morioka, Iwate 020-8550, Japan
Department of Food Science, Obihiro University of Agriculture and Veterinary Medicine, Inada-cho W2-11, Obihiro, Hokkaido 080-8555 Japan

Katsuhito Arai
United Graduate School of Agricultural Sciences, Iwate University, Ueda 3-18-8, Morioka, Iwate 020-8550, Japan
Department of Food Science, Obihiro University of Agriculture and Veterinary Medicine, Inada-cho W2-11, Obihiro, Hokkaido 080-8555 Japan

Nanami Kumada
Department of Food Science, Obihiro University of Agriculture and Veterinary Medicine, Inada-cho W2-11, Obihiro, Hokkaido 080-8555 Japan

Yoshiko Ishida
Department of Food Science, Obihiro University of Agriculture and Veterinary Medicine, Inada-cho W2-11, Obihiro, Hokkaido 080-8555 Japan

Ichiro Tanaka
Department of Food Science, Obihiro University of Agriculture and Veterinary Medicine, Inada-cho W2-11, Obihiro, Hokkaido 080-8555 Japan

Satoru Iwatsuki
Asama Chemical Co. LTD., Nihonbashi-kodenma-cho 20-3, Chuo-ku, Tokyo 103-0001, Japan

Takuji Ohwada
United Graduate School of Agricultural Sciences, Iwate University, Ueda 3-18-8, Morioka, Iwate 020-8550, Japan
Department of Food Science, Obihiro University of Agriculture and Veterinary Medicine, Inada-cho W2-11, Obihiro, Hokkaido 080-8555 Japan

Masao Ohnishi
United Graduate School of Agricultural Sciences, Iwate University, Ueda 3-18-8, Morioka, Iwate 020-8550, Japan
Department of Food Science, Obihiro University of Agriculture and Veterinary Medicine, Inada-cho W2-11, Obihiro, Hokkaido 080-8555 Japan

Yoshihiko Tokuji
United Graduate School of Agricultural Sciences, Iwate University, Ueda 3-18-8, Morioka, Iwate 020-8550, Japan
Department of Food Science, Obihiro University of Agriculture and Veterinary Medicine, Inada-cho W2-11, Obihiro, Hokkaido 080-8555 Japan

Mikio Kinoshita
United Graduate School of Agricultural Sciences, Iwate University, Ueda 3-18-8, Morioka, Iwate 020-8550, Japan
Department of Food Science, Obihiro University of Agriculture and Veterinary Medicine, Inada-cho W2-11, Obihiro, Hokkaido 080-8555 Japan

Luciana Baroni
Primary Care Unit, District No. 4, U.L.S.S. No. 9, Treviso 31100, Italy

Marina Berati
Nutrition Ecology International Center (NEIC), Torino 10100, Italy

Maurizio Candilera
Department of Mathematics, University of Padua, 35121 Padova, Italy

Massimo Tettamanti
Nutrition Ecology International Center (NEIC), Torino 10100, Italy

David H. Kingsley
Food Safety and Intervention Technologies Research Unit, Agricultural Research Service, U.S. Department of Agriculture, James W.W. Baker Center, Delaware State University, Dover, DE 19901, USA

Domenico Meloni
Department of Veterinary Medicine, University of Sassari, Via Vienna 2, 07100, Sassari, Italy

Ellen van der Gaag
Hospital Group Twente, Geerdinksweg 141, 7555 DL Hengelo, The Netherlands

Lisanne Brekhoff
Medical School University of Groningen, Hanzeplein 1, Groningen, The Netherlands

Sebastian E. W. Opitz
Institute of Chemistry and Biological Chemistry, Zurich University of Applied Sciences,
Einsiedlerstrasse 31, CH-8820 Wädenswil, Switzerland

Samo Smrke
Institute of Chemistry and Biological Chemistry, Zurich University of Applied Sciences,
Einsiedlerstrasse 31, CH-8820 Wädenswil, Switzerland

Bernard A. Goodman
State Key Laboratory for Conservation and Utilization of Subtropical Agro-Bioresources,
Guangxi University, Nanning, 530004 Guangxi, China

Marco Keller
Bühler AG, Gupfenstrasse 5, 9240 Uzwil, Switzerland

Stefan Schenker
Bühler AG, Gupfenstrasse 5, 9240 Uzwil, Switzerland

Chahan Yeretzian
Institute of Chemistry and Biological Chemistry, Zurich University of Applied Sciences,
Einsiedlerstrasse 31, CH-8820 Wädenswil, Switzerland

Palmiro Poltronieri
Institute of Sciences of Food Productions, National Research Council, ISPA-CNR, Via Lecce-Monteroni, 73100 Lecce, Italy

Valeria Mezzolla
Institute of Sciences of Food Productions, National Research Council, ISPA-CNR, Via Lecce-Monteroni, 73100 Lecce, Italy

Elisabetta Primiceri
NNL, Institute of Nanoscience-CNR, Via per Arnesano, I-73100 Lecce, Italy
Department of Mathematics and Physics "Ennio De Giorgi", University of Salento, Via per Arnesano, I-73100 Lecce, Italy

Giuseppe Maruccio
NNL, Institute of Nanoscience-CNR, Via per Arnesano, I-73100 Lecce, Italy
Department of Mathematics and Physics "Ennio De Giorgi", University of Salento, Via per Arnesano, I-73100 Lecce, Italy

Qumer Iqbal
Institute of Horticultural Sciences, University of Agriculture, Faisalabad 38000, Pakistan

Muhammad Amjad
Institute of Horticultural Sciences, University of Agriculture, Faisalabad 38000, Pakistan

Muhammad Rafique Asi
Nuclear Institute for Agriculture and Biology (NIAB), Faisalabad 38000, Pakistan

Agustin Ariño
Veterinary Faculty, University of Zaragoza, Zaragoza 50013, Spai

Khurram Ziaf
Institute of Horticultural Sciences, University of Agriculture, Faisalabad 38000, Pakistan

Aamir Nawaz
Faculty of Agricultural Sciences and Technology, Bahauddin Zakariya University, Multan 60000, Pakistan

Tanveer Ahmad
Institute of Horticultural Sciences, University of Agriculture, Faisalabad 38000, Pakistan

Barry K. Hurlburt
Southern Regional Research Center, Agricultural Research Service, United States Department of Agriculture, 1100 Robert E. Lee, Blvd., New Orleans, LA 70124, USA

Jane K. McBride
Southern Regional Research Center, Agricultural Research Service, United States Department of Agriculture, 1100 Robert E. Lee, Blvd., New Orleans, LA 70124, USA

Jacqueline B. Nesbit
Southern Regional Research Center, Agricultural Research Service, United States Department of Agriculture, 1100 Robert E. Lee, Blvd., New Orleans, LA 70124, USA
Current address: Department of Biology, University of New Orleans, New Orleans, LA 70148, USA

Sanbao Ruan
Southern Regional Research Center, Agricultural Research Service, United States Department of Agriculture, 1100 Robert E. Lee, Blvd., New Orleans, LA 70124, USA
Current address: Department of Medicine, Louisiana State University Health Sciences Center, New Orleans, LA 70112, USA

Soheila J. Maleki
Southern Regional Research Center, Agricultural Research Service, United States Department of Agriculture, 1100 Robert E. Lee, Blvd., New Orleans, LA 70124, USA

Caroline Siefarth
Department of Chemistry and Pharmacy, Emil Fischer Centre, Friedrich-Alexander Universität Erlangen-Nürnberg, Schuhstr. 19, Erlangen 91052, Germany
Fraunhofer Institute for Process Engineering and Packaging (IVV), Giggenhauser Str. 35, Freising 85354, Germany

Thi Bich Thao Tran
Fraunhofer Institute for Process Engineering and Packaging (IVV), Giggenhauser Str. 35, Freising 85354, Germany

Peter Mittermaier
Fraunhofer Institute for Process Engineering and Packaging (IVV), Giggenhauser Str. 35, Freising 85354, Germany

Thomas Pfeiffer
Fraunhofer Institute for Process Engineering and Packaging (IVV), Giggenhauser Str. 35, Freising 85354, Germany

Andrea Buettner
Department of Chemistry and Pharmacy, Emil Fischer Centre, Friedrich-Alexander Universität Erlangen-Nürnberg, Schuhstr. 19, Erlangen 91052, Germany
Fraunhofer Institute for Process Engineering and Packaging (IVV), Giggenhauser Str. 35, Freising 85354, Germany

Sencer Buzrul
Tütün ve Alkol Piyasası Düzenleme Kurumu (TAPDK), Ankara 06520, Turkey

Lubna S. Abdalrahman
Department of Biological Science, The University of Tulsa, Tulsa, OK 74104, USA

Mohamed K. Fakhr
Department of Biological Science, The University of Tulsa, Tulsa, OK 74104, USA

Sergio I. Martinez-Monteagudo
Department of Agricultural, Food and Nutritional Science, University of Alberta, Edmonton, AB T6G 2P5, Canada

Marleny D.A. Saldaña
Department of Agricultural, Food and Nutritional Science, University of Alberta, Edmonton, AB T6G 2P5, Canada

Victoria Ferragut
Centre Especial de Recerca Planta de Tecnologia dels Aliments (CERPTA), XaRTA, TECNIO, Departament de Ciència Animal y dels Aliments, Facultat de Veterinària, Universitat Autònoma de Barcelona, Bellaterra 08193, Barcelona, Spain

Dora C. Valencia-Flores
Centre Especial de Recerca Planta de Tecnologia dels Aliments (CERPTA), XaRTA, TECNIO, Departament de Ciència Animal y dels Aliments, Facultat de Veterinària, Universitat Autònoma de Barcelona, Bellaterra 08193, Barcelona, Spain

Marianita Pérez-González
Centre Especial de Recerca Planta de Tecnologia dels Aliments (CERPTA), XaRTA, TECNIO, Departament de Ciència Animal y dels Aliments, Facultat de Veterinària, Universitat Autònoma de Barcelona, Bellaterra 08193, Barcelona, Spain

Joan Gallardo
Centre Especial de Recerca Planta de Tecnologia dels Aliments (CERPTA), XaRTA, TECNIO, Departament de Ciència Animal y dels Aliments, Facultat de Veterinària, Universitat Autònoma de Barcelona, Bellaterra 08193, Barcelona, Spain

Manuela Hernández-Herrero
Centre Especial de Recerca Planta de Tecnologia dels Aliments (CERPTA), XaRTA, TECNIO, Departament de Ciència Animal y dels Aliments, Facultat de Veterinària, Universitat Autònoma de Barcelona, Bellaterra 08193, Barcelona, Spain

Caroline Siefarth
Department of Chemistry and Pharmacy, Emil Fischer Centre, Friedrich-Alexander Universität Erlangen-Nürnberg, Schuhstr. 19, Erlangen 91052, Germany
Fraunhofer Institute for Process Engineering and Packaging (IVV), Giggenhauser Str. 35, Freising 85354, Germany

Thi Bich Thao Tran
Fraunhofer Institute for Process Engineering and Packaging (IVV), Giggenhauser Str. 35, Freising 85354, Germany

Peter Mittermaier
Fraunhofer Institute for Process Engineering and Packaging (IVV), Giggenhauser Str. 35, Freising 85354, Germany

Thomas Pfeiffer
Fraunhofer Institute for Process Engineering and Packaging (IVV), Giggenhauser Str. 35, Freising 85354, Germany

Andrea Buettner
Department of Chemistry and Pharmacy, Emil Fischer Centre, Friedrich-Alexander Universität Erlangen-Nürnberg, Schuhstr. 19, Erlangen 91052, Germany
Fraunhofer Institute for Process Engineering and Packaging (IVV), Giggenhauser Str. 35, Freising 85354, Germany